Healing Herbs for *Child Health Care*

The Authors

Dr. S.K. Sood is Professor and Dean Life Sciences, Himachal Pradesh University, Shimla. He obtained his M.Sc. (1974) and M.Phil. (1976) from Himachal Pradesh University and Ph.D. (1981) from Nagarjuna University, Guntur, specializing in Plant Systematics and Embryology. For his post-doctoral, he underwent specialized training in Ethnobotany. He has had a brilliant academic career (triple gold medalist) followed by over 30 years of teaching and conducting research in Botany at postgraduate as well as research levels. His current major interests are in the theory and practice of systematics, flora of North-West Himalaya, ethnic tribes and conservation research.

Indu Chauhan is a postgraduate from Himachal Pradesh University. At present she is pursuing her doctoral work in the department.

Rakesh Thakur is a UGC-SAP fellow registered for his doctoral degree in the department. His current major interests involve floristics, ethnobotany and management of biodiversity.

Healing Herbs for *Child Health Care*

S.K. Sood
Indu Chauhan
Rakesh Thakur

Department of Biosciences
Himachal Pradesh University
Shimla – 171 005, H.P.

2015
Daya Publishing House®
A Division of
Astral International Pvt. Ltd.
New Delhi – 110 002

Cataloging in Publication Data—DK
Courtesy: D.K. Agencies (P) Ltd. <docinfo@dkagencies.com>

Sood, S. K. (Life scientist), author.
Healing herbs for child health care / S.K. Sood, Indu Chauhan, Rakesh Thakur.
pages cm
Includes bibliographical references (pages).
ISBN 9789351306559 (International Edition)

1. Medicinal plants—India—Classification. 2. Herbs—Therapeutic use. 3. Children—Health and hygiene. 4. Children—Diseases—Alternative treatment. I. Chauhan, Indu, author. II. Thakur, Rakesh, author. III. Title.

DDC 581.6340954 23

Published by : **Daya Publishing House®**
A Division of
Astral International Pvt. Ltd.
– ISO 9001:2008 Certified Company –
4760-61/23, Ansari Road, Darya Ganj
New Delhi-110 002
Ph. 011-43549197, 23278134
E-mail: info@astralint.com
Website: www.astralint.com

Laser Typesetting : **Classic Computer Services**, Delhi - 110 035

Printed at : **Thomson Press India Limited**

PRINTED IN INDIA

Abbreviations

Al	Aluminium
Alc.	Alcohol
α	Alpha
&	And
A.P.	Andhra Pradesh
As	Arsenic
β	Beta
C.	Central
Ca	Calcium
Cl	Chlorine
CNS	Central Nervous System
Co	Cobalt
Cu	Copper
Δ	Delta
δ	Delta
E.	East/Eastern
Eng.	English
et al.	*et alii* (and other authors)
etc.	*et cetera* (so on)
EtOH	Ethanol
Fe	Iron
γ	Gamma
g	Gram
I	Iodine
i.e.	*id est.* (that is)
in vitro	In Glass
in vivo	In Living Tissue
J.&K.	Jammu & Kashmir
K	Potassium
m	Metre
Me	Methyl
mg	Milligram
Mg	Magnesium
ml	Millilitre
Mn	Manganese
M.P.	Madhya Pradesh
N.	North/Northern
Na	Sodium

N.E.	North East
N.W.	North West
OH	Hydroxy
P	Phosphorus
%	Percentage
Ψ	Psi
S	Sulphur
S.	South/Southern
Sans.	Sanskrit
S.E.	South East
sp.	Species
Si	Silicon
Sub sp.	Sub Species
Syn.	Synonymous
tsp.	Tea Spoon
U.P.	Uttar Pradesh
U.S.	United States
var.	Variety
viz.	Vide licet (namely)
Vol. (s)	Volume (s)
W.	West/Western
W.H.O.	World Health Organisation

Preface

There could be no two opinions that healthy children are the wealth of any community. Unfortunately, some of the common ailments like catarrh, constipation, convulsion, cough and cold, dysentery, dyspepsia, fever, measles, mumps, nausea, pneumonia, stomachache, teething, vomiting, wheezing, worms etc. account for high incidence of ill health and mortality in children. Consequently, management of total paediatric health which is still a challenge to medical science has become a vital issue for both the developed and the third world countries. Despite the resurgence of interest in traditional systems of medicine in the last four decades, our knowledge on this aspect is not focused and least documented for a developing country like India. Thus, in the absence of any comprehensive interdisciplinary manual, the need for a valuable ready-reckoner has long been felt by students, researchers, teachers and scientists in traditional Universities, economic botany, ethnobotany, phytomedicine, pharmacology, plant genetic resources, policy planners, health workers, and laymen alike, and necessitated the present compilation. Over 60 photographs, 8 figures, 4 tables and 8 appendices enhance the value of the compilation significantly. It is hoped that this valuable information would be a definitive guide for management of child health care in India and abroad.

Well deserved appreciation is due to Prof. A.D.N. Bajpayee, Vice-Chancellor, Himachal Pradesh University, Prof. T.N. Lakhanpal and Prof. V.K. Mattu for their encouragement and well wishes; to Dr. Kuldeep Dogra, Dr. Romita Sharma, Mr. Ved Prakash, Mr. Ram Kishan and Ms. Poonam for their ungrudging help in critically reading the various drafts of this manuscript, and to Mr. Bal Krishan for his help in various ways. Thanks are also due to authors of books, papers and figures listed in the treatise. Finally, we would like to thank the members of our families for bearing the opportunity cost.

S.K. Sood

Indu Chauhan

Rakesh Thakur

Contents

Abbreviations	*v*
Preface	*vii*
Introduction	1–4
Indigenous People and Plant-Based Medicaments	1–2
Current Status of Phytomedicines Used for Child Health Care	2–3
Enumeration	3–4
Healing Herbs for Child Health Care	5–329
Epilogue	330–340
Literature Cited	341–368
Appendices	
Appendix I: Index to Families, Genera and Species	371–375
Appendix II: Index to Families of Plants	376–383
Appendix III: Index to Plant Part/s Used	384–394
Appendix IV: Index to Ailments and Botanicals Employed	395–414
Appendix V: Index to Botanical Names	415–427
Appendix VI: Index to English Names	428–454
Appendix VII: Index to Hindi Names	455–473
Appendix VIII: Index to Sanskrit Names	474–495

Introduction

Indegenous People and Plant-Based Medicaments

The importance of plants to the living world hardly needs any elaboration. It is an established fact that men's basic necessities of food, shelter, clothing and therapeutic needs have been met by plants over many millennia. Indeed the vegetal kingdom has been the therapeutic arsenal of all the prevalent traditional systems of medicine (Khan and Khanum, 2005). Of the 119 plant based drugs today about 74 per cent are from traditional medicine (Farnsworth and Soejarto, 1991). In recent years, interest in traditional wisdom and in the plants used as ethnomedicine is increasing in both the developing and developed countries. It has been estimated that nearly 80 per cent of the world population rely on traditional medicines for primary health care due to the fact that plant-based medicaments are safer than the costly modern medicines (Tyler, 1986; Wambebe, 1990; WHO 2002). Agenda 21 of the Rio Earth Summit (1992) stated that indigenous people have a vital role in environmental management and development because of their knowledge and traditional practices. In this regard, Maheshwari (1987a, b) states: "Tribal knowledge of plants is important not just for the tribal people themselves, but for the wider world". According to Chandra (1990), "These groups of people are not to be pitied for primitive existence, but honoured and conserved for their richness of human existence in harmony with nature".

Of the estimated 300 million indigenous people all over the world, India with its 67.7 million tribal people belonging to 573 tribal groups with various subsistence patterns covering more than 5000 villages is perhaps the largest country possessing a good treasure of accumulated indigenous/tribal knowledge (Anonymous, 1994; Pushpangadan, 1994) thereby amounting to nearly about 7.7 per cent of India's population (Sinha, 1996). Undeniably, Indian society is a bewildering mosaic of different traditions and cultures. In India, there are over 45,000 species of vascular plants (including 15,000 flowering plants), of which more than 5,000 plants with about 25,000 formulations are being used for treating a variety of ailments in its folk system of medicine whereas the tribal medicine involves the use of over 8,000 wild plants with about 1,75,000 specific preparations/applications (Pushpangadan and George, 2010) from which plant drugs worth Rs 3,400 million per annum are utilized for its various preparations (Gupta, 1986). It is forecasted that the world market for traditional medicine and herbal drugs is expected to grow around $ 5 trillion by the year 2050 (Nanjan, 2010).

Paediatric health care practices have become vital issue for the 3rd world countries due to high neonatal mortality. It is estimated that out of 3.9 million neonatal deaths that occur worldwide, almost 30 per cent occur in India (Black *et al.,* 2003). The primary causes have been due to complications of prematurity (28 per cent), sepsis and pneumonia (26 per cent), birth asphyxia and injuries (23 per cent), tetanus (7 per cent), congenital anomalies (7 per cent) and diarrhoea (3 per cent) (Lawn *et al.,* 2005; Baqui *et al.,* 2007). Therefore, to check high neonatal morbidity and mortality globally, systematic compilation of associated paediatric practices followed by traditional societies is much desired.

Current Status of Phytomedicines Used for Child Health Care

The third world nations of Asia are rich in biodiversity and the indigenous knowledge particularly the traditional ethnomedical practices (Pushpangadan and George, 2010). Glimpses of our knowledge in ethnomedicine are available in Egyptian hieroglyphics, the Bible, the Vedas and Chinese herbals. The earliest credit to document the scientific study of Indian indigenous drugs goes to Van Rheede (1678-1693)- then Dutch Governor of Cochin, a princely state of India, who recorded their use in his book entitled *Hortus Malabaricus* published from Amsterdam in 12 volumes, followed by that of Jones (1799), Fleming (1810), Ainslie (1813), Roxburgh (1832), Royle (1837), O'Shaugnessy (1841), Fluckier and Handburg (1874), Dutta (1877), Dymock (1883), Khory (1887), Dey (1896), Dymock *et al.* (1889-1893), and Watt (1889-1896) in the 18th and 19th centuries. But the study of indigenous drugs gained momentum with the publication of Chopra's book *Indigenous Drugs of India* in 1933. Subsequently, publications of two important directories- *Herbal Drugs Used in Pharmaceutical Industry* (Kapoor and Mitra, 1979), and *Directory of Crude Drugs and Aromatic Plants, Dealers, Producers and Exporters in India* (Shah and Virmani, 1982); two bibliographies, *viz., Bibliography on Investigated Medicinal Plants* (1950-1975) (Iyenger, 1977) and *Bibliography on Pharmacognosy of Medicinal Plants* (Mitra, 1985); and review articles of Mehra *et al.* (1969) on *Research in Pharmacognosy* in India and that of Mitra and Jain (1991) provided a brief history and evoluation of work on indigenous drugs and establishment of various medicinal centres in India. Besides, publications of a large number of medicinal flora of different States of the country, *e.g.*: Arid Zone (Chopra, 1960), A.P. (Raghunathan, 1976; Anonymous, 1999), Garhwal Himalayas (Uniyal, 1989), Gwalior (Singh, 1984), H.P. (Chauhan, 1999), Kachchh-Gujarat (Goyal, 1999), Kerala (Anonymous, 1981; Nambiar, 1985), M.P. (Tiwari, 1998), Nilgiris (Krishnamurthy, 1953), N.E. India (Kumar, 2002), Patalkot (Saxena, 1971),

Saharanpur (Ahuja, 1965), Sikkim Himalayas (Biswas and Chopra, 1982; Rai, 1994), S. India (Sundararaj, 1959), Uttarakhand Himalayas (Uniyal, 1997) and review articles on ethnobotanical aspects (Jain *et al.,* 1984; Jain and Sharma, 2000; Jain, 2002) have also enriched our knowledge on the subject.

Perusal of the aforesaid works revealed that the literature abounds in information on folk medicines and time tested ethnobotanical practices in different parts of India but studies in indigenous phytotherapy for paediatric diseases are scarce and scattered (Radhika and Balasubramanian, 1990; Jain, 1991; Venugopal, 1992; Rao and Henry, 1995; Jain and Mudgal, 1999; Maheshwari, 2000; Parrotta, 2001; Sood and Thakur, 2004; Retnam and Martin, 2005; Sood *et al.,* 2001, 2005, 2009, 2013; Sood and Sudershna, 2008). In this regard, studies pertaining to traditional herbal paediatric practices followed by specific tribe/arboriginals [(Pal *et al.,* 2000), Meitei community, Manipur (Singh and Singh, 2003), Palliyar tribe of Srivilliputhur (Muthukumarswamy *et al.,* 2004), ethnic groups of Purulia district, W. Bengal (Dey and De, 2011)]; and of certain geographical regions like Assam (Borthakur, 1991), Koraput district (Orissa) Srivastava and Rout (1994), Sundargarh district, Orissa (Satapathy and Chand, 2003), Jalgaon district, Maharashtra (Pawar and Patil, 2010), Garhwal Himalayas (Tiwari *et al.,* 2010), S. India (Dixit, 2011), Tilda, Raipur (Lanjewar *et al.,* 2011), Kerala (Beegam and Nayar, 2011), Ahmednagar district, Maharashtra (Mulay and Sharma, 2012) and Jaisinghpur, district Kangra, H.P. (Rawat and Kharwal, 2013) are noteworthy. Based on their studies, Bhattacharya and Patra (2004) examined the significance and potential of medicinal plants for rural women and child health care and suggested the short term and long term strategies whereas nutritional status of mother and under five children among Dhur Gond tribal community of Mahasamund district of Chattisgarh was assessed by Chandraker *et al.* (2009). Recently, Ashraf *et al.* (2010) gave an account of complementary therapies in paediatric age group. Keeping in view the recent growing popularity of plant-based drugs and lack of any exhaustive and multidisciplinary reference work on indigenous phytotherapy for paediatric diseases in India, an effort has been made for the first time to synthesise our current state of knowledge on the subject for the purpose of paediatric care and management.

Enumeration

For collecting data of interest on ethnotherapeutics for child health care in India, scrutiny of outdated classical material, *viz.*, *Vegetable Materia of Western India* (Dymock, 1883), *A Dictionary of Economic Products* (Watt, 1889-1896), *Indigenous Drugs of India* (Dey, 1896), *Materia Medica of Indian and Their Therapeutics* (Katark, 1903), *Vegetable Drugs of India* (Sanyal, 1984), *Materia Medica* (Ainslie, 1926), *Indian Materia Medica* (Nadkarni, 1926), *Commercial Drugs of India* (Dutt, 1928), *Vegetable Drugs of India* (Robert, 1932), *Pharmacopoeia Indica* (Bose, 1932), *Indigenous Drugs of India* (Chopra, 1933), *Indian Medicinal Plants* (Kirtikar and Basu, 1935), *The Wealth of India: A Dictionary of Indian Raw Materials and Industrial Products* (Anonymous, 1948-1976), *Pharmacognosy of Indian Root and Rhizome Drugs* (Dutta, 1949), *Glossary of Indian Medicinal Plants* (Chopra *et al.,* 1956), *Medicinal Plants of Arid Zone* (Chopra, 1960), *Supplement to Glossary of Indian Medicinal Plants* (Chopra *et al.,* 1974), *Dictionary of Economic Products of India* (Maheshwari and Singh, 1965), *Medicinal Plants of India and Pakistan* (Dastur, 1970);

scarcely available standard works (Atkinson, 1982; Singh *et al.,* 1983; Dutta, 1985; Agarwal, 1986; Ambasta, 1986; Singh, 1996; Uphof, 1986; Nayar *et al.,* 1989; Thakur *et al.,* 1989; Jain, 1991; Jain and De Filipps, 1991; Jain *et al.,* 1991, Chatterjee and Pakrashi, 1997; Ambasta *et al.,* 1992; Hussain, 1992; Saklani and Jain, 1994; Warrier *et al.,* 1994-1996; Pandey, 1995; Maheshwari, 1996; Rao and Henry, 1996; Varghese, 1996; Kaul, 1997; Sharma and Sood, 1997; Vedavathy *et al.,* 1997; Rawat and Chaudhary, 1998; Singh and Pandey, 1998; Chauhan, 1999; Kumar, 1999; Joshi, 2000; Maheshwari, 2000; Bhattacharjee, 2001; Sood *et al.,* 2001; Kumar, 2002; Caius, 2003; Sood and Thakur, 2004; Sood *et al.,* 2005 a-c; Prajapati *et al.,* 2006); and a large number of research papers published in various journals on economic botany and ethnobotany was undertaken as per the standered methodology outlined by Schultes (1960), Jain (1987, 1989) and Jain and Mudgal (1999). Wherever possible, data gathered from own field ethnobotanical investigations were also included; and such plants have been marked with asterisk under the present observations. Besides, the information on English, Hindi and Sanskrit names and distribution was largely based upon Hooker (1872-1897), Rastogi and Mehrotra (1998, 1999a,b, 2001, 2002) and Asolkar *et al.* (1992). The relevant informations are recorded in specific proforma, designed by modifying that of Saklani and Jain (1994).

Altogether systematic enumeration of 800 species of plants used for child health care has been presented alphabetically with updated botanical names (nomenclature based upon Bennet, 1986) and synonyms (wherever necessary), followed by the name of the family, English, Hindi and Sanskrit names, distribution, part/s used, utilization and relvant references, colour photographs and 4 tables have been also included to explain the textual matter. Appendices include index to families, genera and species (*Appendix* I), index to families of plants (*Appendix* II), index to plant part/s used (*Appendix* III), index to ailments and botanicals employed (*Appendix* IV), index to botanical names (*Appendix* V), index to English names (*Appendix* VI), index to Hindi names (*Appendix* VII), index to Sanskrit names (*Appendix* VIII).

Healing Herbs for Child Health Care

Abies spectabilis (D. Don.) Spach.

Syn.: *A. densa* Lindl.; *A. webbiana* Lindl.; *Pinus tinctoria* Wall.; *Pinus webbiana* Wall.

Family: Pinaceae

Eng., H. and Sans. Names	:	*Eng.*: East Himalayan Silver Fir, East Himalayan Fir, Indian Silver Fir; *H.*: Talishpatra, Tallispatra; *Sans.*: Talishpatra, Tallispatra.
Distribution	:	Afghanistan-Bhutan, 2,800-4,000m; also found in H.P. between 3,600-4,500m.
Reproductive Cycle	:	April-May.
Part Used	:	Leaf.
Folk Uses	:	Ash of burnt pine boughs sprinkled over skin of young children afflicted with measles, mumps and chickenpox. Infusion of leaves given to infants suffering from fever and chest infection. Leaf juice given to infants in fever during dentition and bronchitis (dose- 5-10 drops in water or in mother's milk).
References	:	Asolkar *et al.* (1992), Chatterjee and Pakrashi (1997), Chauhan (1999), Khare (2004), Watt (1972).

Abrus precatorius L.

Syn.: *A. abrus*(L.) Wight.; *A. minor* Desv.; *A. pauciflorus* Desv.; *Glycine abrus* L.

Family: Fabaceae

Eng., H. and Sans. Names	:	*Eng.*: Abrus, Bread Tree, Crab's eye, Indian Liquorice Root, Rosary Pea; *H.*: Chirmi, Chirmiti, Ghugachi, Ghunchi, Ghunghanchi, Kunch, Rakti, Rati, Ratti; *Sans.*: Aruna, Chataki, Chudala, Gunja, Gunjiku, Kankchchi, Rakta.
Distribution	:	Himalayas-Sri Lanka and Thailand, ascending to 1,200m; commonly occurring in Garhwal region upto a height of 1,200m above the sea level.

Reproductive Cycle	:	August-September (Flowering) and Winters (Fruiting).
Part Used	:	Whole Plant.
Folk Use	:	Decoction of whole plant used in bronchitis of children.
Reference	:	Watt (1972).

Abutilon indicum **(L.) Sweet.**

Syn.: *A. asiaticum* (L.) Sweet.; *A. indicum* (L.) Don.; *Sida indica* L.

Family: Malvaceae

Eng., H. and Sans. Names	:	*Eng.*: Country Mallow, Indian Abutilon;
		H.: Jhampi, Kandhi, Kanghani, Kanghi, Potari, Tepari;
		Sans.: Atibala, Balika, Balya, Bhuribala, Ghanta, Kankati, Shita, Shitapushpa, Vikankata, Vrishyagandha.
Distribution	:	Throughout the tropics and hotter parts of India, ascending to 1,200m.
Reproductive Cycle	:	May-January.
Parts Used	:	Whole Plant and Seed.
Folk Uses	:	Juice of the plant applied as an emollient to relieve soreness of the buttocks in young children. Rectum of children affected with threadworms exposed to smoke emanating from burnt charcoal.
References	:	Asolkar *et al.* (1992), Kirtikar and Basu (1984), Parrotta (2001), Watt (1972).

Abutilon theophrasti **Medik.**

Syn.: *A. avicennae* Gaertn.; *A. theophrasti* var. *chinensis* (Skvorstsov) S. Y. Hu.; *A. theophrasti* var. *nigrum* (Skvorstsov) S. Y. Hu.; *Sida abutilon* L.

Family: Malvaceae

Eng. and Sans. Names	:	*Eng.*: Butterprint, Buttonweed, China Jute, Indian Mallow, Pie-marker, Velvetleaf;
		Sans.: Jaya.
Distribution	:	Found in N.W. India, Sind, Kashmir, Bengal, upto 1,500m.
Reproductive Cycle	:	Flowering during June-October.
Part Used	:	Whole Plant.
Folk Use	:	Whole plant used in fever.
Reference	:	Pal *et al.* (2000).

Acacia caesia **(L.) Willd.**

Syn.: *A. caesia* var.*subnuda* (Craib) I. C. Nielsen.; *A. columnaris* Craib.; *Mimosa caesia* (L.) Willd.

Family: Fabaceae

Eng., H. and Sans. Names	:	*Eng.*: Twisted Acacia; *H.*: Aila, Nikuncham; *Sans.*: Nikunjika.
Distribution	:	Gujarat, Konkan, W. Ghats of Bombay to the Anamalais, Cuddapah, throughout India in dry forests.
Reproductive Cycle	:	September-November.
Parts Used	:	Root and Bark.
Folk Uses	:	Both bark and root used in measles and smallpox.
References	:	Jain (1991), Pande *et al.* (2006).

Acacia catechu **(L.f.) (Pl. 1A)***

Syn.: *A. catechuoides* (Roxb.) Benth.; *A. polyacantha* Willd.; *A. sundra* (Roxb.) Bedd.; *A. wallichiana* DC.; *Mimosa catechu* L.f.

Family: Fabaceae

Eng., H. and Sans. Names	:	*Eng.*: Black Catechu, Catechu, Catechu Nigrum, Cutch, Pegu Catechu; *H.*: Katha, Khair, Khairbabul; *Sans.*: Bahushalya, Balapatra, Balaputra, Balatanaya, Dantadhavana, Homa, Gayatri, Kantaki, Karkati, Khadira, Medhya, Yajnanga.
Distribution	:	Throughout the Himalayas from Punjab-Sikkim, ascending to 1,700m.
Reproductive Cycle	:	November-January.
Part Used	:	Heartwood ('Katha')
Folk Uses	:	Powdered heartwood or its tincture given (dose of 5-10gm) alone or combined with cinnamon or opium in passive diarrhoeas, haemorrhages, tonsillar infections, and as a valuable astringent.
References	:	Chauhan *et al.* (2007), Kirtikar and Basu (1984), Sanjay *et al.* (2007), Sood *et al.* (2012), Watt (1972).

***Acacia farnesiana* (L.) Willd.**

Syn.: *Mimosa farnesiana* L.; *Vachellia farnesiana* Wight and Arn.

Family: Fabaceae

Eng., H. and Sans. Names : *Eng.*: Cassie Flowers, Scented Babul, Sponge Tree;

H.: Diwana Babul, Gandbabul, Guhbabul, Vilayati Babul;

Sans.: Arimaedah, Arimeda, Arimedaka, Kalaskandha, Vita.

Distribution : Throughout the plains of India.

Reproductive Cycle : June-August.

Part Used : Whole Plant.

Folk Use : Whole Plant used in stomachache.

Reference : Pal *et al.* (2000).

***Acacia nilotica* subsp. *indica* (Benth.) Brenan.**

Syn.: *A. arabica* var. *indica* Benth.

Family: Fabaceae

Eng., H. and Sans. Names : *Eng.*: Babul, Babul Acacia, Black Babul, Indian Gum Arabic Tree, Gum Tree;

H.: Babla, Babul, Babur, Kikar;

Sans.: Ajabaksha, Babbula, Babbulaka, Barbura, Bebula, Burbura, Dridhabija, Kantalu, Kashaya, Kashtaki, Pitaka, Vaboula, Varvara, Vavvola.

Distribution : Found from Punjab-Bihar, W. Peninsula and Sri Lanka.

Reproductive Cycle : July-September (Flowering) and April-June (Fruiting).

Parts Used : Leaf, Legumes and Mucilage.

Folk Uses : Bruised leaves applied to sore eyes in children in Rajputana. Powder of tender legumes beneficial in diarrhoea and dysentery; its usefulness further enhanced in combination of some preparation of opium. Mucilage when given as an injection allays rectal irritation in diarrhoea and dysentery,

References : Asolkar *et al.* (1992), Chauhan (1999), Kirtikar and Basu (1984), Watt (1972).

Acacia pennata (L.) Willd.

Syn.: *A. hainanensis* Hayata.; *Albizia tenerrima* de Vriese.; *Mimosa pennata* L.

Family: Fabaceae

Eng., H. and Sans. Names : *Eng.*: Rusty Mimosa;

H.: Aila, Biswul;

Sans.: Ari, Ballikhadira, Khadirapatrika, Sandananika, Silikhadira, Svadikhallari, Uddala.

Distribution : C. and E. Himalayas, ascending to 1,700m; Bihar, Myanmar, E. and W. Peninsula and Sri Lanka.

Reproductive Cycle : February-October.

Parts Used : Whole Plant, Root, Leaf and Bark.

Folk Uses : Whole plant used in measles. 15g of root powder mixed with honey given orally thrice a day in rickets. Leaf juice mixed with milk given to infants in Konkan suffering from indigestion with black stools. Juice of bark pasted on neck in tonsillitis.

References : Augustine and Sivadasan (2004), Farooq (2005), Jha and Varma (1996), Kirtikar and Basu (1984), Parrotta (2001).

Acalypha indica L.

Syn.: *A. canescens* L.; *A. ciliata* L.; *A. spicata* L.

Family: Euphorbiaceae

Eng., H. and Sans. Names : *Eng.*: Indian Acalypha;

H.: Kholi, Kuppi, Rupi;

Sans.: Aritamanjari, Harijan-manjari, Harita-manjari.

Distribution : Throughout the hotter parts of India, ascending the hills of Orissa upto 210m.

Reproductive Cycle : August-September.

Parts Used : Whole Plant, Shoot and Leaf.

Folk Uses : Plant used as a laxative for children. Juice of plant together with a small portion of margosa oil rubbed on tongue of infant for clearing the stomach of viscid phlegm. Juice of leaves used as an emetic for children. Marble-sized paste of fresh leaves introduced into rectum against obstinate constipation of children. Paste of leaves and tender shoot combination within breast milk given to children for catarrh and cough.

References : Anonymous (1948), Dey and Bahadur (1973), Kirtikar and Basu (1984), Prasad *et al.* (1996), Watt (1972).

Achillea millefolium **L.**

Syn.: *A. lanulosa* Nutt.

Family: Asteraceae

Eng., H. and Sans. Names : *Eng.*: Milfoil, Yarrow;

H.: Biranjasif, Gandana, Rajmari;

Sans.: Biranjasipha, Gandana, Shahastaraptri.

Distribution : Himalayas, Kashmir-Kumaon: 1,050 to 3,600m; also found in Bombay and Belgaun areas.

Reproductive Cycle : August-October.

Part Used : Whole Plant.

Folk Use : Whole Plant used in fever.

Reference : Pal *et al.* (2000).

Achyranthes aspera **L.**

Syn.: *Aerva aspera* Spreng.; *Digera muricata* Mart.

Family: Amaranthaceae

Eng., H. and Sans. Names : *Eng.*: Prickly Chaff Flower, Rough Chaff Flower;

H.: Apang, Apmarga, Chirchita, Latjira, Puthkanda;

Sans.: Aghata, Apamarga, Apangaka-Kharaman-Jari.

Distribution : Throughout India, upto 2,100m and in S. Andaman islands.

Reproductive Cycle : June-September.

Parts Used : Whole Plant, Twig, Leaf, Root and Seed.

Folk Uses : Powdered dried plant given to children for colic. Bath with decoction of the whole plant good for children suffering from fever. Paste of twigs with seed oil of *Schleichera oleosa* (3:1) used by Lodha for ear sores of children. Extract of leaves given to children to kill pinworms. Leaf paste along with resin of *Shorea robusta* applied on body for chickenpox. Also, leaves used in tetanus. Slightly warmed root paste (about 10g) given to children once a day on empty stomach as anthelmintic. Ash of seeds along with honey (2-5ml) given three times a day for 4-5 days to children against whooping cough.

References : Asolkar *et al.* (1992), D'Rozario *et al.* (1999), Khan *et al.* (2003), Prajapati *et al.* (2006), Saini (1996), Satapathy (2008), Singh and Kumar (2000), Thakur *et al.* (1989), Watt (1972).

Acmella paniculata **(Wall. *ex* DC.) R. K. Jansen.**

Syn.: *Spilanthes acmella* var. *calva* (DC.).; *S. calva* DC.; *S. lobata* Blanco.; *S. rugosa* Blume *ex* DC.

Family: Asteraceae

Eng. and H. Names : *Eng.*: Panicled Spot Flower;

H.: Pipulka.

Distribution : Throughout India, ascending the Himalayas and other mountains to 1,700m.

Reproductive Cycle : November-December.

Parts Used : Leaf and Flower.

Folk Uses : Leaves used to soothe itching in children during rainy season. Flowers good against stammering in children.

References : Agarwal (2003), Ambasta (1986), Farooq (2005), Kirtikar and Basu (1984), Parrotta (2001), Sood *et al.* (2012), Srivastava *et al.* (2003c).

Aconitum ferox **Wall. *ex* Springs.**

Syn.: *A. dissectum* Stein.; *A. napellus* L.; *A. virosum* D. Don.

Family: Ranunculaceae

Eng., H. and Sans. Names : *Eng.*: Indian Aconite, Monkshood, Wolfsbane;

H.: Bacchang, Bish, Mahoor, Mitha Zahar;

Sans.: Vatsanava, Visha.

Distribution : Wild in the alpine Himalayas of Kashmir and Nepal, ascending to 3,600m.

Reproductive Cycle : July-December.

Parts Used : Whole Plant and Root.

Folk Uses : Whole plant used in curing fever. Roots useful in acute tonsillitis.

References : Ambasta (1986), Asolkar *et al.* (1992), Haridasan *et al.* (2001), Khare (2004), Pal *et al.* (2000).

Aconitum heterophyllum Wallich. *ex* Royle.

Syn.: *A. atees* Royle.; *A. bisma* (Ham.) Rap.; *A. cordatum* Royle.; *A. ovatum* Lindl.; *A. palmatum* D. Don.

Family: Ranunculaceae

Eng., H. and Sans. Names : *Eng.*: Atis Root, Indian Root;

H.: Atis, Ativisha, Atvika, Patis;

Sans.: Amrita, Aruna, Ataicha, Atisaraghni, Ativisha, Bhangura, Bhringi, Kashmira.

Distribution : Common in alpine and sub-alpine belts of the northwestern Himalayan regions, between 1,800-4,500m.

Reproductive Cycle : June-September.

Part Used : Tuberous Roots.

Folk Uses : Powdered tuberous roots given with water or milk for curing abdominal troubles in children. 1-3g root paste with one and a half tsp of *Zingiber officinalis* given twice a day for fever caused by cough and cold. Roots also used in gastro-enteric fever of children. Powdered roots mixed with honey effective for children suffering from cough, coryza, fever and vomiting. Roots also prescribed as tonics for children.

References : Anonymous (1976b), Arya and Prakash (2000), Beigh *et al.* (2003), Chatterjee and Pakrashi (1997), Kaul (1997), Khare (2004), Kirtikar and Basu (1984), Singh (1999), Thakur *et al.* (1989).

Aconitum lethale Griff.

Syn.: *A. balangrense* Lauener.; *A. balfourii* Stapf.; *A. deltoideum* Lauenar.

Family: Ranunculaceae

H. Names : *H.*: Mitha- Bis, Vatsanabh.

Distribution : Kumaon and Garhwal, in the shady slopes and forest edges, between 300-4,200m.

Reproductive Cycle : June-October.

Part Used : Root.

Folk Use : Used in tonsillitis.

References : Jain (1991), Samant *et al.* (2001b).

Aconitum violaceum **Stapf.**

Syn.: *A. multifidum* Royle.; *A. nepellus* var. *dissectum* Duthie.

Family: Ranunculaceae

Eng., H. and Sans. Names : *Eng.*: Indian Aconite;

H.: Dudhia-Bis, Mitha Telia, Telia Kachang, Tilla;

Sans.: Prativishaa, Shyaamkandaa.

Distribution : Slopes of alpine and sub-alpine meadows of cold arid region from Kashmir-Uttaranchal Himalayas: 3,400-4,500m.

Reproductive Cycle : July-November.

Parts Used : Root and Flower.

Folk Uses : Root powder mixed with mother's milk given to infants in cold, cough and stomachache. A flower given twice a day to asthmatic and coughing children in Himalayas.

References : Bhattacharjee (2001), Kaul (1997), Sharma (2003).

Acorus calamus **L.**

Family: Acoraceae

Eng., H. and Sans. Names : *Eng.*: Calamus, Flag Root, Myrtle Root, Sea Sedge, Sweet Cinnamon, Sweet Flag;

H.: Bach, Bash, Bhoja, Ghorbach, Gorbach, Vacha;

Sans.: Bhadra, Galani, Golomi, Jalaja, Jatila, Kunga, Ugra.

Distribution : Cultivated in damp marshy places in India and Myanmar ascending to 2,000m; exceedingly common in Manipur and Naga hills.

Reproductive Cycle : April-October.

Parts Used : Whole Plant, Leaf, Rhizome and Root.

Folk Uses : Whole plant used for deworming, fever and toothache. Also, the plant cytoplasm given to kids for severe sore throats and mouth sores. Poultice of small pieces of root tied around the neck to cure Kwashiorkor disease in children. Two to three spoons of fresh root extract given thrice a day to children suffering from diarrhoea and dysentery. Rhizome used in rickets and for teething problems of children. Root paste applied on chest to treat pneumonia in children. Juice of rhizome with a little sugar or honey and black pepper given once daily with half a glass of milk for three four days in dyspepsia with abdominal pain in chronic liver disorders. Also, root paste given to children in morning on empty stomach for flatulence and headache. An amulet of three or five small pieces of rhizome worn around the neck of children to expel worms and for checking cough and cold. Paste of small piece of rhizome with fruit of 'Jaiphal' and 'Rada' given with mother's milk to children suffering from cold, cough and fever. Rhizome powder applied externally on the abdomen to expel flatus and in tympanitis. Paste of rhizome and leaves in mustard oil makes good lotion for joint pains and body massage of children.

References : Ahmed and Borthakur (2005), Bhattacharjee (2001), Bhuyan (2003), Borthakur *et al.* (2004), Dey and Bahadur (1973), Gopal and Chandra (2003), Islam (1996), Jain (1991), Kaul (1997), Pal *et al.* (2000), Prasad *et al.* (1996), Sharma *et al.* (2003a,b), Singh (1999), Sood *et al.* (1991), Thakur *et al.* (1989), Watt (1972).

***Acranthera grandiflora* Bedd.**

Family: Rubiaceae

Distribution : Moist forests throughout India, ascending to 1,000m.

Parts Used : Leaf and Flower.

Folk Use : 1g paste of equal quantity of leaves and its flowers consumed with cow's milk twice a day for three days to cure skin diseases of children.

Reference : Viswanathan *et al.* (2006).

***Acronychia pedunculata* (L.) Miq.**

Syn.: *A. barberi* Gamble.; *A. laurifolia* Bl.; *Cyminosma pedunculata* (L.) DC.; *Jambolifera pedunculata* L.

Family: Rutaceae

Eng. and Sans. Names : *Eng.*: Clawflowered Laurel, Laka Wood;

Sans.: Kattukanni.

Distribution : Sikkim Himalayas (900-1,200m), Khasia mountains (0-1,200m), Assam, Chittagong, W. Peninsula- on the Ghats from Concan-Travancore.

Reproductive Cycle : June-December.

Part Used : Leaf.

Folk Use : Leaves burnt near smallpox patients for its effective cure.

Reference : Kirtikar and Basu (1984).

Adansonia digitata **L.**

Syn.: *A. bahobab* L.; *A. integrifolia* Raf.; *Ophelus sitularius* Lour.

Family: Bombacaceae

Eng., H. and Sans. Names : *Eng.*: African Baobab Tree, Baobab, Cream of Tartar Tree, Monkey Bread Tree;

H.: Gorakh Amli, Hathi Kathian;

Sans.: Gorakshi.

Distribution : Gardens in the plains of India; common in A.P.

Reproductive Cycle : February-October.

Part Used : Whole Plant.

Folk Use : Whole plant used to cure fever in children.

Reference : Pal *et al.* (2000).

Adiantum incisum **Forsk.**

Syn.: *A. caudatum* L.

Family: Adiantaceae

Eng., H. and Sans. Names : *Eng.*: Tailed Maiden Hair;

H.: Marshikha;

Sans.: Nilkantha Shikha.

Distribution : Lower slopes of hills, old walls and shady places of Maharashtra, Rajasthan, Punjab, Tamil Nadu, and W. Bengal.

Reproductive Cycle : June- December.

Part Used : Whole Plant.

Folk Use : Whole Plant good for fever in children.

Reference : Pal *et al.* (2000).

Adiantum lunulatum **L.**

Syn.: *A. philippense* L.

Family: Adiantaceae

Eng., H. and Sans. Names : *Eng.*: Walking Maiden Hair Fern;

H.: Hansapadi, Hansapagi, Kalijhamp, Kalijhant, Paresiyavasan;

Sans.: Brahmadani, Chitrapada, Dharttarashtrapadi, Gondhangri, Hansaghri, Hansapadi, Kirramata, Kitamari, Padangi, Raktapadi, Shitangi, Suvaka.

Distribution : Throughout N. India, in moist places; very general on W. side in the plains and lower slopes of hills in S. India.

Parts Used : Whole Plant and Root.

Folk Uses : Plant beneficial in febrile affections of children. Decoction of roots used for treating throat infection and febrile conditions in children.

References : Kirtikar and Basu (1984), Pal *et al.* (2000), Parrotta (2001), Watt (1972).

Aeginetia indica **L.**

Family: Orobanchaceae

Eng. and H. Names : *Eng.*: Forest Ghost Flower, Ye gu;

H.: Aankuri Bankuri.

Distribution : Throughout India.

Reproductive Cycle : July-December.

Part Used : Rhizome.

Folk Use : Juice of rhizome used for relief from mumps.

References : Lalramnghinglova (1996, 2003).

Aegle marmelos **L. Correa *ex* Roxb.***

Syn.: *Belou marmelos* (L.) Lyons.; *Bilacus marmelos* (L.) Kuntze.; *Crataeva religiosa* Ainslie; *Feronia pellucida* Roth.

Family: Rutaceae

Eng., H. and Sans. Names : *Eng.*: Bael Tree, Bengal Quince, Golden Apple, Holy Fruit Tree, India Quince;

H.: Bel, Bili, Siriphal, Sirphal;

Sans.: Asholam, Bilva, Mangalya, Putivata, Sadaphala, Shalatu, Shalya, Sriphalah, Sivadrumah, Tripatra.

Distribution : Throughout India, ascending to 1,200m in W. Himalayas.

Reproductive Cycle : February-March and September-December.

Parts Used : Whole Plant, Root, Root Bark and Fruit.

Folk Uses : Plant good against fever. Decoction of roots given with sugar and fried rice for checking diarrohea and gastric irritability in infants. Aqueous extract of root bark (3/4g) administered with barley pulp and sugar for curing vomiting and emaciation in children. Fruits used for diarrhoea and dysentery in children.

References : D' Rozario *et al.* (1999), Dey and Bahadur (1973), Kirtikar and Basu (1984), Pal *et al.* (2000), Prajapati *et al.* (2006).

Aerva lanata Juss. *ex* Schultes.

Syn.: *Achyranthes lanata* L.; *A. elegans* Moq.; *A. floribunda* Wight.; *Illecebrum lanatum* (L.) Muer.

Family: Amaranthaceae

Eng., H. and Sans. Names : *Eng.*: Javanese Wood Plant, Mountain Knot Grass;
H.: Chaya, Gorkhabundi, Kapurijadi, Papurijadi;
Sans.: Astmabayday, Bhadra.

Distribution : Throughout tropical India, up to 900m.

Reproductive Cycle : August-April.

Parts Used : Whole Plant and Root.

Folk Uses : Bath with decoction of whole plant good for babies to improve their complexion; also used as a vermifuge for children. Dried roots worn as pendant in chronic fever. Decoction of roots along with tuberous root of *Cyperus rotundus* and fennel given to cure diarrhoea in children in southern Orissa.

References : Ansarali and Sivadasan (2009), Kirtikar and Basu (1984), Mohanty *et al.* (1996), Parrotta (2001), Prusti (2007).

Aeschynanthus parviflorus (D. Don) Spreng.

Syn.: *A. parviflorus* C. B. Clarke.; *A. sikkimensis* (C. B. Clarke) Stapf.

Family: Gesneriaceae

Eng. Name : *Eng.*: Sikkim Blushwort.

Distribution : Temperate Himalayas from Nepal eastwards-Assam and Meghalaya; frequent in Mizoram particularly in tropical semi-evergreen forests.

Reproductive Cycle : August-December.

Parts Used : Root and Flower.

Folk Uses : 5ml decoction of roots taken as febrifuge, anodyne and for all kinds of ailments in children. Flowers used against tonsillitis.

References : Lalramnghinglova (1996, 2003), Lalramnghinglova and Jha (1999).

***Aethusa cynapium* L.**

Syn.: *A. cynica* Dulac.; *A. toxicaria* Salisb.

Family: Apiaceae

Eng. Names	:	*Eng.*: Dog's Parsley, Fool's Parsley.
Distribution	:	Native to the U.K., also found in Europe, New England; cultivated in district Saharanpur in U.P. in India.
Reproductive Cycle	:	July-September.
Part Used	:	Whole Plant.
Folk Use	:	Helps in indigestion during teething in children.
Reference	:	Prajapati *et al.* (2006).

Ageratum conyzoides* L. (Pl. 1B)

Syn.: *A. cordifolium* Roxb.

Family: Asteraceae

Eng., H. and Sans. Names	:	*Eng.*: Appa Grass, Bastard Agrimony, Floss Flower, Goat Weed, White Weed ; *H.*: Koobhi, Nil Phool; *Sans.*: Visamustih.
Distribution	:	Throughout India, ascending Himalayas to 1,700m.
Reproductive Cycle	:	January-December.
Parts Used	:	Whole Plant and Leaf.
Folk Uses	:	Plant used for skin diseases and preventing tetanus. Extract of leaves crushed in week-long stored water given on Sunday only to control bed wetting in children during sleep.
References	:	Asolkar *et al.* (1992), Pal *et al.* (2000), Sahoo and Mudgal (1995), Upadhyay and Singh (2005).

***Agrimonia eupatoria* L.**

Syn.: *A. bracteosa* E. Mey.

Family: Rosaceae

Eng. Names	:	*Eng.*: Agrimony, Cocklebur, Stickle Wart.
Distribution	:	Temperate Himalayas, Muree, Kashmir, Sikkim; N. Bengal, ascending to 3,000m; Nagaland and Khasia Hills in Meghalaya: 1,200-2,400m.
Reproductive Cycle	:	April-September.
Parts Used	:	Aerial parts.

Folk Use	:	Aerial parts used against diarrhoea of children.
Reference	:	Karnick (1994a).

***Ailanthus altissima* (Mill.) Swingle.**

Syn.: *A. giraldii* Dode.; *A. glandulosa* Desf.; *Rhus cacodendron* Ehrh.

Family: Simaroubaceae

Eng. Names	:	*Eng.*: Ailanto, Chinese Sumach, Japan Varnish Tree, Stinking Cedar, Tree of the Gods.
Distribution	:	Cultivated in hills of Punjab; planted as a shade tree in N. India.
Reproductive Cycle	:	June-October.
Part Used	:	Bark.
Folk Use	:	Decoction of bark used for curing dysentery in infants.
Reference	:	Kirtikar and Basu (1984).

***Ailanthus excelsa* Roxb.**

Syn.: *Pongelion wighttii* Van Tiegh.

Family: Simaroubaceae

Eng., H. and Sans. Names	:	*Eng.*: Tree Of Heaven; *H.*: Limbado, Maha Nim, Maharukha; *Sans.*: Aralu, Atarusha, Madala, Mahanimba, Maharakha, Pisasha.
Distribution	:	Found in N.W. India, Bihar, W. Peninsula.
Reproductive Cycle	:	February-April.
Parts Used	:	Whole Plant and Leaf.
Folk Uses	:	Plant known to cure fever in children. Warmed poultice of two or three leaves tied on the chest to cure bronchitis and cough in children. Leaf juice used as a vermifuge in children.
References	:	Bhatt and Jain (2009), Kirtikar and Basu (1984), Kothari and Londhe (2000), Pal *et al.* (2000).

***Ailanthus triphysa* (Dennst.) Alston.**

Syn.: *Adenanthera triphysa* Dennst; *Ailanthus malabarica* DC.

Family: Simaroubaceae

Eng., H. and Sans. Names	:	*Eng.*: White Bean, White Palle, White Siris; *H.*: Guggul Dhup, Matti, Muddhe Dhup, Perumaram, Pongilyam;

		Sans.: Atisarahita, Mahanimbu.
Distribution	:	Konkan, Deccan, W. Ghats from N. Kanara and Mysore-Travancore up to 1,000m, Sri Lanka, Eastern slopes of the Pegu yoma in Myanmar; Often planted in S. India.
Reproductive Cycle	:	February-May.
Part Used	:	Whole Plant.
Folk Use	:	Plant used for curing fever in children.
Reference	:	Pal *et al.* (2000).

***Ajuga integrifolia* Buch.-Ham.**

Syn.: *A. alba* (Gurke) Robyns.; *A. bracteosa* Wall. *ex* Benth.

Family: Lamiaceae

Eng., H. and Sans. Names	:	*Eng.*: Bracted Bungleweed;
		H.: Kauributi;
		Sans.: Nilkanthi.
Distribution	:	Native to N.W. India; Kashmir, Nepal, ascending to 2,300m; common in H.P. in fields, bushes, barren places and particularly shady slopes.
Reproductive Cycle	:	March-December.
Part Used	:	Leaf.
Folk Uses	:	Leaves used for removing deposition on tongues of children suffering from stomach complaints and fever.
References	:	Bhattacharjee (2001), Chauhan (1999), Pal *et al.* (2000).

***Alangium salviifolium* (L.f.) Wang.**

Syn.: *A. decapetalum* Lamk.; *A. lamarckii* Thw.; *A. latifolium* Miq.; *A. tomentosum* Lamk.

Family: Alangiaceae

Eng., H. and Sans. Names	:	*Eng.*: Sage- Leaved Alangium;
		H.: Akhaul, Akol, Akola, Ankora, Dhera, Kwel;
		Sans.: Ankola, Bodha, Ghalanta, Kalhora, Madana, Pita, Vamaka.
Distribution	:	Throughout India, from N.W. Himalayas-Sri Lanka.
Reproductive Cycle	:	April-June.
Parts Used	:	Whole Plant and Seed.

Folk Uses : Plant used for curing fever and eye- sore in children. Oil from seeds applied for in-born defective limbs in children.

References : Pal *et al.* (2000), Panda and Das (2000).

Albizia amara **Boiv.**

Syn.: *Mimosa amara* Roxb.

Family: Fabaceae

Eng., H. and Sans. Names : *Eng.*: Bitter Albizia, Oil Cake Tree, Silk Plant;

H.: Krishnasiris, Lallei;

Sans.: Krishnasirisha.

Distribution : Drier areas of Indian Peninsula, also found in M.P.

Reproductive Cycle : March-October.

Part Used : Leaf.

Folk Uses : Medicinal oil obtained by boiling its tender leaves grounded along with *Tamarindus indica*, *Capparis zeylanica* in coconut oil, applied all over the body of a child to cure cough and fever.

Reference : Murugesan *et al.* (2011).

Alcea rosea **L.**

Syn.: *A. biennis* Winterl.; *Althaea chinensis* Wall.; *A. rosea* (L.) Cav.

Family: Malvaceae

Eng. Names : *Eng.*: Hollyhock, Indian Spring.

Distribution : Commonly cultivated in Indian gardens.

Reproductive Cycle : August-September.

Part Used : Whole Plant.

Folk Use : Plant used for curing fever in children.

Reference : Pal *et al.* (2000).

Alhagi maurorum **Medik.**

Syn.: *A. camelorum* Fisch.; *A. pseudalhagi* (M. Bieb) Fisch.; *Hedysarum pseudalhagi* M. Bieb.

Family: Fabaceae

Eng., H. and Sans. Names : *Eng.*: Arabian Manna Plant, Camel Thorn, Persian Manna Plant;

H.: Javansa, Javasa, Jawasa, Junwasa, Juwasa, Yavasa;

		Sans.: Ananta, Balapatra, Duralambha, Durlabha, Gandhari, Kantaki, Rodanika, Samudranta.
Distribution	:	Widely spread in the Ganges valley and acrid tracts in Punjab, Rajasthan, Gujarat and Maharashtra, ascending to 900m.
Reproductive Cycle	:	April-July.
Part Used	:	Whole Plant.
Folk Uses	:	Whole plant used as a laxative; its decoction beneficial in cough and in smallpox eruptions.
References	:	Chatterjee and Pakrashi (1997), Dey and Bahadur (1973), Khare (2004), Kirtikar and Basu (1984), Watt (1972).

Allium ascalonicum **L.**

Syn.: *A. carneum* Willd.; *A. fissile* Gray.; *Cepa ascalonica* (L.) Garsault.

Family: Liliaceae

Eng. and H. Names	:	*Eng.*: Eschallot, Shallot, Shalot, Spanish Garlic;
		H.: Ek-Kanda-Lasun, Gandana.
Distribution	:	Cultivated in Indian Gardens.
Part Used	:	Whole Plant.
Folk Use	:	Plant used for curing earache.
Reference	:	Pal *et al.* (2000).

Allium cepa **L. (Pl. 1C)***

Syn.: *A. angolense* Baker.; *A. aobanum* Araki.; *Ascalonicum sativum* P. Renault.

Family: Liliaceae

Eng., H. and Sans. Names	:	*Eng.*: Bulb Onion, Garden Onion, Onion, White Onion;
		H.: Palandu, Piyaj;
		Sans.: Basl, Dirghapatra, Mahakanda, Palandu, Rajapriya.
Distribution	:	Cultivated all over India.
Reproductive Cycle	:	Plant grown in cold season and crop harvested in April-May.
Part Used	:	Bulb.
Folk Uses	:	Children suffering from nasal bleeding asked to inhale smell of bulbs of the plant. Massage with medicinal oil obtained by boiling equal amounts of bulbs of *Allium*

cepa, cloves of *Allium sativum*, fresh rhizomes of *Cyperus rotundus* and fresh leaves of *Dactyloctenium aegyptium* and mustard oil good for chest, backbone, forehead, back of pinna, neck etc. and also to get relief in vomiting and diarrhoea.

References : Sen and Behera (2007), Singh and Singh (2003a), Watt (1972).

Allium sativum L.*

Syn.: *A. arenarium* Sadler *ex* Rchb.; *A. pekinense* Prokh.; *Porrum sativum* (L.) Rchb.

Family: Liliaceae

Eng., H. and Sans. Names : *Eng.*: Churl's Treacle, Garlic, Poor Man's Treacle;

H.: Lahsan, Lasan, Rasun;

Sans.: Arishtha, Bhutabhna, Dirghapatraka, Grinjana, Katukanda, Lashuna, Mahakanda, Rahutrishta, Rasona, Vatari, Yavaneshta.

Distribution : Widely cultivated throughout India.

Reproductive Cycle : March-April.

Part Used : Bulb.

Folk Uses : Bulbs good against antirickettsial activity, dysentery and fever. Poultice of bulb applied to chest and spine in infantile convulsions. Macerated mass of garlic kept for a while on sting's (bees and wasp) spot to cures the reactions. Gargles of ½ spoon of 'lassun' powder in hot water check tonsillitis. Garlic oil used in bronchitis.

References : Asolkar *et al.* (1992), Kirtikar and Basu (1984), Pal *et al.* (2000), Singh and Singh (2003a), Sood *et al.* (2012), Watt (1972).

Allium stracheyi Baker.

Family: Liliaceae

H. and Sans. Names : *H.*: Pharna;

Sans.: Jambu.

Distribution : W. Himalayas (Kashmir-Kumaon): 3,300-4,000m.

Reproductive Cycle : May-November.

Part Used : Whole Plant.

Folk Uses : Tincture of plant given for preventing worms and colic condition in children.

Reference : Kapur (1996a).

***Allium tuberosum* Rottl. *ex* Spreng.**

Syn.: *A. argyi* H. Lev.; *A. clarkei* Hook.f.; *A. yezoense* Nakai.; *Nothoscordium sulvia* (Buch.-Ham. *ex* D. Don) Kunth.

Family: Liliaceae

Eng. and H. Names : *Eng.*: Chinese Leeks, Fragrant Onion, Garlic Chives; *H.*: Bungandana.

Distribution : E. India and W. Himalayas, Khasia mountains: 1,700-2,000m.

Reproductive Cycle : July-September.

Part Used : Leaf.

Folk Uses : Poultice of crushed leaves on abdomen and soles of young children substantially lower the body temperature in high fever.

Reference : Sharma *et al.* (2003a).

***Allophylus cobbe* (L.) Raeusch.**

Syn.: *A. serratulatus* Radlk.; *Aporetica gemella* DC.; *Picrodendron arboreum* (Mill.) Planch.

Family: Sapindaceae

Eng. and Sans. Names : *Eng.*: Diverse Trifoliate Sumach, Indian Allophylus; *Sans.*: Triputah.

Distribution : Found in Peninsular India from W. and C. Maharashtra-Kerala and in Karnataka, Tamil Nadu, A.P., Orissa and from W. Bengal to Nagaland: ascending to height of 1,000m.

Reproductive Cycle : Throughout the year.

Part Used : Leaf.

Folk Use : Decoction of leaves given for reducing fever in infants and children.

References : Asolkar *et al.* (1992), Parrotta (2001).

***Alocasia macrorrhizos* (L.) G. Don.**

Syn.: *A. cordifolia* (Bory) Cordem.; *Arum rapiforme* Roxb.; *Calla badian* Blanco.; *Colocasia peregrina* (L.) Raf.

Family: Araceae

Eng. Names : *Eng.*: Elephant Ear Taro, Giant Taro.

Distribution : Throughout the tropical parts of India.

Reproductive Cycle : Throughout the Year.

Part Used : Petiole.

Folk Uses : Dose of 5-15ml preparation comprising an equal volume of crushed juice of *Alocacia* petiole and *Polygonum* leaves mixed with a required amount of honey prescribed orally three-four times daily for 3 days for sore mouth and tongue in children. Juice of petiole used as ear drops for children suffering from otorrhoea, toothache and cough.

References : Kumar (2002), Singh and Singh (2003a).

Aloe officinalis **Forssk.**

Syn.: *Aloe vera* var. *officinalis* Forssk. Baker.

Family: Aloaceae

Eng. and H. Names : *Eng.*: Barbados Aloe;

H.: Ghikawar.

Distribution : Semi-wild condition in Bengal and N.W. regions.

Part Used : Whole Plant.

Folk Uses : Whole plant used as intestinal dewormer in children. Warmed poultice of plant over a betel leaf applied to the belly of a child to act as an aperient. Inspissated juice in combination with gum asafoetida applied as a warm plaster in colic and pneumonia of infants.

Reference : Watt (1972).

Aloe vera **(L.) Burm.f.***

Syn.: *A. barbadensis* Mill.; *A. chinensis* Steud. *ex* Baker.; *A. indica* Royle.; *A. perfoliata* var. *vera* L.; *A. vulgaris* Lam.

Family: Aloaceae

Eng., H. and Sans. Names : *Eng.*: Curacao Aloe, Indian Aloe, Jaffarabad Aloe;

H.: Ghikumar, Ghikumari;

Sans.: Ghartkumari, Kumari.

Distribution : Cultivated or seen in drier parts of India; completely naturalized especially in hot dry valleys of N.W. Himalayas.

Reproductive Cycle : March-April.

Part Used : Leaf.

Folk Use : Leaves good for helminthiasis in children.

Reference : Prajapati *et al.* (2006).

Alpinia galanga (L.) Willd.

Syn.: *Amomum galanga* (L.) Lour.

Family: Zingiberaceae

Eng., H. and Sans. Names : *Eng.*: Blue Ginger, Greater Galangal, Java Galangal, Siamease Ginger;

H.: Kulanjan, Kulinjan, Sugandhbach;

Sans.: Aruna, Dhumala, Koraja, Kulanjan, Sugandhavacha.

Distribution : E. Himalayas, S.W. India and W. Ghats.

Reproductive Cycle : May-November.

Parts Used : Whole Plant and Rhizome.

Folk Uses : Whole plant used for curing fever and skin diseases. Rhizome powder given orally for curing rheumatism and respiratory troubles in children.

References : Anonymous (1948), Jain (1994), Kumar (2002), Pal *et al.* (2000), Singh *et al.* (2003).

Alpinia officinarum Hance.

Family: Zingiberaceae

Eng., H. and Sans. Names : *Eng.*: Alpinia Chinensis of Chemist, The Lesser Galangal;

H.: Shitta-rattai;

Sans.: Rasna.

Distribution : Grassland slopes, wood or thickets; cultivated in N. India.

Reproductive Cycle : April-September (Flowering) and May-November (Fruiting).

Part Used : Whole Plant.

Folk Uses : Paste of powdered plant made with oil or water removes freckles. Plant also used for making infants to talk early.

Reference : Watt (1972).

Alstonia scholaris (L.) R. Br.

Syn.: *Echites pala* Ham. Spreng.; *E. scholaris* L.

Family: Apocynaceae

Eng., H. and Sans. Names : *Eng.*: Devils Tree, Ditta Bark Tree, Scholar's Tree;

H.: Chhatim, Saittan Ka Jat, Saptarna, Sataparna, Satwin, Shaitan;

		Sans.: Ayugmaparna, Devavriksha, Palagaruda, Saptaparna, Vishalatvaka.
Distribution	:	Drier forests of India, in tropical region of the W. Himalayas from Jumna eastwards to Assam and southwards to Sri Lanka, Singapore and Penang.
Reproductive Cycle	:	December-July.
Parts Used	:	Whole Plant. Seed. Bark.
Folk Uses	:	Whole plant used as an anthelmintic and for curing fever in children. Both bark and seeds used for curing pneumonia in children.
References	:	Pal *et al.* (2000), Tewari *et al.* (2001).

Alternanthera sessilis **(L.) R. Br. *ex* DC.**

Syn.: *Achyranthes sessilis* (L.) Besser.; *Alternanthera triandra* Roxb.; *Gomphrena sessilis* L.

Family: Amaranthaceae

Eng., H. and Sans. Names	:	*Eng.*: Sessile Alternanthera, The Light Galangal;
		H.: Gudrisag, Kanchari, Lancheri;
		Sans.: Giajjhra, Matsyaksi, Patturah.
Distribution	:	Moist places throughout India, ascending to 2,100m.
Reproductive Cycle	:	Throughout the year.
Part Used	:	Whole Plant.
Folk Uses	:	Plant used for curing stomach pain, fever, and as an anthelmintic.
Reference	:	Pal *et al.* (2000).

Althaea officinalis **L.**

Syn.: *A. multiflora* Rchb. *ex* Regel.; *Malva althaea* E. H. L. Krause.; *M. maritima* Salisb.

Family: Malvaceae

Eng., H. and Sans. Names	:	*Eng.*: Bread and Cheese, Bullseyes, Guimauve, Marsh Mallow, White Mallow Wymote;
		H.: Gulkhaira, Khaira, Khaira-Kajhor, Khitmigajhar;
		Sans.: Khatmi.
Distribution	:	Naturalized in India, Occurring from Punjab-Kashmir.
Reproductive Cycle	:	Flowering in autumn and summer.
Part Used	:	Whole Plant.

Folk Use	:	Whole plant used as demulcent for children.
References	:	Bhattacharjee (2001), Kirtikar and Basu (1984).

***Alysicarpus vaginalis* L.**

Syn.: *Alysicarpus vaginalis* var. *nummularifolius* (Willd.) Miq.

Family: Fabaceae

Eng. and H. Names	:	*Eng.*: Alyce Clover, Sheathed Alysicarpus; *H.*: Chauli, Sauri.
Distribution	:	N.W. regions, ascending to 1,350m; Himalayas-Malacca and Sri Lanka.
Reproductive Cycle	:	September-October.
Part Used	:	Root.
Folk Use	:	Amulet of roots pulled out in one breath hung around neck to cure hiccups in children.
Reference	:	Hembrom (1996).

Amaranthus spinosus* L.

Syn.: *Galliaria spinosa* (L.) Nieuwl.; *G. spitosa* Nieuwl.

Family: Amaranthaceae

Eng., H. and Sans. Names	:	*Eng.*: Prickly Amaranth, Spiny Pigweed, Thorny Amaranth; *H.*: Cholai, Choli, Janumarak, Kantabhaji, Kantenatia; *Sans.*: Alpamarisha, Bahuvirya, Bhandira, Granthila, Kandera, Sushaka, Tandula, Vira.
Distribution	:	Wastelands and cultivated fields throughout India.
Reproductive Cycle	:	March-October.
Parts Used	:	Whole Plant, Root and Leaf.
Folk Uses	:	Plant used for curing nightblindness. Boiled leaves and roots given to children as a laxative. Root (2 tsp twice daily, for a week) with a pinch of common salt given as vermicide against roundworms. Small children made to take bath in root extract to keep hair clean and for dandruff.
References	:	Gogoi and Borthakur (2001), Kirtikar and Basu (1984), Kshirsagar and Singh (2003, 2007), Mishra (2008), Pal *et al.* (2000), Parrotta (2001), Upadhyay and Singh (2005).

Amaranthus viridis **L.**

Family: Amaranthaceae

Eng., H. and Sans. Names : *Eng.*: Green or Wild Amaranth, Green Pigweed;
H.: Cholayi, Karund;
Sans.: Tanduliya, Vishaghna.

Distribution : Throughout India.

Reproductive Cycle : May-June.

Part Used : Whole Plant.

Folk Use : Extract of the entire plant grounded with seeds of *Opium somnifera* administered orally to promote the growth of retarded children.

Reference : Hosagoudar and Henry (1996c).

Amomum aromaticum **(Roxb.)**

Syn.: *Geocallis fasciculata* Horan.; *Renealmia fasciculata* Rosc.

Family: Zingiberaceae

Eng., H. and Sans. Names : *Eng.*: Bengal Cardamom, Jalpaiguri Cardamom;
H.: Banillachi, Barielachi, Elachi, Morang-elachi;
Sans.: Ela, Upakunchika.

Distribution : Found in entire N.E. region.

Reproductive Cycle : May-September.

Part Used : Fruit.

Folk Use : Fruits used to cure small-pox in Meghalaya by Garo tribes.

References : Asolkar *et al.* (1992), Jain (1991, 1995), Kumar (2002), Kumar *et al.* (1980), Tripathi and Goel (2001).

Amorphophallus sylvaticus **(Roxb.)**

Syn.: *A. zeylanicus* Blume.; *Arum sylvaticum* Kunth.; *Pythonium sylvaticum* (Roxb.) Wight.

Family: Araceae

Eng. Name : *Eng.*: Sylvaticus Voodoo Lily.

Distribution : Deccan Peninsula, from northern Circars to the Konkan and southwards to Sri Lanka.

Reproductive Cycle : September-October.

Part Used : Whole Plant.

Folk Use : Plant used for curing fever in children.

Reference : Pal *et al.* (2000)

Ampelocissus barbata **(Wall.) Planch.**

Syn.: *Vitis barbata* Wall.

Family: Vitaceae

H. Name : *H.*: Jangli Angur.

Distribution : Hills of Assam and Meghalaya up to 1,200m.

Reproductive Cycle : June-October.

Part Used : Whole Plant.

Folk Use : Plant good against eczema in children.

Reference : Pal *et al.* (2000).

Amphineurion marginatum **(Roxb.) D. J. Middleton.**

Syn.: *Aganosma marginata* (Roxb.) G. Don.; *Echites apoxys* Voigt.; *Holarrhena procumbens* Blanco.; *Ichnocarpus velutinus* (A. DC.) Fern. -Vill.

Family: Apocynaceae

Eng. Name : *Eng.*: Khruea Sai Tan.

Distribution : From Silhet and Chittagong to Malacca and Singapore.

Part Used : Whole Plant.

Folk Use : Plant used for curing fever in children.

Reference : Pal *et al.* (2000).

Anacardium occidentale **L.**

Family: Anacardiaceae

Eng., H. and Sans. Names : *Eng.*: Cashew, Cashew Apple, Cashew Nut;
H.: Kaju, Kajubadam;
Sans.: Agnikrita, Kajutaka, Parvati.

Distribution : Cultivated in coastal areas of the country: Maharashtra, Goa, Karnataka, Kerala, Tamil Nadu, A.P., Orissa and W. Bengal.

Reproductive Cycle : March-August.

Part Used : Seed.

Folk Use : Resinous juice contained in the seeds good for loss of memory following recovery from smallpox.

Reference : Parrotta (2001).

Anacolosa crassipes **Kurz.**

Family: Olacaceae

Distribution : Endemic to N.E. India; fairly frequent in tropical wet evergreens in Mara autonomous district, S. Mizoram.

Part Used : Leaf.

Folk Use : Decoction of leaves used for bathing children suffering from measles and skin eruptions.

References : Lalramnghinglova (2003), Lalramnghinglova and Jha (1999).

Anacyclus pyrethrum **DC.**

Syn.: *A. depressus* Maire.; *A. freynii* Porta and Rigo.; *Anthemis pyrethrum* Hayne.

Family: Asteraceae

Eng., H. and Sans. Names : *Eng.*: Pellitory, Spanish Pellitory, The Pellitory of Spain;

H.: Akarkara, Akorkara;

Sans.: Agragrahi, Akara, Akarakarabhah, Karava.

Distribution : Indigenous to N. Africa; grown on an experimental scale at 900m at Katra (Jammu and Kashmir).

Reproductive Cycle : Flowering from Spring-the early Summer.

Part Used : Root.

Folk Uses : Decoction of roots useful in tonsillitis. Roots given to infants in the Deccan and Konkan to enable them to talk.

References : Chatterjee and Pakrashi (1997), Prajapati *et al.* (2006), Watt (1972).

Ananas comosus **(L.) Merr.**

Syn.: *A. sativa* Lindl.; *A. sativus* Schult. f.; *Bromelia comosa* L.; *B. rubra* Schult. and Schult.f.

Family: Bromeliaceae

Eng., H. and Sans. Names : *Eng.*: Ananas, Pineapple;

H.: Ananas, Anannas;

Sans.: Ama, Anannasa, Kautukasanjaka, Paravati.

Distribution : Throughout India.

Reproductive Cycle : January-March.

Part Used : Leaf.

Folk Uses : About two spoonfuls juice of young leaves mixed with molasses given on empty stomach in early morning as a vermifuge for 3 consecutive days. Paste of basal portion of tender leaves of Ananas, mature leaves of *Musa balbisiana*, leaves of *Erythrina variegata* and bark of *Pongamia pinnata* in equal quantities given to expel roundworms in children.

References : Borthakur (1993), Gogoi and Borthakur (2001).

Anaphalis margaritacea (L.) Benth. and Hook.f.

Syn.: *A. cinnamomea* (DC.) C. B. Clarke.

Family: Asteraceae

Eng. Names : *Eng.*: Pearly Everlasting, Western Pearly Everlasting.

Distribution : Temperate Himalayas, Kashmir (1,300-3,000m) to Sikkim and Bhutan (1,700-3,500m), Khasia mountains (1,300-2,000m), Sri Lanka, Central regions (1,700-2,700m).

Reproductive Cycle : August-December.

Part Used : Leaf.

Folk Uses : Slightly warmed roll of leaves kept on the head of children for about 10 minutes at an interval of 1-2 minutes for 3-5 days to cure intestinal disorders (locally termed as 'jug') and to kill ascaris manifestation.

Reference : Arya and Prakash (2000).

Andrographis echioides Nees.

Syn.: *Indoneesiella echioides* L.; *Justicia echioides* L.

Family: Acanthaceae

Eng. Name : *Eng.*: False Waterwillow.

Distribution : Drier districts in the tropical India; also found in Sri Lanka.

Reproductive Cycle : March-December.

Part Used : Whole Plant.

Folk Use : Plant beneficial against fever in children.

Reference : Pal *et al.* (2000).

Andrographis laxiflora (Blume) Lindau.

Syn.: *A. glomerulifera* Bremek.; *A. tenuiflora* T. Anders.; *Justicia laxiflora* Blume.

Family: Acanthaceae

Distribution : Eastern frontier: 170-1,000m, from Assam-Tavoy, frequent, Andamans.

Reproductive Cycle : August-March.

Part Used : Leaf.

Folk Use : Fresh leaves eaten against tonsillitis.

Reference : Lalramnghinglova (1996).

Andrographis paniculata (Burm.f.) Wallich *ex* Nees.

Syn.: *Justicia paniculata* Roxb.

Family: Acanthaceae

Eng., H. and Sans. Names : *Eng.*: Chiretta, King of Bitter, The Creat;

H.: Chirayata, Kalmegh, Kirata, Kiravat, Kirayat, Kiryato, Kulujnath;

Sans.: Bhunimba, Bhunimbah, Kirata, Kiratatikah.

Distribution : Throughout India, from H.P.-Assam and Mizoram and all over S. India.

Reproductive Cycle : August-February.

Parts Used : Whole Plant, Shoot, Leaf and Root.

Folk Uses : Whole plant used for loss of appetite, flatulence, diarrhoea, fever and as an anthelmintic. Decoction of shoots (10g) given twice daily for 4-5 days in acute jaundice due to hepatitis. Young leaves used as a liver tonic. Paste of leaves given to arrest sneezing. Leaves also used in infantile jaundice, kalazar, and general debility of children. Pills made out of juice of leaves in combination with cardamom, clove and cinnamom found beneficial for digestive disorders in children. One tsp of root with rhizome of *Zingiber zerumbentin* in equal parts by weight given thrice daily for 15 days in whooping cough. 3g powdered roots given with water to children once a day as anthelmintic among the Kondhs in S.W. Orissa.

References : Anonymous (1948), Asolkar *et al.* (1992), Banerjee and Ghora (1996), Borthakur (1993), Borthakur *et al.* (2004), Chatterjee and Pakrashi (1997), Dey and Bahadur

(1973), Girach (1992), Gogoi *et al.* (2003), Jain (1994), Khare (2004), Pal (2000), Pal *et al.* (2000), Parrotta (2001), Patil (2007), Rawat and Chowdhury (1998), Satapathy and Brahmam (2000), Watt (1972).

Aneilema scapiflorum Wight.

Syn.: *A. tuberosum* Buch.; *Commelina scapiflora* Roxb.; *C. edulis* Stokes.; *Murdannia tuberosa* Royle.

Family: Commelinaceae

Eng. and H. Names : *Eng.*: Black Musala;
H.: Siyahmusli.

Distribution : Temperate and tropical Himalayas, Gangetic plains, W. Peninsula.

Reproductive Cycle : September-November.

Part Used : Root Bark.

Folk Use : Root bark used in infantile convulsions.

References : Agarwal (2003), Kirtikar and Basu (1984), Watt (1972).

Anemone pulsatilla L.

Syn.: *A. acutipetala* Schleich.; *A. collina* Salisb.; *Pulsatilla vulgaris* Mill.

Family: Ranunculaceae

Eng. Names : *Eng.*: European Pasque Flower, Pasque Flower, Wind Flower.

Distribution : Introduced in W. Himalayas.

Reproductive Cycle : March-May.

Part Used : Whole Plant.

Folk Use : Plant used in measles.

Reference : Bhattacharjee (2001).

Anemone vitifolia Buch. Ham. *ex* DC.

Syn.: *A. elegans* Decne.; *Eriocapitella elegans* (Decne.) Nakai.

Family: Ranunculaceae

Eng. and H. Names : *Eng.*: Grape Leaf Anemone, Vine-leaved Anemone;
H.: Agali.

Distribution : Throughout the temperate Himalayas from Kashmir-Arunachal Pradesh, ascending to 2,700m.

Reproductive Cycle : Flowering during August-September.

Part Used	:	Leaf.
Folk Use	:	A band of crushed leaves tightened on the hands of children (age 5-10 years) for 2 hours to create a permanent burning spot on the outer surface of skin and considered a vaccination by local tribes against diseases like chickenpox and smallpox.
Reference	:	Arya *et al.* (1999).

Anethum graveolens L. (Pl. 1D)

Syn.: *A. graveolens* L. var. *sowa* Roxb.; *A. sowa* Roxb. *ex* Flem.; *Peucedanum graveolens* L.; *P. sowa* Roxb.

Family: Apiaceae

Eng., H. and Sans. Names	:	*Eng.*: Anet, Dilly, Indian Dill, Sowa; *H.*: Soa, Soi, Sowa, Soya, Sutopsha; *Sans.*: Bahala, Chhatr, Ghosha, Karavi, Madhavi, Misi, Satapushpi, Shalini, Vajana.
Distribution	:	Throughout tropical and sub-tropical India.
Reproductive Cycle	:	January-March.
Parts Used	:	Whole Plant, Fruit and Seed.
Folk Uses	:	Plant increases milk production and prevent colic in their babies, when taken regularly by nursing mothers. Fruits beneficial in flatulent pains in infants. Decoction of seeds used as a stomachic medicine for infants. Oil from seed used as an important ingredient of gripe water for infants for curing flatulence in children.
References	:	Agarwal (2003), Ambasta (1986), Anonymous (1948), Chatterjee and Pakrashi (1997), Dey and Bahadur (1973), Farooq (2005), Jain (1991), Karnick (1994a), Thakur *et al.* (1989).

Angelica glauca Edgew.

Family: Apiaceae

Eng., H. and Sans. Names	:	*Eng.*: Angelica, Long Wort; *H.*: Canda, Chora, Choru; *Sans.*: Khatmi.
Distribution	:	Kashmir-Shimla, 2,700-3,500m.
Reproductive Cycle	:	July-August.
Parts Used	:	Whole Plant and Root.

Folk Uses	:	Plant used as an intellect promoting tonic for children. Roots administered along with warm water for checking stomach ailments and vomiting in children.
References	:	Asolkar *et al.* (1992), Khare (2004), Maikhuri *et al.* (2001).

Anisochilus carnosus **(L.f.) Wall.**

Syn.: *A. crassus* Benth.; *Lavandula carnosa* L.f.; *Plectranthus strobiliferus* Roxb.

Family: Lamiaceae

Eng., H. and Sans. Names	:	*Eng.*: Thick- leaved Lavender; *H.*: Panjirikapat, Sitakipanjiri; *Sans.*: Ajapada, Ulapalabhenda.
Distribution	:	W. Himalayas, Bengal, C. India, N. Circars, Deccan, Carnatic, Sri Lanka, Java.
Reproductive Cycle	:	October-November.
Parts Used	:	Whole Plant, Stem and Leaf.
Folk Uses	:	Stem along with whole plant used to cure cough. Juice of leaves mixed with sugar and human milk considered a popular remedy in Mysore for cough in children.
References	:	Kirtikar and Basu (1984), Watt (1972).

Anisomeles indica **O. Kuntze.**

Syn.: *A. ovata* R. Br.; *Ballota distichia* L.; *Nepata indica* L.

Family: Lamiaceae

Eng., H. and Sans. Names	:	*Eng.*: Indian Catmint, Malabar Catmint; *H.*: Gobara, Gopali, Kala Bhangra; *Sans.*: Kala Bhangra.
Distribution	:	Throughout India, up to 1,800m.
Reproductive Cycle	:	June-October.
Parts Used	:	Leaf and Flower.
Folk Uses	:	1-2 tsp extract of tender leaves and flowers applied externally on head and tongue of a child to cure pneumonia. Powdered leaves (7-10g) given along with cow's milk twice a day before sunrise and after sunset for a period of 2-3 weeks for treating rickets.
References	:	Jadhav (2009), Maliya (2007).

***Anisomeles malabarica* R. Br. *ex* Sims.**

Syn.: *Ajuga fructicosa* Roxb.; *A. intermedia* Wight.; *Nepeta malabarica* L.

Family: Lamiaceae

Eng., H. and Sans. Names : *Eng.*: Malabar Mint;

H.: Codhara;

Sans.: Mahadronah, Oshthaphala, Vaikunthah.

Distribution : Throughout S. India and Sri Lanka.

Reproductive Cycle : Flowering in rainy season.

Part Used : Leaf.

Folk Uses : Infusion of leaves beneficial in fever, colic, and dyspepsia arising during teething in children. One tsp of juice extracted from 10g of fresh leaves with water given to children three times a day for a period of two days to arrest vomiting.

References : Agarwal (2003), Ambasta (1986), Anonymous (1948), Kirtikar and Basu (1984), Prajapati *et al.* (2006), Subramanian (2000), Subramanian *et al.* (2003).

***Annona squamosa* L.**

Syn.: *A. asiatica* L.; *A. asiatica* Vahl.; *Xylopia glabra* L.

Family: Annonaceae

Eng., H. and Sans. Names : *Eng.*: Custard Apple, Sugar Apple, Sweet Apple;

H.: At, Atasitaphal, Sharifah, Shariphal, Sitaphal;

Sans.: Agrimakhya, Atripya, Krishnabija, Sitaphalam, Subha, Suda.

Distribution : Throughout India, naturalized especially in W. Peninsula.

Reproductive Cycle : September-January.

Part Used : Leaf.

Folk Use : Infusion of leaves considered efficacious in prolapsus ani.

References : Prajapati *et al.* (2006), Watt (1972).

***Anthemis odontostephana* Boiss.**

Syn.: *A. gayana* Boiss.; *A. scaposa* J. Gay.

Family: Asteraceae

Distribution : Punjab at Peshawar and Afghanistan.

Part Used	:	Whole Plant.
Folk Use	:	Plant used to cure fever in children.
Reference	:	Pal *et al.* (2000).

Antiaris toxicaria **Lesch.**

Syn.: *A. dubia* Span.; *A. rufa* Miq.; *Ficus challa* Schweinf.

Family: Moraceae

Eng., H. and Sans. Names	:	*Eng.*: Antiaris, Bark cloth tree, False Iroko, False Mvule Sacking Tree, Upas Tree;
		H.: Jangli Lakuch, Jasund;
		Sans.: Upas, Valkala.
Distribution	:	Deccan Peninsula, on the Ghats from Concan southwards, virgin tropical forests of India.
Reproductive Cycle	:	March-June.
Part Used	:	Whole Plant.
Folk Use	:	Plant used to cure fever in children.
Reference	:	Pal *et al.* (2000).

Antidesma acidum **Retz.**

Syn.: *A. diandrum* (Roxb.) Roth.; *Stilago diandra* Roxb.; *S. lanceolaria* Roxb.

Family: Euphorbiaceae

Eng. and H. Names	:	*Eng.*: Black Currant, Chinese Laurel;
		H.: Aburok, Amrul, Khetua, Matta-ara.
Distribution	:	Tropical Himalayas from Garhwal eastwards and southwards to Travancore and Tenasserim, ascending to 500m.
Reproductive Cycle	:	May-November.
Part Used	:	Whole Plant.
Folk Uses	:	Plant good for curing diarrhoea and dysentery.
References	:	Hembrom (1996), Pal *et al.* (2000).

Apocynum cannabinum **L.**

Syn.: *A. hypercifolium* Ait.; *A. pubescens* Mitch. *ex* R.Br.; *A. sibiricum* Jacq.

Family: Apocynaceae

Eng. Names	:	*Eng.*: Amy Root, Black Indian Hemp, Dogbane, Hemp Dogbane.

Distribution	:	Open wooded areas, ditches and hillsides and moist places in temperate regions.
Reproductive Cycle	:	June-November.
Part Used	:	Root.
Folk Use	:	Roots used in snuffles in children.
Reference	:	Prajapati *et al.* (2006).

Aporosa cardiosperma **(Gaertn.) Merr.**

Syn.: *A. lindleyana* (Wight) Baill.; *A. sphaerocarpa* Mull. Arg.; *Croton cardiospermum* Gaertn.; *Scepa lindleyana* Wight.

Family: Euphorbiaceae

Eng. Name	:	*Eng.*: Lindley's Aporosa.
Distribution	:	W. Peninsula, throughout W. Ghats.
Part Used	:	Whole Plant.
Folk Use	:	Plant used to cure fever in children.
Reference	:	Pal *et al.* (2000).

Aquilaria malaccensis **Lam.**

Syn.: *A. angallocha* Roxb.; *A. ovata* Cav.

Family: Thymelaeaceae

Eng., H. and Sans. Names	:	*Eng.*: Agallochum, Agar Wood, Aloe Wood, Eagle Wood; *H.*: Agar; *Sans.*: Agaru, Agurah, Asara, Krsnaguruh.
Distribution	:	Forests of Assam, Bhutan and Myanmar.
Reproductive Cycle	:	May-August.
Part Used	:	Wood.
Folk Use	:	External application of essential oil from wood useful in vomiting in children.
Reference	:	Prajapati *et al.* (2006).

Arachis hypogaea **L.**

Syn.: *A. nambyquarae* Hoehne.

Family: Fabaceae

Eng., H. and Sans. Names	:	*Eng.*: Chinese Almond, Groundnut, Monkey nut, Peanut; *H.*: Mung-phali, Sheng, Sheng-Dana, Vilayeti Mung; *Sans.*: Bhucanakah, Mandapi, Tribija.

Distribution : Throughout India.

Reproductive Cycle : August-October.

Part Used : Nut.

Folk Use : Peanut oil used for infantile paralysis.

Reference : Ambasta (1986).

Archidendron clypearia **(Jack) I. C. Nielsen.**

Syn.: *Pithecellobium angulatum* (Benth) Kosterm.; *P. cypearia* (Jack) Benth.

Family: Fabaceae

Eng. Names : *Eng.*: Grasshopper Tree, Greater Grasshopper Tree, Monkey-Pod.

Distribution : E. Himalayas, Sikkim, Khasi, Mishmi and Lushai hills of Assam, up to 1,200m.

Reproductive Cycle : December-March.

Part Used : Leaf.

Folk Uses : Leaf paste applied on chickenpox and smallpox.

References : Ambasta (1986), Anonymous (1969).

Ardisia solanacea **Roxb.**

Syn.: *A. humilis* Vahl.; *A. litoralis* Adr.; *A. polycephula* Wight.; *A. umbellata* Roth.

Family: Myrsinaceae

Eng. and H. Names : *Eng.*: Duck's eye, Ellliptical-Leaf Ardisia, Shoe Button Ardesia;

H.: Banjan, Bisi.

Distribution : Throughout India, alt. 0-1,700m; from Himalayas-Sri Lanka and Singapore.

Reproductive Cycle : April-June.

Part Used : Leaf.

Folk Use : Leaf paste mixed with coconut oil used externally for curing mumps.

Reference : Sinha *et al.* (1996).

Areca catechu **L.***

Syn.: *A. cathechu* Burm.f.; *A. faufel* Gaertn.; *A. hortensis* Lour.

Family: Arecaceae

Eng., H. and Sans. Names : *Eng.*: Aerica Nut, Areca Palm, Betal Nut, Betel Nut Palm, Pinang Palm;

		H.: Kasail, Supari, Supyari; *Sans.*: Gubak, Guvaka, Kramuka, Phalum, Poag, Pooga, Poogi.
Distribution	:	Cultivated in coastal regions of India; Bengal, Assam.
Reproductive Cycle	:	Flowering during August-January and Fruiting occurs about a year later.
Part Used	:	Whole Plant.
Folk Use	:	Plant used in smallpox.
References	:	Asolkar *et al.* (1992), Jain (1991).

Argemone mexicana **L. (Pl. 2A)**

Syn.: *A. spinosa* Moench.; *A. vulgaris* Spach.; *Papaver mexicanum* (L.) E.H.L. Krause.

Family: Papaveraceae

Eng., H. and Sans. Names	:	*Eng.*: Mexican Poppy, Mexican Prickly Poppy, Prickly Poppy, Yellow Thistle; *H.*: Bharband, Bhatbhamt, Brahmaduni, Dhattura, Kandiari, Kateli, Pharamgi, Satyanasi; *Sans.*: Brahmadani, Hemashikha, Kshirini, Srigalakanta, Svarnaksiri.
Distribution	:	Throughout India, specially from Bengal-Punjab and Shimla.
Reproductive Cycle	:	January-May.
Parts Used	:	Whole Plant, Fruit and Latex.
Folk Uses	:	Plant used against skin diseases, scabies, ringworm, and eye complaints. Juice of fruits prescribed in tetanus. One or two drops of latex applied on tongue, once everyday for 3-4 months in lisping.
References	:	Banerjee (2000a), Borthakur (1993), Pal *et al.* (2000).

Argyreia nervosa **(Burm.f.) Boj.**

Syn.: *A. speciosa* Sweet.; *Convolvulus speciosus* L.; *Lettsomia nervosa* Roxb.

Family: Convolvulaceae

Eng., H. and Sans. Names	:	*Eng.*: The Elephant Creeper, Woolly Morning Glory; *H.*: Samandar Ka Pat, Samandarphaind, Samandarsof, Samandarsokh; *Sans.*: Bastantri, Samudrapalaka, Vrddhadarukah.

Distribution	:	Throughout India, ascending to 300m from Assam-Belgaum and Mysore; frequent in Bengal and W. India.
Reproductive Cycle	:	August-November.
Part Used	:	Whole Plant.
Folk Uses	:	Plant used for curing fever and diarrhoea in children. Juice of the plant mixed with an equal quantity of gingelly oil and a little powdered 'dill' (*Anethum graveolens*) seed used for external application in smallpox, scabies and other skin diseases of children.
References	:	Asolkar *et al.* (1992), Jain (1991), Pal *et al.* (2000), Watt (1972).

***Aristolochia bracteolata* Lamk.**

Syn.: *A. abyssinica* Klotzsch.; *A. bracteata* Retz.; *A. crenata* Ehrenb. *ex* Duch.

Family: Aristolochiaceae

Eng., H. and Sans. Names	:	*Eng.*: Bracteated Birthwort, Worm Killer;
		H.: Kidamari, Kiramar, Gandan, Gandati;
		Sans.: Bhringi, Dhumapatra, Gridhrani, Gridhrapatra, Kitamari, Kitari, Krimighni, Sulabha.
Distribution	:	Bengal, Upper Gangetic Plains, W. Peninsula; Haryana, W. Bengal and southwards to Tamil Nadu and Kerela.
Reproductive Cycle	:	August-September.
Part Used	:	Leaf.
Folk Uses	:	Bruised leaves mixed with castor oil applied in navel region to move the bowels. Juice of leaves mixed with mother's milk given orally to increase appetite in children.
References	:	Anonymous (1948), Chatterjee and Pakrashi (1997), Kirtikar and Basu (1984), Kumar and Pullaiah (2000), Watt (1972).

***Aristolochia fontanesii* Boiss. and Reut.**

Syn.: *A. longa* L.

Family: Aristolochiaceae

Eng., H. and Sans. Names	:	*Eng.*: Long-Rooted Birthwort;
		H.: Isarmul, Isvarmul;
		Sans.: Ahiganda, Arkamula, Rudrajata, Sunanda.

Distribution	:	Algeria, Morocco, Libya, France, Germany, Italy, Egypt.
Part Used	:	Root.
Folk Uses	:	Roots used against indigestion and bowel complaints.
Reference	:	Watt (1972).

Aristolochia indica **L.**

Syn.: *A. lanceolata* Wight.; *A. pandurata* Wall. *ex* Duch.

Family: Aristolochiaceae

Eng., H. and Sans. Names	:	*Eng.*: Indian Birthwort; *H.*: Isharmul, Ishermul; *Sans.*: Ahigandha, Arkamula, Ishvara, Ishvari Mul, Jata, Nakuli, Sunanda.
Distribution	:	Found throughout low hills and plains of India.
Reproductive Cycle	:	September-March.
Parts Used	:	Whole Plant, Root, Leaf, Seed and Bark.
Folk Uses	:	Plant given to cure fever. Roots used for minor ailments such as flatulence and dyspepsia. Juice of fresh leaves useful in treating cough by inducing vomiting without depression. Fresh juice of leaves and of bark used in bowel complaints. Seeds beneficial in dyspnoea of children.
References	:	Anonymous (1948), Asolkar *et al.* (1992), Chatterjee and Pakrashi (1997), Khare (2004), Kirtikar and Basu (1984), Pal *et al.* (2000), Parrotta (2001).

Aristolochia saccata **Wall.**

Syn.: *A. cathcartii* Hook.f.; *Siphisia saccata* (Wall.) Klotzsch.

Family: Aristolochiaceae

Eng. Name	:	*Eng.*: Birthwort.
Distribution	:	Sikkim Himalayas and Khasia mountains, ascending to 1,000m.
Reproductive Cycle	:	November-January.
Part Used	:	Root.
Folk Use	:	Roots used for smallpox by Garos in Meghalaya.
References	:	Asolkar *et al.* (1992), Jain (1991).

***Artemisia abrotanum* L.**

Syn.: *A. altissima* Ehrh. *ex* DC.; *A. tenuissima* Spreng. *ex* Besser.

Family: Asteraceae

Eng. Names	:	*Eng.*: Boy's Love, Garderobe, Lover's Plant, Old Man, Oldman Wormwood, Sitherwood, Southernwood.
Distribution	:	Cultivated in Indian gardens.
Reproductive Cycle	:	July-September.
Part Used	:	Aerial Parts.
Folk Use	:	Infusion of aerial parts given to children as a dewormer.
Reference	:	Prajapati *et al.* (2006).

***Artemisia absinthium* L.**

Syn.: *Absinthium officinale* Brot.; *A. vulgare* (L.) Lam.

Family: Asteraceae

Eng., H. and Sans. Names	:	*Eng.*: Absinthi, Ajenjo, Green Ginger, Maddar Wood, Madderwort, Worm Wood;
		H.: Afsantin, Vilayatiafranthin;
		Sans.: Amar, Aphasantin, Damara, Indhana.
Distribution	:	Kashmir, ascending to 2,350m.
Reproductive Cycle	:	May-June.
Parts Used	:	Whole Plant. Leaf.
Folk Uses	:	Plant used for curing fever. Decoction of leaves given to children for expelling worms from abdomen.
References	:	Beigh *et al.* (2003), Kapur *et al.* (1996), Pal *et al.* (1992).

***Artemisia capillaris* Thunb.**

Syn.: *Oligosporus capillaris* (Thunb.) Poljakov.

Family: Asteraceae

Eng. and H. Names	:	*Eng.*: Chinese Worm Wood;
		H.: Sita bani.
Distribution	:	Throughout India, ascending to 3,657m in Himalayas.
Reproductive Cycle	:	July-September.
Part Used	:	Whole Plant.
Folk Use	:	Plant good against earache in children.
Reference	:	Pal *et al.* (2000).

Artemisia dracunculus **L.**

Syn.: *Achillea dracunculus* Hort. *ex* Steud.; *Artemisia changaica* Krasch.

Family: Asteraceae

Eng. Names : *Eng.*: Dragon Mugwort, Estragon, Tarragon.

Distribution : W. Himalayas: 4,200-4,800m.

Reproductive Cycle : May-October.

Part Used : Whole Plant.

Folk Use : Plant good for curing fever in children.

Reference : Pal *et al.* (2000).

Artemisia indica **Willd.**

Syn.: *A. nilagirica* (Clarke) Pamp.; *A. vulgaris* auct. non L.

Family: Asteraceae

Eng., H. and Sans. Names : *Eng.*: Fleabane, Indian Worm Wood, Motherwort, Mugwort;

H.: Dauna, Davana, Dona, Gathivana, Majtari, Mastaru, Nagdana, Nugduna;

Sans.: Barha, Barhipushpa, Damanah, Damanakah, Granthika, Kukura, Nagadamani, Shuka.

Distribution : Throughout the mountainous regions of India, ascending to 3,048m in W. Himalayas; up to 2,438m in Sikkim and Khasia Hills and in Mount Abu and W. Ghats.

Reproductive Cycle : August-October.

Parts Used : Whole Plant, Flowering Tops and Leaf.

Folk Uses : Plant juice applied to the head of young children for prevention of convulsions. Plant also used against measles in Sikkim. Leaves and flowering tops used against measles. Leaves beneficial in convulsive diseases of children.

References : Asolkar *et al.* (1992), Chatterjee and Pakrashi (1997), Jain and Saklani (1992), Kirtikar and Basu (1984), Kumar (2002), Kumar and Mathur (2000), Prajapati *et al.* (2006), Srivastava *et al.* (1987), Watt (1972).

Artemisia japonica **Thunb.**

Syn.: *Oligosporus japonicus* (Thunb.) Poljakov.

Family: Asteraceae

Eng. and H. Names : *Eng.*: Japanese Wormwood;

H.: Pati, Ptee.

Distribution	:	Found in temperate Himalayas.
Reproductive Cycle	:	September-December.
Part Used	:	Whole Plant.
Folk Use	:	Plant good for convulsions in children.
References	:	Jain (1991), Sood *et al.* (2012).

***Artemisia maritima* L.**

Syn.: *A. pseudogallica* (Rouy) A.W. Hill.; *A. salina* Willd.

Family: Asteraceae

Eng., H. and Sans. Names	:	*Eng.*: Sea Worm Seed, Worm Seed; *H.*: Ajavayana, Chhuari; *Sans.*: Chauhara, Gandha, Parasi, Yavania.
Distribution	:	W. Himalayas, (Kashmir – Kumaon): 2,350-3,500m, coasts of Europe and saline tracts of N. Asia.
Reproductive Cycle	:	May-August.
Parts Used	:	Whole Plant and Flower Heads.
Folk Uses	:	Plant used for curing fever. Flower heads used for children suffering from worms with great irritability.
References	:	Pal *et al.* (2000), Prajapati *et al.* (2006).

***Artemisia persica* Boiss.**

Syn.: *A. subspinescens* Boiss.

Family: Asteraceae

H. Name	:	*H.*: Davana.
Distribution	:	Himalayas, Karakorum, Hindukush: 3,000-4,650m.
Reproductive Cycle	:	July-October.
Part Used	:	Whole Plant.
Folk Use	:	Plant used for curing fever in children.
Reference	:	Pal *et al.* (2000).

***Artemisia sievarsiana* Willd.**

Family: Asteraceae

Eng., H. and Sans. Names	:	*Eng.*: Sage Brush; *H.*: Dauna; *Sans.*: Agnidamanaka, Damana, Damani, Dandi, Danta, Gandhotkata, Jatila, Patri, Pavitraka, Tapasapatri.

Distribution	:	W. Himalayas; Kashmir-Lahaul (2,700-3,500m).
Reproductive Cycle	:	July-September.
Part Used	:	Whole Plant.
Folk Use	:	Plant good for curing fever in children.
Reference	:	Pal *et al.* (2000).

Artocarpus heterophyllus **Lam.**

Syn.: *A. integra* auct.; *A. integrifolia* auct. non L.; *Polyphema jaca* Lour.; *Sitodium cauliforum* Gaertn.

Family: Moraceae

Eng., H. and Sans. Names	:	*Eng.*: Jaca Tree, Jack Fruit, Jack Wood, Orange Wood; *H.*: Cakki, Chakki, Kanthal, Kathal, Kathar; *Sans.*: Ashaya, Panasa, Panasah, Pansa, Phanasa.
Distribution	:	Cultivated throughout the hotter parts of India, mainly in Kerela, Bihar and S. India. Assam being the largest producer of Jack fruit.
Reproductive Cycle	:	Flowering during winter and Fruiting in rainy season.
Parts Used	:	Whole Plant and Leaf.
Folk Uses	:	Plant used for curing smallpox. Leaves used in fast breathing of children.
References	:	Asolkar *et al.* (1992), Jain (1991), Khanna (2002b).

Artocarpus lakoocha **Roxb.**

Syn.: *A. ficifolius* W.T. Wang.

Family: Moraceae

Eng., H. and Sans. Names	:	*Eng.*: Monkey Jack; *H.*: Badahara, Barhal, Dahu, Dhao, Dhava, Dhou, Lakuch; *Sans.*: Airawata, Amlaka, Dahu, Karshya, Kashayi, Likucha, Nikucha, Shalashura.
Distribution	:	Sub-Himalayan tracts and outer hills, Khasia, Myanmar, Malay, W. Peninsula, Sri Lanka.
Reproductive Cycle	:	February-September.
Part Used	:	Seed.
Folk Use	:	Seeds valued as good purgative for children.
Reference	:	Kirtikar and Basu (1984).

***Asclepias curassavica* L.**

Family: Asclepiadaceae

Eng., H. and Sans. Names	:	*Eng.*: Blood Flower, Curssavian Shallow-Wort, False Ipecae, Milk Weed, Silk Weed, West Indian Ipecacuanha; *H.*: Kakatundi, Kauri dodi; *Sans.*: Bharadwaji, Kaktundi.
Distribution	:	Naturalized throughout Himalayas, ascending to 1,500m.
Reproductive Cycle	:	May-July.
Part Used	:	Whole Plant.
Folk Use	:	Plant used in smallpox.
Reference	:	Pal *et al.* (2000).

***Asclepias speciosa* Torr.**

Syn.: *A. giffordii* Eastw.

Family: Asclepiadaceae

Eng. Names	:	*Eng.*: Greek Milkweed, Showy Milkweed.
Distribution	:	Native of N. America.
Reproductive Cycle	:	May-October.
Part Used	:	Root.
Folk Use	:	A tea made by boiling its roots useful for treating measles.
Reference	:	Bhattacharjee (2001).

***Asparagus adscendens* Roxb.**

Syn.: *A. adscendens* (Roxb.) Kunth.; *A. satawur* James A. Murray.; *A. volubilis* Wall.

Family: Asparagaceae

Eng., H. and Sans. Names	:	*Eng.*: Asparagus; *H.*: Hazarmuli, Safedmusli, Satavar, Satawar; *Sans.*: Safed Musli, Shatawari.
Distribution	:	Throughout tropical and sub-tropical parts of India.
Reproductive Cycle	:	September-December.
Part Used	:	Whole Plant.
Folk Use	:	Plant considered beneficial in poliomyelitis.
Reference	:	Pal *et al.* (2000).

Asparagus filicinus **Buch.-Ham. *ex* D.Don. Prodr.**

Syn.: *A. qinghaiensis* Y. Wan.; *Protasparagus filicinus* (Buch.-Ham. *ex* D.Don) Kamble.

Family: Asparagaceae

Eng., H. and Sans. Names	:	*Eng.*: Fern Asparagus; *H.*: Chiriya-Kanda, Sahasimuli; *Sans.*: Satavari.
Distribution	:	Temperate and tropical Himalayas from Kashmir-Bhutan, Khasia hills, Assam, Myanmar, China.
Reproductive Cycle	:	Flowering during May-July.
Part Used	:	Twig.
Folk Use	:	Twig kept in the hands of patients as a curative measure against smallpox in Kanawar.
Reference	:	Kirtikar and Basu (1984).

Asparagus gonoclados **Baker**

Syn.: *Protasparagus gonoclados* (Baker) Kamble.

Family: Asparagaceae

H. and Sans. Names	:	*H.*: Shakakul, Shakaloo; *Sans.*: Abheerupatri.
Distribution	:	Peninsular India from Konkan southwards in the Ghats between 1,200-1,800m and in the plains; also found in Bihar and Gujarat.
Reproductive Cycle	:	June-September.
Part Used	:	Root.
Folk Use	:	An amulet of pieces of root hung around neck to cure fever in children.
Reference	:	Hembrom (1996).

Asparagus racemosus **Willd. (Pl. 2B)**

Syn.: *Asparagopsis abyssinica* Kunth.; *A. acerosa* Kunth.; *A. floribunda* Kunth.; *A. javanica* Kunth.; *Asparagus acerosus* Roxb.

Family: Asparagaceae

Eng., H. and Sans. Names	:	*Eng.*: Indian Asparagus, Wild Asparagus; *H.*: Bojhidan, Kurilo, Sataavar, Satavare, Satawar, Satmuli, Shakakul; *Sans.*: Abhiru, Aheru, Atmagupta, Bahusutaa, Indivari, Satavari.

Distribution : Common at low altitudes in shade and tropical climate throughout India.

Reproductive Cycle : August-December.

Parts Used : Whole Plant, Stem, Leaf, Tuber and Root.

Folk Uses : Plant used for curing dysentery and poliomyelitis. Stem paste rubbed on body to cure rickets by Jenukurubas. Leaves used against smallpox. Pounded tuber applied in umbilicus infection of new born babies. Orange coloured decoction of mixture of its tuberous roots with small pieces of stem of *Byttneria pilosa* given to new-born babies in colic pain, vomiting and stomach troubles. Pieces of roots kept in the bed of children up to 1 year of age for inducing sleep and preventing confluence of smallpox.

References : Ahmed and Borthakur (2005), Farooq (2005), Hembrom (1996), Kshirsagar and Singh (2007), Pal *et al.* (2000), Watt (1972).

***Asparagus sarmentosus* L.**

Syn.: *Asparagopsis sarmentosa* (L.) Kunth.; *Protasparagus sarmentosus* (L.) Kamble.

Family: Asparagaceae

Eng., H. and Sans. Names : *Eng.*: Hundred Husbands;

H.: Satavari, Shakakul, Shaqaqul;

Sans.: Sahasravirya, Satavari, Wari.

Distribution : A climber found in upper India; common in Konkans and the Deccan.

Part Used : Root.

Folk Uses : Root juice boiled with ghee given to children in debility and emaciation as a cooling and nourishing medicine.

Reference : Watt (1972).

***Asplenium ruta-muraria* (L.)**

Syn.: *A. cryptolepis* Fernald; *Phyllitis ruta-muraria* (L.) Moench.

Family: Aspleniaceae

Eng. Names : Tentwort, Wall Rue, White Maiden Hair.

Distribution : Kashmir, N.E. Himalaya, C. India, Rajasthan, Sikkim, S. India.

Reproductive Cycle : June-October.

Part Used : Whole Plant.

Folk Use : Plant considered beneficial in curing rickets.

References : Ambasta (1986), Das (2003), Kirtikar and Basu (1984), Singh and Viswanathan (1996).

Asystasia gangetica **(L.) T. Anders.**

Syn.: *A. coromandeliana* Nees.; *A. violacea* Dalz.; *Justicia gangetica* L.

Family: Acanthaceae

Eng., H. and Sans. Names : *Eng.*: Chinese Violet, Creeping Foxglove, Ganges Primerose;

H.: Lavanavalli;

Sans.: Lavana-valli.

Distribution : Peninsular India, Orissa and W. Bengal; cultivated as an ornamental plant.

Reproductive Cycle : November-April.

Part Used : Whole Plant.

Folk Uses : Juice of plant administered to children suffering from swelling and rheumatism.

References : Ambasta (1986), Anonymous (1948).

Athyrium pectinatum **(Wall. *ex* Mett.) T. Moore.**

Syn.: *Asplenium pectinatum* Wall. *ex* Mett.; *Athyrium pectinatum* C. Presl.

Family: Woodsiaceae

Distribution : Mount Abu of Rajasthan.

Part Used : Whole Plant.

Folk Use : Generally given by tribals as a strong vermifuge to their children.

References : Bhattacharjee (2001).

Atropa belladonna **C.B. Clarke.**

Syn.: *A. acuminata* Royle.

Family: Solanaceae

Eng., H. and Sans. Names : *Eng.*: Belladona, Deadly Nightshade;

H.: Angurshefa, Luckmuna, Sag-angur;

Sans.: Suchi.

Distribution : Shimla-Kashmir: 2,000-4,000m; found wild in Kanawar at 2,800m.

Reproductive Cycle	:	Flowering during autumn and Fruiting in winters.
Parts Used	:	Root and Leaf.
Folk Use	:	Regarded good for tetanus.
Reference	:	Watt (1972).

Atylosia scarabaeoides **Benth.**

Syn.: *Cajanus scarabaeoides* (L.) Thouars.; *Centharospermum pauciflorum* Wight. and Arn.; *Dolichos scarabaeoides* L.

Family: Fabaceae

Eng., H. and Sans. Names	:	*Eng.*: Wild Kuthi;
		H.: Jangli Tur, Kulthi, Wakollu;
		Sans.: Masaparni, Vana-Kulathha.
Distribution	:	Throughout India, up to 1800m on the W. Himalayas; Malaya, China, Mauritius and Madagascar.
Reproductive Cycle	:	May-September.
Part Used	:	Whole Plant.
Folk Use	:	Plant used for curing smallpox.
References	:	Asolkar *et al.* (1992), Jain (1991).

Averrhoa bilimbi **L.**

Syn.: *A. abtusangulata* Stokes.; *A. obtusangula* Stokes.

Family: Oxalidaceae

Eng., H. and Sans. Names	:	*Eng.*: Cucumber Tree;
		H.: Belumbu.
		Sans.: Amlee, Karamaranga, Karmaraka, Karukha, Krama, Siralam.
Distribution	:	Throughout hotter parts of India; also naturalized as escape.
Reproductive Cycle	:	April-October.
Part Used	:	Leaf.
Folk Use	:	Poultice of paste of leaves applied on mumps.
Reference	:	Chatterjee and Pakrashi (1997).

Averrhoa carambola **L.**

Family: Oxalidaceae

Eng., H. and Sans. Names	:	*Eng.*: Carambola Apple, Coromandel Gooseberry;

		H.: Kamarakh, Kamaramga, Karmal;
		Sans.: Brihaddala, Karmarangah, Karukah, Rujakara.
Distribution	:	Throughout hotter parts of India.
Reproductive Cycle	:	June-November.
Parts Used	:	Whole Plant, Shoot, Leaf and Fruit.
Folk Uses	:	Plant used for curing fever. Shoot valued for treatment of chickenpox. Leaves used by Garos of Meghalaya against measles. Juice of fruit with or without macerated mass applied at the area of burns and scalds.
References	:	Jain (1991), Pal *et al.* (2000), Parrotta (2001), Rao and Shanpru (1981), Singh and Singh (2003a).

Avicennia officinalis **L.**

Syn.: *A. obovata* Griff.; *A. oepata* Buch.-Ham.; *Halodendrum thouarsii* Roem. and Schult.

Family: Verbenaceae

Eng., H. and Sans. Names	:	*Eng.*: Indian Mangrove, White Mangrove;
		H.: Bina;
		Sans.: Sagarodhurga, Tuvara, Tuvaraka.
Distribution	:	Throughout India in salt marshed and tidal creeks, Baluchistan, Sri Lanka, Malaya, shores of the Indian and Pacific oceans.
Reproductive Cycle	:	March-August.
Part Used	:	Seed.
Folk Use	:	Seeds regarded good for smallpox in Chennai.
Reference	:	Kirtikar and Basu (1984).

Azadirachta indica **A. Juss.**

Syn.: *Melia azadirachta* L.; *M. indica* (A. Juss.) Brandis.

Family: Meliaceae

Eng., H. and Sans. Names	:	*Eng.*: Indian Lilac, Margosa Tree, Neem Tree;
		H.: Nim, Nimb;
		Sans.: Nimbah, Prabhadrah.
Distribution	:	Everywhere in Indian plains including some parts of arid zone.
Reproductive Cycle	:	March-July.
Parts Used	:	Whole Plant, Bud, Leaf and Gum.

Folk Uses : Plant valued for curing smallpox, measles and other skin diseases. A little amount of powdered young buds taken daily for 15-20 days especially during September and November months to prevent attacks of measles, chickenpox and smallpox. Paste of powdered fresh leaves and turmeric rhizome rubbed on body to prevent infection of smallpox. Leaf paste applied to cure chickenpox. Dried leaf powder applied locally to the anus of children suffering from intestinal worms. Leaves also used in bronchitis and other skin diseases. 2-5g fresh leaves boiled in rice-starch water administered twice a day for blood purification. Froth obtained by putting powdered dry leaves in water applied on the body once daily to cure fever in infants. 2-3g gum dissolved in rice water administered twice daily to check diarrhoea.

References : Asolkar *et al.* (1992), Bondya *et al.* (2006), D' Rozario *et al.* (1999), Das and Sharma (2003), Dhiman (2007), Jain (1991), Kalpesh *et al.* (2004), Kirtikar and Basu (1984), Mohanty *et al.* (1996), Pal *et al.* (2000), Pande *et al.* (2006), Parrotta (2001), Rajendran *et al.* (2003), Rao and Henry (1996), Saren *et al.* (2000), Satapathy and Brahman (2000), Sharma (2002), Singh and Singh (2003a), Sinha (1986, 1987), Sood *et al.* (2012), Verma *et al.* (1995), Watt (1972).

***Azima tetracantha* Lam.**

Syn.: *Fagonia montana* Miq.; *Monetia barlerioides* L'Herit.

Family: Salvadoraceae

Eng., H. and Sans. Names : *Eng.*: Mistletoe Berrythorn;

H.: Katagurkamai;

Sans.: Kantangur, Kundali, Trikantajata.

Distribution : Deccan Peninsula, Sri Lanka, Bengal, Orissa, and Peninsular India.

Reproductive Cycle : April May.

Part Used : Leaf.

Folk Uses : Leaves used for ulcers and smallpox.

References : Kirtikar and Basu (1984), Watt (1972).

***Baccaurea ramiflora* Lour.**

Syn.: *B. sapida* (Roxb.) Muell.-Arg.; *B. wrayi* King *ex* Hook.f.; *Pierardia sapida* Roxb.

Family: Euphorbiaceae

Eng. and H. Names : *Eng.*: Burmese Grape;
H.: Leteku.

Distribution : Base of the E. Himalayas, Assam, Silhet, Myanmar, the Malay Peninsula and Andaman islands.

Reproductive Cycle : March-August.

Part Used : Sap.

Folk Use : About 100cc of sap given in early morning as tonic.

Reference : Agarwal (2003).

***Bacopa monnieri* (L.) Penell.**

Syn.: *B. monnieria* (L.) Wettst.; *Gratiola monnieria* L.; *Herpestris monniera* (L.) Kunth.; *Lysimachia monnieria* L.

Family: Scrophulariaceae

Eng., H. and Sans. Names : *Eng.*: Thyme Leaved Gratiola, Water Hyssop;
H.: Brahmi, Jalnim;
Sans.: Eiendri, Nira Brahmi, Sarasvati.

Distribution : Common in moist places throughout India, ascending to 1,320m.

Reproductive Cycle : July-April.

Parts Used : Whole Plant and Leaf.

Folk Uses : Plant used for promoting intellect in children. A poultice of boiled plant placed on chest of children to check cough. Leaf juice given to infants in bronchitis, catarrh and diarrhoea. Juice of leaves along with ginger juice, sugar and bark extract of *Moringa oleifera* given to children in stomach disorders. Leaves also consumed by children to increase memory power.

References : Ansarali and Sivadasan (2009), Asolkar *et al.* (2992), Borthakur *et al.* (2004), Chatterjee and Pakrashi (1997), Gogoi *et al.* (2003), Islam and Jha (2003), Jain (1994), Khan *et al.* (2003), Khare (2004), Kirtikar and Basu (1984), Kothari and Londhe (2000), Pal (2000), Parrotta (2001), Watt (1972).

Balanites aegyptiaca **(L.) Delile.**

Syn.: *Balanites roxburghii* Planch.; *Ximenia aegyptiaca* Juss.

Family: Balanitaceae

Eng., H. and Sans. Names : *Eng.*: Desert Date, Zachum Oil Plant;

H.: Hingan, Hingol, Hingot, Hingu;

Sans.: Angarvriksha, Angulidala, Gaurtvaka, Hinguputra, Ingudee, Jalajantu.

Distribution : Drier parts of India from Kanpur-Sikkim, Bihar, Gujarat, Deccan and Myanmar.

Reproductive Cycle : April-May.

Parts Used : Fruit and Seed.

Folk Uses : Fruit pulp mixed with goat's milk rubbed on chest to treat pneumonia in children. Seed paste given to children in colic and stomachache.

References : Bhatt *et al.* (1999, 2003), Chatterjee and Pakrashi (1997), Khare (2004).

Baliospermum solanifolium **(Burm.) Suresh.**

Syn.: *B. axillare* Blume.; *B. indicum* Decne.; *Croton roxburghii* Wall.; *Jatropha montana* Willd.; *Ricinus montanus* (Willd.) Wall.

Family: Euphorbiaceae

Eng., H. and Sans. Names : *Eng.*: Red Physic Nut, Wild Castor, Wild Croton;

H.: Danti;

Sans.: Danti, Dirgha, Erandhaphala, Nikumbha, Rechani, Shigra, Udumbaraparni.

Distribution : N.E. India from eastern U.P.-Assam, Bangladesh and Mayanmar, Chotta Nagpur, M.P.; also found in dry and scrub forests in western and southern regions of Peninsular India.

Reproductive Cycle : January-May.

Parts Used : Whole Plant and Root.

Folk Uses : Whole plant used to cure ringworm, diarrhoea, dysentery and bowel complaints, and for blindness of children. Root juice taken by nursing mothers to act as anthelmintic for babies.

References : Aminudin and Girach (1996), Pal *et al.* (2000), Parrotta (2001).

Bambusa arundinacea **(Retz.) Willd.**

Syn.: *Arundo bambos* L.; *Bambusa bambos* Druce.; *B. orientalis* Nees.; *B. spinosa* Roxb.

Family: Bambusaceae

Eng., H. and Sans. Names : *Eng.*: Bamboo, Spiny Bamboo, Thorny Bamboo;

H.: Bans, Bhans, Kalangbans, Kantabans, Kattang;

Sans.: Bahupallava, Brihattrina, Dhatushya, Kitali, Tajana, Vansa, Vansh, Venul.

Distribution : Throughout India, in areas up to 2,100m.

Reproductive Cycle : Flowers gregariously once in 30-45 years.

Part Used : Whole Plant.

Folk Use : Plant used in poliomyelitis.

Reference : Pal *et al.* (2000).

Bambusa tulda **Roxb.**

Syn.: *Dendrocalamus tulda* Voigt.

Family: Bambusaceae

Eng. and H. Names : *Eng.*: Calcutta Cane, Common Bamboo of Bengal;

H.: Deobans.

Distribution : Arborescent in N.E. India, wild and cultivated. Wild in E. India, hills of A.P. and Orissa, up to 600m.

Reproductive Cycle : March-July (it flowers only once in life and die after blooming).

Parts Used : Whole Plant and Leaf.

Folk Uses : Plant used in tetanus. Leaves prescribed in smallpox and chickenpox in Manipur.

References : Asolkar *et al.* (1992), Sinha (1987).

Baphicacanthus cusia **(Nees) Bremek.**

Syn.: *Strobilanthes flaccidifolius* (Nees) Imlay.

Family: Acanthaceae

Eng. Names : *Eng.*: Assam Indigo, Assam Rum.

Distribution : In low hills; often cultivated in India, Assam, China, Myanmar, Indonesia.

Reproductive Cycle : Flowering in Autumn.

Parts Used : Leaf and Root.

Folk Use : Both valued for tonsillitis.

Reference : Prajapati *et al.* (2006).

Baptisia tinctoria **(L.) Vent.**

Syn.: *B. gibbesii* Small.; *Podalyria tinctoria* (L.) Willd.; *Sophora tinctoria* L.

Family: Fabaceae

Eng. Names : *Eng.*: American Indigo, Wild Indigo, Yellow Broom, Yellow Indigo.

Distribution : All over India in plains and hills.

Reproductive Cycle : June-August.

Parts Used : Leaf and Root.

Folk Use : Both root and leaves used in tonsillitis.

References : Karnick (1994a, 1994b), Prajapati *et al.* (2006).

Barleria prionitis **L.**

Syn.: *B. coriacea* Oberm.; *B. echinata* St.-Lag.; *B. spicata* Roxb.

Family: Acanthaceae

Eng., H. and Sans. Names : *Eng.*: Yellow-nail-dye plant;

H.: Jhinti, Katsareya, Vajradanti;

Sans.: Ananta, Bana, Bhindi, Dasi, Jhinti, Karunta, Kinkirata, Pura, Vira.

Distribution : Throughout the hotter parts of India, also commonly grown as a hedge plant in gardens.

Reproductive Cycle : September-December (Flowering) and January-April (Fruiting).

Parts Used : Whole Plant and Leaf.

Folk Uses : Plant used to cure toothache. Leaves used for cough and diarrhoea. Leaf juice mixed with honey given in catarrhal infection and fever.

References : Agarwal (2003), Ambasta (1986), Anonymous (1976b), Chatterjee and Pakrashi (1997), Khare (2004), Kirtikar and Basu (1984), Mishra (2008), Pal *et al.* (2000), Parrotta (2001), Srivastava *et al.* (2003a), Watt (1972).

Barringtonia acutangula (L.) Gaertn. subsp. *spicata* (Bl.) Payens.

Syn.: *B. edaphocarpa* Gagnep.; *B. pedicellata* Ridley.; *B. spicata* Bl.; *Stavadium acutangulum* (L.) Miers.

Family: Lecythidaceae

Eng., H. and Sans. Names : *Eng.*: Indian Oak, Small Indian Oak;

H.: Hijgal, Ijal, Ingar, Jujar, Neora, Panniari, Samundarphal;

Sans.: Abdhiphal, Ambudhiphala, Ambuja, Dhatriphala, Hijjala, Ijjala, Nichula, Nishula, Shosha, Udadhiphala, Yaha.

Distribution : Throughout India from Himalayas-Sri Lanka.

Reproductive Cycle : April-October.

Parts Used : Leaf and Seed.

Folk Uses : Paste of leaves along with turmeric, applied all over body to prevent insect bites like mosquito, bed bug, ant etc. Powder of seeds mixed with milk given to children for enlarged spleen.

References : Bhattacharjee (2001), Chatterjee and Pakrashi (1997), Kirtikar and Basu (1984), Padhy and Dash (2003), Parrotta (2001), Watt (1972).

Basella alba L.

Syn.: *B. cordifolia* Lam.; *B. rubra* L.

Family: Basellaceae

Eng., H. and Sans. Names : *Eng.*: Indian Spinach, Malabar Night Shade;

H.: Lalbaclu, Poi;

Sans.: Potaki, Putaki, Upodika.

Distribution : Throughout India, cultivated in N.E. states, sometimes growing wild near hedges.

Reproductive Cycle : March-December.

Parts Used : Leaf and Stem.

Folk Uses : 25-50ml juice of leaves given once daily for 4-7 days to relieve constipation, catarrhal affections and malnutrition in children. Stem and leaf both prescribed as a laxative.

References : Ambasta (1986), Anonymous (1948), Bondya *et al.* (2006), Chatterjee and Pakrashi (1997), Kumar (2002), Mishra (2008), Prajapati *et al.* (2006), Watt (1972).

***Basilicum polystachyon* (L.) Moench.**

Syn.: *Moschosma polystachyon* Benth.; *Ocimum polystachyon* L.; *Perxo polystachyon* (L.) Raf.; *Plectranthus micranthus* Spreng.

Family: Lamiaceae

Eng. and Sans. Names	:	*Eng.*: Musk Basil, South African Basil, Swamp Basil; *Sans.*: Ajagandhi, Surabhi, Tungi.
Distribution	:	Moist regions of E. and S. India.
Reproductive Cycle	:	Flowering and Fruiting in August.
Part Used	:	Leaf.
Folk Use	:	Leaf extract applied to check headache among children.
References	:	Agarwal (2003), Anonymous (1962).

***Bauhinia purpurea* L.**

Syn.: *B. coromandeliana* DC.; *B. triandra* Roxb.; *Phanera purpurea* (L.) Benth.

Family: Fabaceae

Eng., H. and Sans. Names	:	*Eng.*: Bauhinia, Butterfly Tree, Geranium Tree, Mountain Ebony, Orchid Tree; *H.*: Barial, Gurial, Gwiar, Kachnar, Kaniar, Padrian; *Sans.*: Ashmantaka, Chamrika, Gandhari, Girija, Vanaraj.
Distribution	:	Throughout India, from foot of W. Himalayas and Khasia mountains (alt. 1,350m)-Sri Lanka and Penang.
Reproductive Cycle	:	September-January.
Parts Used	:	Bark and Flower.
Folk Use	:	Bark and flowers applied on sores of smallpox.
References	:	Jain (1991), Jain and R. De (1966), Pande *et al.* (2006), Sharma (2004), Srivastava *et al.* (2003a).

***Bauhinia racemosa* Lam.**

Syn.: *B. parviflora* Vahl.; *Piliostigma racemosum* (Lam.) Benth.

Family: Fabaceae

Eng., H. and Sans. Names	:	*Eng.*: Bidi-Leaf Tree; *H.*: Katmauli; *Sans.*: Yamalapatrakah, Yugmapatra.
Distribution	:	Native of N.E. India, grows best in sandy or pebbly, loamy soil that is moist.

Reproductive Cycle	:	Flowering during February-May.
Part Used	:	Whole Plant.
Folk Use	:	Plant used in stomach pain and stomach heat.
Reference	:	Pal *et al.* (2000).

Bauhinia variegata **L. (Pl. 2C)***

Syn.: *B. candida* Roxb.; *B. decora* Uribe.; *Phanera variegata* (L.) Benth.

Family: Fabaceae

Eng., H. and Sans. Names	:	*Eng.*: Bauhinia, Buddhist, Mountain Ebony, Orchid Tree; *H.*: Barial, Kachnar, Kandan, Rakta; *Sans.*: Chamari, Chamarika, Kovidara.
Distribution	:	Throughout India, ascending to 1,300m in Himalayas; from foot of W. Himalayas and Sikkim, 1,300m.
Reproductive Cycle	:	February-July.
Parts Used	:	Whole Plant, Flower Buds and Bark.
Folk Uses	:	Plant used for curing stomach pain and stomach heat. Decoction of young flower buds given to children after measles to prevent stomach disorders. About 5g bark powder given once a day with lukewarm water to treat mumps. Decoction of bark given in pox.
References	:	Khanna *et al.* (1996b), Khare (2004), Pal *et al.* (2000), Pande and Joshi (2001).

Benincasa hispida **(Thunb.) Cogn.**

Syn.: *B. cerifera* Savi.; *Cucurbita hispida* Thunb.

Family: Cucurbitaceae

Eng., H. and Sans. Names	:	*Eng.*: Ashgourd, Ash Pumpkin, Tallow Gourd, White Gourd; *H.*: Gol Kaddu, Pani Kumhra, Petha; *Sans.*: Kooshmanda, Kushmanda, Parvaphala, Phalashaka.
Distribution	:	Cultivated in tropical and sub-tropical parts of India.
Reproductive Cycle	:	June-October.
Part Used	:	Fruit.
Folk Use	:	Fruits used in marasmus.
Reference	:	Chatterjee and Pakrashi (1997).

Berberis aristata DC. (Pl. 2D)*

Syn.: *B. bussmul* K. Koch *ex* Miq.; *B. gracilis* Lindl.; *B. undulata* K. Koch.

Family: Berberidaceae

Eng., H. and Sans. Names	:	*Eng.*: Indian Barberry, Tree Turmeric; *H.*: Chitra, Chotra, Darhald, Kashmal, Rasuat; *Sans.*: Daruharidra, Darvi, Kata, Pitadaru, Suvarnavarna.
Distribution	:	Temperate Himalayas between 2,000-3,350m; extending from Bhutan-Kanawar, Nilgiri hills, Sri Lanka.
Reproductive Cycle	:	August-October.
Parts Used	:	Root and Fruit.
Folk Uses	:	Dried extract of roots used as a purgative for children. Fruits also regarded as mild laxative for children.
References	:	Chatterjee and Pakrashi (1997), Farooq (2005), Parmar and Kaushal (1982), Watt (1972).

Berberis asiatica Roxb. *ex* DC.

Syn.: *B. hypoleuca* Lindl.

Family: Berberidaceae

Eng., H. and Sans. Names	:	*Eng.*: Queen of the shores; *H.*: Daru Haldi, Rasant; *Sans.*: Daruharidra, Darupita, Hemakanti, Karnavati, Parjani, Pitadru, Sthirraga.
Distribution	:	Dry outer Himalayas (700-2,800m) and Assam dry valley of the Himalayas (1,000-2,500m) from Bhutan-Garwhal.
Reproductive Cycle	:	March-June.
Part Used	:	Fruit.
Folk Use	:	Fruits given to children as a mild laxative.
References	:	Ambasta (1986), Anonymous (1948), Asolkar *et al.* (1992), Kumar (2002).

Berberis lycium Royle.*

Syn.: *B. insignis* Hook. f.; *B. parkeriana* C.K. Schn.

Family: Berberidaceae

Eng., H. and Sans. Names	:	*Eng.*: Raisin Berberry; *H.*: Chitra Kushmal; *Sans.*: Daruharidra.

Distribution : Dry hot places in W. Himalayas ascending to 3,000m (Garwhal-Hazara).

Reproductive Cycle : March-October.

Part Used : Whole Plant.

Folk Uses : Plant good for constipation, toothache and dysentery in children.

References : Chauhan (1999), Pal *et al.* (2000).

Berchemia lineata **(L.) DC.**

Syn.: *B. poiretiana* DC.; *Rhamnus lineata* L.

Family: Rhamnaceae

Eng. Name : *Eng.*: Supple -Jack.

Distribution : N.W. Himalayas, Khazan valley, ascending to 2,350m; Sikkim, at 3,350m.

Reproductive Cycle : July-November.

Part Used : Whole Plant.

Folk Use : Plant known to cure fever in children.

Reference : Pal *et al.* (2000).

Bergenia ciliata **(Haw.) Sternb. (Pl. 3A)***

Syn.: *B. ligulata* var. *ciliata* (Royle) Engl.

Family: Saxifragaceae

Eng., H. and Sans. Names : *Eng.*: Rockfoil;

H.: Pakhanabheda, Pakhanbed;

Sans.: Ashmaghna, Bhimayojini, Giribhita, Shveta.

Distribution : Temperate Himalayas (Kashmir – Bhutan), Khasia hills, ascending to 1,350m.

Reproductive Cycle : March-November.

Part Used : Whole Plant.

Folk Use : Plant used to cure fever in children.

Reference : Pal *et al.* (2000).

Bergenia pacumbis **(Buch.-Ham. *ex* Don.) C.Y.Wu and J.T.Pan.**

Syn.: *B. ligulata* Engl.

Family: Saxifragaceae

Eng. Name : *Eng.*: Bergenia.

Distribution : Grows usually in rocky areas and cliffs between 1,800-5,100m in the Himalayas.

Reproductive Cycle : April-August.

Part Used : Whole Plant.

Folk Use : Used against fever.

Reference : Pal *et al.* (2000).

Bidens pilosa **auct. (non L.) Hook. f. (Pl. 3B)***

Syn.: *B. biternata* (Lour.) Merr. and Sherff.; *B. chinensis* Willd.

Family: Asteraceae

Eng. and H. Names : *Eng.*: Black Jack, Broomstick, Devil's Needle, Spanish Needle;

H.: Chirchitta.

Distribution : Throughout India, ascending Himalayas and other mountains to 2,000m.

Reproductive Cycle : August-October.

Parts Used : Whole plant and Capitulum.

Folk Uses : Plant valued against earache. Decoction of capitulae used for treating illness that affects children.

References : Keller *et al.* (2010), Pal *et al.* (2000).

Bischofia javanica **Blume.**

Syn.: *Andrachne trifoliata* Roxb.; *Bischofia cummingiana* Decne.

Family: Euphorbiaceae

Eng. and H. Names : *Eng.*: Bishop Wood, Cedar, Vinegar Wood;

H.: Bhillar, Irum, Kotsemba, Paniala, Pankain.

Distribution : Moderately common from sea-level-mid-montane in primary or secondary forests, forest edges, grassy slopes and thickets.

Reproductive Cycle : January-July.

Part Used : Stem.

Folk Use : Liquid from stem used to treat children having skin rashes.

Reference : Prajapati *et al.* (2006).

Bistorta macrophylla **(D. Don.) Sojak.**

Syn.: *Polygonum macrophyllum* D. Don.

Family: Polygonaceae

Eng. Names : *Eng.*: Large Leaved knotweed, Red knotweed.

Distribution : Himalayas (Garwhal-Bhutan): 3,300-4,600m.

Reproductive Cycle : July-September.

Part Used : Rhizome.

Folk Use : Paste of sweetish rhizomatic rootstock given to children and infants having stomach problems.

Reference : Chauhan (1999).

Bixa orellana L.

Syn.: *B. americana* Poir.; *B. katangensis* Delpierre.; *B. odorata* Ruiz. and Pav. *ex G.* Don.

Family: Bixaceae

Eng., H. and Sans. Names : *Eng.*: Arnatto, Chinese Dye Tree, Lipstick Tree, Monkey Tree, Roucou;

H.: Kesari, Latkan, Vatkana;

Sans.: Raktabija, Sinduri, Trivapushpi.

Distribution : Largely cultivated throughout India; indigenous to tropical America.

Reproductive Cycle : April-September.

Part Used : Whole Plant.

Folk Use : Plant used to cure fever in children.

Reference : Pal *et al.* (2000).

Blechnum orientale L.

Syn.: *Blechnopsis orientale* (L.) J. Presl.

Family: Blechnaceae

Eng. and H. Names : *Eng.*: Fish Fern, Shield Fern;

H.: Haththazori.

Distribution : Throughout India.

Reproductive Cycle : August-December.

Part Used : Rhizome.

Folk Use : Inner soft portion of rhizome (15g) grounded with 7 black peppers given to nursing mothers to cure boils in babies.

Reference : Brahman and Saxena (1990).

Blepharispermum sub-sessile **DC.**

Family: Asteraceae

Distribution	:	Bastar, M.P., The Concan, Canara, and Mysore; also on the Ghats.
Reproductive Cycle	:	August-October.
Part Used	:	Whole Plant.
Folk Use	:	Whole plant used for cold rhinitis in children.
Reference	:	Jain (1991).

Blumea balsamifera **(L.) DC.**

Syn.: *Baccharis balsamifera* Stokes.; *B. gratissima* Blume. *ex* DC.; *Blumea grandis* DC.

Family: Asteraceae

Eng., H. and Sans. Names	:	*Eng.*: Ngai Camphor; *H.*: Kakoranda, Kukronda; *Sans.*: Kukundara, Tamracuda.
Distribution	:	Tropical Himalayas; Nepal and Sikkim, 300-1,350m. Assam, Khasia mountains, Chittagong, Pegu, Myanmar and the E. Peninsula to Singapur and Penang.
Reproductive Cycle	:	Flowering-year round.
Part Used	:	Leaf.
Folk Use	:	One tsp juice of young leaves given once a day till recovery to babies suffering from bowel complaints due to mother's defective milk.
References	:	Asolkar *et al.* (1992), Borthakur (1993), Jain (1991).

Blumea lacera **(Burm.f.) DC.**

Syn.: *Baccharis ariculata* Wall. *ex* DC.; *Blumea subcapitata* DC.; *Conyza lacera* Burm.f.

Family: Asteraceae

Eng., H. and Sans. Names	:	*Eng.*: Blumea, Malay Blumea; *H.*: Jangli Muli, Kukaraundha, Kukronda; *Sans.*: Kukkuradry, Kukundarh, Kukuradru.
Distribution	:	Throughout the plains of India, from N.W. ascending to 700m in the Himalayas-Travancore and Singapore and in Sri Lanka.
Reproductive Cycle	:	September-October.

Part Used	:	Leaf.
Folk Uses	:	Warm leaf juice applied on chest of children suffering from pneumonia and to cure cramp disease. Bruised leaves also introduced into rectum to kill threadworms and pinworms.
References	:	Khan *et al.* (2003), Khanna *et al.* (1996c), Pandey (2003).

***Blumea membranacea* DC.**

Syn.: *B. balansae* Gagnep.; *B. paucifolia* DC.

Family: Asteraceae

Distribution	:	Common throughout India.
Reproductive Cycle	:	March-May.
Part Used	:	Leaf.
Folk Use	:	Extract of leaves applied on anus in children to kill pin-worms.
Reference	:	Saini (1996).

***Blumeopsis flava* (DC.) Gagnep.**

Syn.: *Blumea flava* DC.; *Laggera flava* (DC.) Benth. Hook.f.

Family: Asteraceae

Distribution	:	Almost throughout India.
Reproductive Cycle	:	Flowering almost throughout the year.
Part Used	:	Leaf.
Folk Use	:	Fresh leaf extract applied externally on forehead and back of children to reduce high body temperature.
References	:	Salam *et al.* (2011), Singh and Singh (2003a).

***Boerhavia diffusa* L.**

Syn.: *B. adscendens* Willd.; *B. caespitosa* Ridl.

Family: Nyctaginaceae

Eng., H. and Sans. Names	:	*Eng.*: Spreading Hogweed; *H.*: Biskhafra, Gadepurna, Punernava, Sani; *Sans.*: Punarnava.
Distribution	:	Throughout India, ascending to 2,300m in the hot Himalayan valleys.
Reproductive Cycle	:	Throughout the year.
Parts Used	:	Whole Plant and Shoot.

Folk Uses	:	Paste of 20-25g shoot given daily before meals for a week in jaundice associated with loss of appetite in children. Plant used as haematinic and growth promoter in children.
References	:	Asolkar *et al.* (1992), Borthakur *et al.* (2004), Gogoi *et al.* (2003).

Bombax ceiba **L.***

Syn.: *B. malabarica* DC.; *Salmalia malabarica* (DC.) Schott and Endl.

Family: Bombacaceae

Eng., H. and Sans. Names	:	*Eng.*: Indian Bombax, Red Cotton Tree, Silkcotton Tree; *H.*: Kantisenbal, Pagun, Ragatsemal, Ragatsenbal, Semal, Semul, Semur, Shembal; *Sans.*: Moca, Mocha, Salmali.
Distribution	:	Throughout India, scattered up to 1,400m in Himachal Pradesh, tropical E. Himalaya and hotter forest regions of India to Myanmar and Sri Lanka.
Reproductive Cycle	:	January-February.
Parts Used	:	Whole Plant, Root, Stem Bark, Seed and Gum.
Folk Uses	:	Whole plant valued for smallpox and poliomyelitis. Root administered with rice for one week on empty stomach in morning to cure bed-wetting syndrome in children. Juice of crushed stem bark good against tonsillitis. Seed paste applied over the body in smallpox and chickenpox. 20-30 grains of gum with equal parts of sugar prescribed in diarrhoea. Gum applied on the body of chickenpox patients for quick relief, and also given to children as a laxative.
References	:	Asolkar *et al.* (1992), Chauhan (1999), Ghosh and Sensarma (1997), Jain (1991), Katewa and Arora (1997), Lalramnghinglova (1996), Pal *et al.* (2000), Parrotta (2001), Sharma (2003), Sood *et al.* (2012), Upadhyay and Chauhan (2003), Watt (1972).

Bombax insigne **Wall.**

Syn.: *B. scopulorum* Dunn.; *B. tenebrosum* Dunn.

Family: Bombacaceae

Eng., H. and Sans. Names	:	*Eng.*: Showy Silk Cotton Tree, Silk Cotton Tree; *H.*: Bagun; *Sans.*: Kutasalmali.

Distribution	:	Myanmar and Malay Peninsula, Assam, Andamans and Manipur, very rare in Tripura; common in Mizoram, particularly in tropical mixed deciduous forests.
Reproductive Cycle	:	December-April.
Part Used	:	Bark.
Folk Use	:	Juice of decorticated bark considered a effective remedy against tonsillitis.
Reference	:	Lalramnghinglova (2003).

Breonia chinensis **(Lam.) Capuron.**

Syn.: *Anthocephalus chinensis* (Lam.) Hassk.; *A. indicus* A. Rich.; *Cephalanthus chinensis* Lam.; *Nauclea citrifolia* Poir.

Family: Rubiaceae

Eng., H. and Sans. Names	:	*Eng.*: Wild Cinchona; *H.*: Kadam, Kadamba; *Sans.*: Ashokari, Dasa, Deva, Haripriya, Jala, Jirnaparna, Kadamba, Nipa, Priya, Priyaka, Surabhi, Vrittapushpa.
Distribution	:	Himalayas-Sri Lanka and Malacca; wild or cultivated.
Reproductive Cycle	:	May-September.
Parts Used	:	Whole Plant, Leaf, Fruit and Bark.
Folk Uses	:	Whole plant good against fever in children. Extract of leaves or powdered leaves with water given to children against intestinal worms. Fruit juice along with cumin ('jeera') and sugar also given in gastric irritability. Fresh juice of bark applied to the head of infants in Konkan at the time of sinking of fontanelle.
References	:	Chatterjee and Pakrashi (1997), Khare (2004), Kirtikar and Basu (1984), Mahato and Mahato (1996), Pal *et al.* (2000).

Breynia vitis-idaea **(Burm.f.) C.E.C. Fisch.**

Syn.: *B. accrescens* Hayata.; *B. rhamnoides* (Retz.) Mull. Arg.; *Phyllanthus rhamnoides* Retz.

Family: Euphorbiaceae

Eng., H. and Sans. Names	:	*Eng.*: Indian Snowberry; *H.*: Surasaruni, Tikhar; *Sans.*: Aruni.

Distribution : Found in tropical Asia from India and Sri Lanka-Malaysia, China and the Philippines. Found chiefly in deciduous forests in Peninsular India.

Reproductive Cycle : July-February.

Part Used : Leaf.

Folk Uses : Leaf paste applied on cheeks to get relief from pain during mumps. Powdered dried leaves smoked like tobacco in parts of Bihar to get relief from tonsils.

References : Agarwal (2003), Ambasta (1986), Dash and Misra (2000), Nayak *et al.* (2003a), Parrotta (2001).

Bridelia retusa (L.) A. Juss.

Syn.: *B. cambodiana* Gagnep.; *B. pierrei* Gagnep.; *B. spinosa* (Roxb.) Willd.

Family: Euphorbiaceae

Eng., H. and Sans. Names : *Eng.*: Spinous Kino Tree;

H.: Ekdania, Gauli, Gondai, Kaj, Karka, Kasai, Kassi, Khoja;

Sans.: Asana, Ekadivi, Ekavira, Mahavira.

Distribution : Secondary deciduous and semi-deciduous forests throughout India excepting very dry regions, up to 1,800m.

Reproductive Cycle : May-January.

Part Used : Fruit.

Folk Use : Fruits used for checking vomiting in children.

Reference : Parrotta (2001).

Bridelia stipularis (L.) Blume.

Syn.: *B. dasycalyx* Kurz.; *B. scandens* (Roxb.) Willd.; *Clutia stipularis* L.

Family: Euphorbiaceae

H. and Sans. Names : *H.*: Kangi abel, Khaj, Khaji;

Sans.: Ghonta.

Distribution : Throughout tropical and sub-tropical India, Bangladesh, Myanmar including Himalayan foothills from Kashmir-Bhutan.

Reproductive Cycle : October-April.

Parts Used : Leaf and Bark.

Folk Uses : Leaves used for cleaning sores on the scalp of children. Decoction of bark given to children to cure cough, fever, asthma and mouth sores (as a gargale).

References : Asolkar *et al.* (1992), Nisha and Sivadasan (2007), Parrotta (2001).

Bryophyllum pinnatum **(Lam.) Oken.**

Syn.: *B. calycinum* Salisb.; *Cotyledon pinnata* Lam.; *Kalanchoe pinnata* (Lam.) Pers.

Family: Crassulaceae

Eng., H. and Sans. Names : *Eng.*: Air Plant, Tree of Life;

H.: Hamsagar, Zakhm-e-Hayat;

Sans.: Asthibhaksha, Hamsagar, Parnabija, Parnapeejam.

Distribution : Throughout the warm and moist parts of India, especially abundant in W. Bengal and hilly areas of S. India, up to 1,000m.

Reproductive Cycle : February-June.

Part Used : Leaf.

Folk Use : Leaves used as a lotion for smallpox.

References : Anonymous (1959), Khare (2004).

Buchanania cochinchinensis **(Lour.) M. R. Almeida.**

Syn.: *B. lanzan* Spreng.; *B. latifolia* Roxb.

Family: Anacardiaceae

Eng., H. and Sans. Names : *Eng.*: Almondette Tree, Cheranjee, Cuddapah Almond;

H.: Achar, Chironji, Piyal, Piyar;

Sans.: Chara, Piyala, Priyala, Priyalah.

Distribution : Native to India and Myanmar, commonly occurring at Rater of Chhindwara district of M.P. and in drier parts of India.

Reproductive Cycle : December-June.

Part Used : Stem Bark.

Folk Use : 3g mixture of powdered stem bark with *Syzygium cumini* (jamun) bark given twice a day for infantile diarrhoea.

References : Girach and Aminuddin (1995), Parrotta (2001).

Bulbophyllum fuscopurpureum **Wight.**

Syn.: *Phyllorkis fuscopurpurea* (Wight.) Kuntze.

Family: Orchidaceae

Distribution : Nilgiri hills, at the Kaitia falls.

Reproductive Cycle : February-April.

Part Used : Whole Plant.

Folk Uses : Paste of whole plant applied all over children's body followed by bath after 1 hour to cure diarrhoea and dysentery.

Reference : Murugesan *et al.* (2011).

Butea monosperma **(Lam.) Taub.**

Syn.: *B. frondosa* Koen. *ex* Roxb.; *Erythrina monospermia* Lam.; *Palso monosperma* Kuntze.

Family: Fabaceae

Eng., H. and Sans. Names : *Eng.*: Bal Kino, Bastard Teak, Bengal Kino, Flame of the forest;

H.: Dhaak, Dhak, Palaas, Palas, Tesu;

Sans.: Bijasneha, Karaka, Palaasha, Palasha.

Distribution : Deciduous forests throughout India, up to 1,200m and in open areas.

Reproductive Cycle : March-May.

Parts Used : Whole Plant, Leaf, Flower and Seed.

Folk Uses : Plant given to cure dysentery. Leaf juice used for heat eruptions. Flower juice given in fever and cold. Seed powder mixed with milk given daily to children as a tonic. Santals give seed oil to children suffering from intestinal worms and skin diseases.

References : Bhandary *et al.* (1996), Bhatt and Jain (2009), Bhatt *et al.* (2000), D'Rozario *et al.* (1999), Jain (1994), Pal *et al.* (2000).

Butea superba **Roxb.**

Family: Fabaceae

Eng., H. and Sans. Names : *Eng.*: Creeping Butea, Red Kwao Kruva;

H.: Bela Palas, Bodha, Chihumt, Palas Lata;

Sans.: Latapalas, Latapalasha.

Distribution : Throughout C. and S. India.

Reproductive Cycle	:	March-July.
Parts Used	:	Whole Plant and Leaf.
Folk Uses	:	Plant used against dysentery. Leaf juice mixed with curd and *Curcuma aromatica* beneficial for heat eruptions in children.
References	:	Chatterjee and Pakrashi (1997), Pal *et al.* (2000), Parrotta (2001).

Cadaba trifoliata **Wight and Arn.**

Syn.: *C. triphylla* Wight.; *Desmocarpus missionis* Wall.

Family: Capparaceae

Eng. and Sans. Names	:	*Eng.*: Indian Cadaba; *Sans.*: Balaya.
Distribution	:	Southern parts of the Carnatic, Karnataka, Sri Lanka.
Reproductive Cycle	:	October-November.
Part Used	:	Whole Plant.
Folk Use	:	Plant used for indigestion of children.
Reference	:	Kirtikar and Basu (1984).

Caesalpinia crista **L.**

Syn.: *C. bonduc* (L.) Roxb.; *C. bonducella* (L.) Fleming.; *C. crista* L.; *C. joyabo* Naza.; *C. nuga* Ait.f.; *Guilandia bonduc* L.; *G. bonducella* L.

Family: Fabaceae

Eng., H. and Sans. Names	:	*Eng.*: Bonduc Nut, Fever Nut, Physic Nut; *H.*: Karanju, Kat-karanja; *Sans.*: Kantakikaranjah, Kuberakshi, Lata-karanj, Putikaranja.
Distribution	:	Throughout hotter parts of India particularly along the sea coast and up to 830m, W. Bengal and S. India; cultivated as hedge plant in villages.
Reproductive Cycle	:	July-January.
Parts Used	:	Whole Plant, Leaf, Root, Seed and Bark.
Folk Uses	:	Plant used for curing fever and earache. Root, bark, leaves and seeds used in smallpox. 10-20ml of leaf extract given twice a day to expel intestinal worms. Infusion of seeds used in infantile convulsions.
References	:	Kirtikar and Basu (1984), Nayak *et al.* (2003a), Pal *et al.* (2000), Parrotta (2001), Sharma and Singh (2001).

***Cajanus cajan* (L.) Millsp.**

Syn.: *C. indicus* Spreng.; *C. obcordifolia* V. Singh.; *Cystisus cajan* L.

Family: Fabaceae

Eng., H. and Sans. Names	:	*Eng.*: Congopea, Pigeon Pea, Red Grass; *H.*: Arhar, Tur, Tuvar, Tuver; *Sans.*: Adhaki, Tuvari, Tuwari.
Distribution	:	Cultivated as pulse crop throughout India.
Reproductive Cycle	:	June-October.
Part Used	:	Shoot.
Folk Use	:	5 drops of juice mixed with a tsp of honey given daily for 3 days in acute viral hepatitis.
References	:	Borthakur *et al.* (2004), Gogoi *et al.* (2003).

***Cajanus crassus* (King) Maesen.**

Syn.: *Atylosia crassa* King.; *A. mollis* Benth.; *A. volubilis* (Blanco) Gamble.; *Cantharospermum volubile* (Blanco) Merr.

Family: Fabaceae

H. Name	:	*H.*: Jungli Baler.
Distribution	:	Almost throughout India.
Part Used	:	Root.
Folk Use	:	Roots used against mumps by inhabitants of Orissa.
References	:	Jain (1991), Saxena *et al.* (1981).

***Cajanus scarabaeoides* (L.) Thouars.**

Syn.: *Atylosia scarabaeoides* (L.) Benth.; *Cantharospermum pauciflorum* Wight and Arn.; *Dolichos minutus* Wight and Arn.

Family: Fabaceae

Eng. Names	:	*Eng.*: Peanut Grass, Showy Pigeon, Tropical Clover.
Distribution	:	Usually spread in the plains throughout India, ascending to 2,000m in W. Himalayas.
Reproductive Cycle	:	Flowering during June-October.
Part Used	:	Seed.
Folk Use	:	Seeds used against worms in children.
Reference	:	Topno and Ghosh (2000).

Calamus rotang **L.**

Syn.: *C. roxburaghii* Griff.

Family: Arecaceae

Eng., H. and Sans. Names	:	*Eng.*: Choir-bottomcane, Common Rattan, Rattan Chair, Slender Rattan, Water Rattan; *H.*: Bemt, Bet, Chachibet; *Sans.*: Kalana, Latavamsa, Nichula, Ratha, Shita, Vanira, Verta, Vetasa, Vidula.
Distribution	:	Dry tract in C. and S. India, up to 450m.
Reproductive Cycle	:	Flowering with a peak during March-April, Fruiting throughout the year.
Part Used	:	Whole Plant.
Folk Use	:	Whole plant used to cure fever in children.
Reference	:	Pal *et al.* (2000).

Calamus travancoricus **Bedd. *ex* Becc.**

Family: Arecaceae

Eng. Names	:	*Eng.*: Cane, Rattan.
Distribution	:	Deccan Peninsula from Malabar-Travancore.
Reproductive Cycle	:	October-December (Flowering), April-July (Fruiting).
Part Used	:	Whole Plant.
Folk Use	:	Plant used in earache.
Reference	:	Pal *et al.* (2000).

Calendula officinalis **L. (Pl. 3C)**

Syn.: *C. eriocarpa* DC.

Family: Asteraceae

Eng., H. and Sans. Names	:	*Eng.*: African Marigold, Calenduls, Common Marigold, Garden Marigold, Marigold, Pot Marigold, Scotch Marigold; *H.*: Gul-e-Asharfi, Sadberg, Zergul; *Sans.*: Zergul.
Distribution	:	Ornamental plant throughout India.
Reproductive Cycle	:	February-December.
Part Used	:	Whole Plant.

Folk Use : Whole plant considered useful for scrofulous children.

Reference : Kirtikar and Basu (1984).

***Callicarpa longifolia* Lam.**

Syn.: *C. attenuata* Wall. *ex* Walp.; *C. roxburghiana* Schult.

Family: Lamiaceae

Eng. Name : *Eng.*: Long Leaved Beauty Berry.

Distribution : Assam, Khasia hills.

Reproductive Cycle : Flowering during May-August.

Part Used : Whole Plant.

Folk Use : Plant used for curing fever in children.

Reference : Pal *et al.* (2000).

***Callicarpa tomentosa* (L.) L.**

Syn.: *C. lanata* L.; *C. villosa* Vahl.; *C. wallichiana* Walp.; *Tomex tomentosa* L.

Family: Lamiaceae

Eng., H. and Sans. Names : *Eng.*: Fever Nut, French Mulberry of Western Ghats, Nickar Bean, Perfumed Cherry;

H.: Bastra;

Sans.: Priyangu.

Distribution : Konkan, N. Kanara, W. Ghats of Bombay, Tamil Nadu.

Reproductive Cycle : July-September.

Part Used : Whole Plant.

Folk Use : Plant used for curing fever in children.

Reference : Pal *et al.* (2000).

***Calotropis gigantea* (L.) R. Br. *ex* Ait.**

Syn.: *Ascelepias herbacea* Roxb.

Family: Asclepiadaceae

Eng., H. and Sans. Names : *Eng.*: Bow String Hemp;

H.: Ak, Akada, Madar;

Sans.: Aditya, Aharbandhava, Arka, Bhanu, Divakar, Jambhala, Ravi, Tulaphala.

Distribution : Throughout India, in warm and dry places, ascending to 1,000m in Himalayas.

Reproductive Cycle : November-January.

Parts Used : Whole Plant, Leaf and Root.

Folk Uses : Plant valued for curing fever and smallpox. Poultice of castor smeared leaves applied for relief from pain in abdomen of children due to constipation. Decoction of roots used by Santals in infantile convulsions and smallpox.

References : Asolkar *et al.* (1992), Chauhan (1999), Goel and Rajendran (2000), Jain (1991), Kalpesh *et al.* (2004), Pal *et al.* (2000), Watt (1972).

Calotropis procera **(Aiton) Dryand. (Pl. 3D)**

Syn.: *C. hamiltonii* W.

Family: Asclepiadaceae

Eng., H. and Sans. Names : *Eng.*: Gigantic Swallow-Wort, Madar Milk Weed;

H.: Aak, Aakada, Ag, Madar, Safedak;

Sans.: Alarka, Ark, Bimbora, Ganrupaka, Mandara, Vasuka.

Distribution : More or less throughout India in warm dry places from Punjab-Western, C. and S. India, abundant in the sub-Himalayan tract and adjacent plains in N. India.

Reproductive Cycle : Most part of the year.

Parts Used : Whole Plant, Twig and Leaf.

Folk Uses : Plant used against smallpox. Twig of the plant used for 'jhada' (tantaric activity) by Bhopas for general debility amongst kids. Leaves used in gastric troubles of children. Toasted fresh leaves coated with castor oil kept on mumps.

References : Bhattacharyya (1996), Jain (1991), Sharma (2002), Sen and Batra (1997), Venkataratnam and Raju (2004).

Calvatia cyathiformis **(Bosc) Morgan.**

Syn.: *Bovista lilacina* Berk. *et* Mont.; *Calvatia lilacina* (Berk. *et* Mont.) Lloyd.; *Lycoperdon lilacinum* (B. *et* M.) Massee.; *L. novae-zelandiae* Lev.

Family: Lycoperdaceae

Eng. Name : *Eng.*: Purple Spored Puffball.

Distribution : Temperate Himalayas.

Part Used : Spores.

Folk Use : Spores used for healing mouth sores and navel of recently born children.

Reference : Keller *et al.* (2011).

Cananga odorata **(Lam.) Hook.f. and Thoms.**

Syn.: *C. mitrastigma* (F. Muell.) Domin.; *Canangium odoratum* (Lam.) King.

Family: Annonaceae

Eng., H. and Sans. Names : *Eng.*: Kenanga, Ylang-Ylang;

H.: Ban champak;

Sans.: Lanji, Vanachampaka.

Distribution : Cultivated or naturalized in forests, slopes or gullies from sea-level-mid-montane throughout India.

Reproductive Cycle : April-August (Flowering) and December-March of the following year (Fruiting).

Part Used : Leaf.

Folk Use : Leaves used for treating infantile diarrhoea.

References : Asolkar *et al.* (1992), Prajapati *et al.* (2006).

Canna edulis **Ker.-Gawl. (Pl. 4A)***

Syn.: *C. chinensis* Willd.; *C. floribunda* Bouche.; *C. hifida* Roem.; *C. indica* L.; *Xyphostylis lutea* (Mill.) Raf.

Family: Cannaceae

Eng., H. and Sans. Names : *Eng.*: Indian Bead, Indian Reed, Indian Shot;

H.: Keli, Sabbajaya, Silarumba;

Sans.: Devakili, Sarvajaya, Silarumba.

Distribution : Native of C. and S. America, naturalized in Asia and West Indies; common all over India and Sri Lanka as ornamental flowering plant.

Reproductive Cycle : Throughout the year.

Part Used : Root.

Folk Use : A few spoons of local drink 'Rokshi' prepared from roots of plant and corns of *Zea mays* given to children for cough and cold.

Reference : Kumar and Rao (2001).

Cannabis sativa L. (Pl. 4B)*

Syn.: *C. gigantea* E.; *C. indica* Lamk.

Family: Cannabinaceae

Eng., H. and Sans. Names : *Eng.*: Congo Tobacco, Indian Hemp, Marijuma;

H.: Bhang, Charas, Ganja, Kinnab, Siddhi;

Sans.: Bhanga, Ganjka, Indrasana, Shivpriya, Ujaya, Vijaya.

Distribution : Native of C. and W. Asia, naturalized in the sub-Himalayan tract and abundant in wastelands from Punjab eastwards to Bengal and Bihar extending southwards.

Reproductive Cycle : March-May.

Parts Used : Whole Plant, Leaf and Flower.

Folk Uses : Plant considered useful in infantile convulsions. A small quantity of leaves mixed with other drugs and spices forms an useful compound in diarrhoea and indigestion of children. Medicated oil obtained by boiling its dried leaves in mustard oil and camphor massaged over chest to treat pneumonia in children. Powdered mixture obtained by roasting tobacco leaves of about 2-3 in number, about 50ml of honey or concentrated cane juice, 10-20g of processed 'ganja' of hemp plant and 2-3 earthworms obtained from banana plants applied for relief from scratching inflammation of skin. Flowering tops are used in infantile leucorrhoea and in tetanus.

References : Khanna *et al.* (1996a), Singh and Singh (2003a), Watt (1972).

Capparis deciduas (Forsk) Edgew.

Syn.: *C. aphylla* Roth.; *Sodaela decidua* Forsk.

Family: Capparaceae

Eng., H. and Sans. Names : *Eng.*: Berry, Caper;

H.: Karel, Karer, Karil, Karu, Ker;

Sans.: Granthila, Gudhapatra, Karaka, Karira.

Distribution : Dry areas throughout India.

Reproductive Cycle : March-June.

Part Used : Whole Plant.

Folk Use : Plant used for toothache in children.

Reference : Pal *et al.* (2000).

Capparis sepiaria L.

Syn.: *C. incanescens* DC.

Family: Capparaceae

Eng., H. and Sans. Names : *Eng.*: Indian Caper;
H.: Hiun, Kanthari;
Sans.: Kakadani, Kanthari.

Distribution : Throughout dry regions of India including Andamans.

Reproductive Cycle : February-June.

Part Used : Root.

Folk Use : Paste of roots crushed along with ginger and asafoetida applied externally to cure mumps by Bhils, Dhanka, Dubada and Nayaka tribes of Khedbrahma region in N. Gujarat.

References : Bhatt and Sabins (1987), Jain (1991), Parrotta (2001).

Capsicum annum L. (Pl. 4C)*

Syn.: *C. frutescens* Sensu Cl.

Family: Solanaceae

Eng., H. and Sans. Names : *Eng.*: Guinea Pepper, Red Chilli, Sweet Pepper;
H.: Gramirch, Lal Mirch, Wangru;
Sans.: Lanka, Marich.

Distribution : Throughout India.

Reproductive Cycle : July-September.

Parts Used : Fruit and Seed.

Folk Uses : Infusion of fruit powder (1g) of *Capsicum annum* and shoot (25g) of *Leucas plukenetii* prescribed for gargling twice a day in tonsillitis. Macerated chili kept for a while on sting's (bee and wasp) spot to cure reaction in children. Seeds also used in tonsillitis.

References : Asolkar *et al.* (1992), Dhiman (2003), Dutta and Nath (2000), Singh and Singh (2003a).

Caralluma aucheriana (Decne.) N. E. Br.

Syn.: *Boucerosia aucheriana* Decne.; *Crenulluma aucheriana* (Decne.) Plowes.

Family: Asclepiadaceae

Distribution : Dry hills in the W. Punjab.

Part Used : Whole Plant.

Folk Use	:	Plant used for curing fever in children.
Reference	:	Pal *et al.* (2000).

***Cardiospermum helicacabum* L.**

Syn.: *C. microcarpum* Kunth.

Family: Sapindaceae

Eng., H. and Sans. Names	:	*Eng.*: Ballon Vine, Blister Creeper, Heart Pea, Winter cherry;
		H.: Kanphuti, Kapalphoti;
		Sans.: Indravalli, Jyotir, Sakralata.
Distribution	:	Throughout Indian plains.
Reproductive Cycle	:	August-January.
Parts Used	:	Whole Plant and Seed.
Folk Uses	:	One or 2 tsp extract of fresh plant fed to infants in flatulence. Plant also used to cure earache. A necklace of its seeds worn to prevent contagious diseases affecting children.
References	:	Barua *et al.* (2000), Pal *et al.* (2000), Patil (2004), Punjani (2002).

***Carex baccans* Nees.**

Syn.: *C. curvirostris* Kuntze.; *C. javanica* Boeckeler.; *C. recurvirostra* Kuntze.

Family: Cyperaceae

Eng. Name	:	*Eng.*: Crimson Seeded Sedge.
Distribution	:	Sikkim and Khasia to the Naga hills up to 2,350m; Malabar Ghats-Sri Lanka: 1,000-2,000m.
Reproductive Cycle	:	Blooms in April-May.
Part Used	:	Root.
Folk Use	:	Root paste applied on boils among children.
Reference	:	Brahman and Saxena (1990).

***Careya arborea* Roxb.**

Syn.: *C. coccinea* Sensu Alston.

Family: Myrtaceae

Eng., H. and Sans. Names	:	*Eng.*: Patana Oak, Wild Guava;
		H.: Kamba, Kuba, Kumbhi, Vakamba;
		Sans.: Bhadrendrani, Girikarnika, Vishaghnika.

Distribution	:	Throughout India, Sri Lanka, Malay Peninsula and Siam.
Reproductive Cycle	:	March-May.
Parts Used	:	Stem and Bark.
Folk Uses	:	Stem fibre tied to the waist to cure 'dhuda' (a form of diarrhoea in case of infants). Bark paste applied all over the body of children for healthy growth. Also, it acts as an antipyretic and antipruritic in eruptive fevers, particularly in smallpox.
References	:	Anonymous (1950), Hosagoudar and Henry (1996a), Pal *et al.* (2000), Parrotta (2001), Sen and Behera (2007).

***Carica papaya* L. (Pl. 4D)**

Syn.: *Papaya carica* Gaertn.

Family: Caricaceae

Eng., H. and Sans. Names	:	*Eng.*: Melon Tree, Papaw, Paw-Paw, Papaya Tree; *H.*: Anda, Kharbuja, Papita; *Sans.*: Chirbita, Eranda, Naikadala, Parisha.
Distribution	:	Commonly cultivated and more or less naturalized in India.
Reproductive Cycle	:	Throughout the year.
Parts Used	:	Fruit and Latex.
Folk Uses	:	Fruits used in tonsillitis. Unripe fruits valued for excretory troubles of children. Latex smeared on body of children to cure rickets. Also, infusion of latex and honey given to children for expelling roundworms.
References	:	Asolkar *et al.* (2000), Bhattacharjee (2001), Kshirsagar and Singh (2003, 2007), Sood *et al.* (2012).

***Carissa carandas* L.**

Syn.: *Arduina carandas* (L.) Baill.; *Capparis carandas* (L.) Burm.f.

Family: Apocynaceae

Eng., H. and Sans. Names	:	*Eng.*: Caranda, Christ's Thorn, Jasmine Flowered Carrisa, Karaunda; *H.*: Karaunda; *Sans.*: Avighnah, Karamardah.
Distribution	:	Rajasthan, Gujarat and W. Ghats from Maharashtra southwards.

Reproductive Cycle : April-June.

Part Used : Whole Plant.

Folk Use : Plant used for curing fever in children.

Reference : Pal *et al.* (2000).

Carthamus tinctorius **L.**

Syn.: *Carduus tinctorius* Ehrh.; *Carthamus glaber* Burm.f.

Family: Asteraceae

Eng., H. and Sans. Names : *Eng.*: African Saffron, American Saffron, Bastard Saffron, Dyer's Saffron, Safflower, Wild Saffron;

H.: Barre, Karrah, Kasumba, Kussum;

Sans.: Agnishikha, Gramyakunkuma, Kamlottama, Kusumbha, Lohita, Pita, Rakta.

Distribution : Throughout India.

Reproductive Cycle : Autumn and Winter.

Parts Used : Whole Plant and Flower.

Folk Uses : Whole plant including infusion of flower given to children and infants in measles.

References : Agarwal (2003), Anonymous (1976b), Bhattacharjee (2001), Khare (2004), Kirtikar and Basu (1984), Sharma (2003), Singh *et al.* (1996), Watt (1972).

Carum carvi **L.**

Syn.: *Bunium carvi* (L.) M. Bieb.; *Carum aromaticum* Salisb.

Family: Apiaceae

Eng., H. and Sans. Names : *Eng.*: Caraway, Common Cari;

H.: Kala Zira, Shia-Jira, Zira;

Sans.: Bhedini, Hridya, Jarana, Krishna, Krishnajaji, Nila, Ruchya, Sugandha, Sushavi, Varshakali.

Distribution : Wild in N. Himalayas; cultivated in plains, Kashmir, Kumaon, Garhwal, Chamba.

Reproductive Cycle : July-November.

Part Used : Fruit.

Folk Use : Fruits used for flatulent colic in children.

Reference : Karnick (1994a).

Cascabela thevetia **(L.) Lippold.**

Syn.: *Cerbera peruviana* Pers.; *Thevetia peruviana* (Pers.) Schum.

Family: Apocynaceae

Eng., H. and Sans. Names	:	*Eng.*: Bastard Oleander, Exile Oleander, Lucky Nut Tree, Yellow Oleander;
		H.: Kananir, Pila;
		Sans.: Ashvamaraka, Nakharotila, Sidhapushpa.
Distribution	:	Planted throughout India.
Reproductive Cycle	:	July-September.
Parts Used	:	Whole Plant, Leaf and Latex.
Folk Uses	:	Plant used for curing skin diseases. Leaves used for children's ailment by Khasi and Jantia tribes in Meghalaya. Plant latex with salt and turmeric powder smeared on swellings due to mumps.
References	:	Kharkonger and Joseph (1981), Pal *et al.* (2000), Viswanathan (1997).

Casearia tomentosa **Roxb.**

Syn.: *Anavigna lanceolata* Lam.; *Casearia anavinga* Dals. and Gibs.; *C. canziala* Ham.; *C. elliptica* Willd.

Family: Samydaceae

Eng., H. and Sans. Names	:	*Eng.*: Toothed Leaf Chilla;
		H.: Bairi, Beri, Bheri, Chilara, Chilla, Kirchi;
		Sans.: Chilhaka, Kampillaka.
Distribution	:	Moist, hilly locations up to 900m from the foothills of the Himalaya southwards and in the Andaman islands throughout India.
Reproductive Cycle	:	February-June.
Part Used	:	Root Bark.
Folk Use	:	Root bark applied on enlarged scrotum in children.
Reference	:	Kothari *et al.* (2003).

Cassia fistula **L.**

Syn.: *C. rhombifolia* Roxb.; *Cathartocarpus fistula* Pers.

Family: Fabaceae

Eng., H. and Sans. Names	:	*Eng.*: Golden Shower, Indian Laburnum, Purging Fistula;

		H.: Amaltas, Girimala, Kirala, Sinar;
		Sans.: Aragvadha, Krtamalah, Suvarnaka.
Distribution	:	Wild in many parts of India; also cultivated as an ornamental plant for attractive yellow blossoms in pendant racemes in C. and E. Himalayas-Sri Lanka and Malacca, ascending to 1,000m.
Reproductive Cycle	:	Flowering during April-June and Fruiting throughout the year.
Parts Used	:	Whole Plant, Root, Leaf, Stem Bark and Fruit.
Folk Uses	:	Plant used for fever, poliomyelitis, stomach pain, dysentery and as a moderate laxative for children. Roots used in misperistalsis in infants. Young leaves or its extracts given to children for eradication of intestinal worms. Extract of stem bark administered orally to check vomiting sensation in children. Fruit pulp given to children above 12 years of age as a purgative for chronic constipation; also commonly applied round navel to produce motions in flatulent colic. Pulp also given to children to relieve disorders of liver and biliousness.
References	:	Balasubramanian and Prasad (1996), Khare (2004), Pal *et al.* (2000), Parrotta (2001), Sanjay *et al.* (2007), Watt (1972).

***Cassia italica* (Mill.) Lam. *ex* F. W. Andrews.**

Syn.: *Senna italica* Mill. Gard.

Family: Fabaceae

Eng., H. and Sans. Names	:	*Eng.*: Alexandria Senna, Country Senna, Italian Senna, Senegal Senna;
		H.: Chhotataroda;
		Sans.: Bhumyahuli, Bhutalapota.
Distribution	:	Deccan, Gujarat, Punjab.
Reproductive Cycle	:	August-January.
Part Used	:	Leaf.
Folk Use	:	Leaves used in tonsillitis; also used as a purgative for children below 10 years.
Reference	:	Kalpesh *et al.* (2004).

Cassine glauca **(Rottb.) Kuntze.**

Syn.: *Barringtonia sphaerocarpa* C. A. Gardner.; *Elaeodendron glaucum* (Rottb.) Pers.; *E. paniculatum* Wight and Arn.; *Mangifera glauca* Rottb.

Family: Celastraceae

Eng., H. and Sans. Names	:	*Eng.*: Ceylon Tea; *H.*: Bakra, Chouri, Dhebri, Jamrassi; *Sans.*: Bhutapala.
Distribution	:	Throughout India.
Reproductive Cycle	:	April-August.
Parts Used	:	Whole Plant and Leaf.
Folk Uses	:	Plant good against diarrhoea, dysentery and blood dysentery. Paste of crushed leaves applied on children's body to treat cramp diseases followed by bath with fresh water.
References	:	Khanna *et al.* (1996a), Pal *et al.* (2000).

Cassytha filiformis **L.**

Syn.: *C. americana* Meissn.

Family: Lauraceae

Eng., H. and Sans. Names	:	*Eng.*: Filiformis Cassytha, Sita's Yarn; *H.*: Amarbel, Amarbeli; *Sans.*: Akashavalli, Akasvalli.
Distribution	:	Found throughout the greater parts of India, particularly nearby the sea coasts.
Reproductive Cycle	:	January-December.
Parts Used	:	Whole Plant and Stem.
Folk Uses	:	An amulet of portion of plant worn around neck, arm and ankles by the Santals as a cure for rickets. Medicated oil obtained by boiling its stem together with larvae of tiger beetle (found in cow dung) and mustard oil used for massaging weak legs of children.
References	:	Banerjee (2000a), Dhiman (2003), Kapur and Nanda (1996).

Catharanthus roseus **G. Don. (Pl. 5A)**

Syn.: *Ammocallis rosia* (L.) Small.; *Lochnera rosea* (L.) Reichb.; *Vinca rosea* L.

Family: Apocynaceae

Eng., H. and Sans. Names : *Eng.*: Old Maid, Pink Periwinkle, Red Periwinkle, Rosy-flowered Indian Periwinkle;

H.: Sadabahar, Sadaphul, Sadasuhagen;

Sans.: Nityakalyani, Rasna, Sadampuspa.

Distribution : Cultivated on a large scale in some areas of Tamil Nadu.

Reproductive Cycle : During the rainy season.

Parts Used : Whole Plant, Leaf and Flower.

Folk Uses : Plant considered useful in ringworm and dysentery. Decoction of flowers with a few drops of alcohol used as an eye wash for infants. Leaves good for treating acute leukaemia.

References : Bhattacharjee (2001), Jha and Kumar (2003), Pal *et al.* (2000), Sharma (2003).

Catunaregam spinosa **(Thunb.) Tirveng.**

Syn.: *Randia brandisii* Gamble.; *R. dumetorum* (Retz.) Poir.; *R. tomentosa* Wight and Arn.; *Xeromphis spinosa* (Thunb.) Keay.

Family: Rubiaceae

Eng., H. and Sans. Names : *Eng.*: Common Emetic Nut, Emetic Nut Tree;

H.: Arar, Madan, Mainphal;

Sans.: Bastishodhana, Madana, Ramachhardanaka.

Distribution : Moist and dry deciduous forests throughout India up to 1,350m in Himalayas and 1,200m in W. Ghats.

Reproductive Cycle : April-July.

Part Used : Fruit.

Folk Use : Fruits useful as nervine, calmative and antispasmodic remedy for the teething ailments of the children.

References : Anonymous (1969), Chatterjee and Pakrashi (1997), Chauhan (1999), Dhiman (2003), Jha and Varma (1996), Kirtikar and Basu (1984), Parrotta (2001), Watt (1972).

Cedrus deodara (Roxb. *ex* D. Don) G. Don.*

Syn.: *Abies deodara* Lindl.; *Cedrus libani* var. *deodara* Hook.f.; *Pinus deodara* Roxb.

Family: Pinaceae

Eng., H. and Sans. Names : *Eng.*: Deodar, Himalayan Cedar, True Cedar;

H.: Dedwar, Deodar, Devdarru;

Sans.: Devadaru, Indravriksha, Kilima, Mastadaru, Shatapadapa.

Distribution : Temperate W. Himalayas: 1,500-3,000m.

Reproductive Cycle : June-July.

Part Used : Whole Plant.

Folk Use : Useful for fever in children.

Reference : Pal *et al.* (2000).

Ceiba pentandra (L.) Gaertn.

Syn.: *Bombax orientale* Spreng.; *B. pentandrum* L.; *Eriodendron anfractuosum* DC.; *E. orientale* Kostel.

Family: Malvaceae

Eng., H. and Sans. Names : *Eng.*: Sinhaporekapok, True Kopok Tree, White Cotton Tree;

H.: Hattian, Saphed Simal;

Sans.: Chirayu, Kutasalmali, Moch, Svetasalmali.

Distribution : Throughout the hotter parts of India.

Reproductive Cycle : October-December.

Part Used : Gum.

Folk Use : Gum used in incontinence of urine in children.

Reference : Watt (1972).

Celastrus paniculatus Willd.

Syn.: *Ceanothus paniculatus* Roth.; *Celastrus alnifolia* Don.; *C. dependens* Wall.; *C. multiflora* Roxb.; *C. nutans* Roxb.; *C. rothiana* Roem.

Family: Celastraceae

Eng., H. and Sans. Names : *Eng.*: Climbing Staff Tree, Intellect Tree;

H.: Malkangani, Malkunki;

Sans.: Jyotishmati, Malkangni, Pitataila.

Distribution	:	Tropical and sub-tropical Himalayas ascending up to 1,400m from Punjab – Assam, Bihar, U.P. and S. India.
Reproductive Cycle	:	April June.
Parts Used	:	Whole Plant, Root, Leaf, Seed and Bark.
Folk Uses	:	Plant useful in scabies. Leaf paste applied on infant's head to strengthen the skull bones. Bark or root paste applied on forehead to cure boils in children. Seed oil used for allergy of children.
References	:	Brahman and Saxena (1990), Pal *et al.* (2000), Topno and Ghosh (2000), Yasodamma *et al.* (2009).

Celosia argentea L.

Syn.: *C. cristata* L.

Family: Amaranthaceae

Eng., H. and Sans. Names	:	*Eng.*: Cock's Comb, Quail Grass; *H.*: Debkoti, Kokan, Lalmurghka, Safed Murgha, Sufaidmurga, Survari; *Sans.*: Barhichuda, Kekishikha, Madhuchhada, Shikha, Shikhini, Vitunna.
Distribution	:	Throughout India, ascending to 1,350m.
Reproductive Cycle	:	May-October.
Part Used	:	Seed.
Folk Use	:	Seed oil massaged over the body of children to cure cramp disease.
Reference	:	Khanna *et al.* (1996b).

Centaurea cyanus L.

Syn.: *C. concinna* (Boiss. and Huet.) Trautv.; *C. cyaneum* St.-Lag.; *C. lanata* Roxb.; *Cyanus cyanus* Hill.

Family: Asteraceae

Eng. Names	:	*Eng.*: Bluebottle, Cornflower, Devil's Flower, Hurtlesickle, Red Campion.
Distribution	:	Native to the near east, it grows wild in all temperate regions often in cornfields (N.W. India, Punjab, Kashmir and Mysore).
Reproductive Cycle	:	June-October.
Part Used	:	Seed.
Folk Use	:	Seeds used as a mild laxative for children.
Reference	:	Prajapati *et al.* (2006).

Centella asiatica* (L.) Urban.

Syn.: *Hydrocotyle asiatica* L.; *H. erecta* L.f.

Family: Apiaceae

Eng., H. and Sans. Names	:	*Eng.*: Indian Pennywort, Thick-leaved Pennywort; *H.*: Bemgsag, Brahmamanduki, Brahmi, Khulakhudi; *Sans.*: Bhekaparni, Brahmi, Divya, Mandukparni.
Distribution	:	Throughout marshy places up to 1,828m in India.
Reproductive Cycle	:	March-April.
Parts Used	:	Whole Plant and Leaf.
Folk Uses	:	Plant considered good for smallpox and measles. Paste of leaves and turmeric used externally for curing eczema in young children. Roasted leaves given in bowel complaints. Three-four fresh leaves with 1-2 black pepper given to children to strengthen memory. Juice of leaves of *Centella asiatica* and *Euphorbia hirta* given to children in dysentery.
References	:	Chauhan (1999), Chowdhury *et al.* (1996), Dhiman (2007), Gogoi and Das (2003), Kirtikar and Basu (1984), Kshirsagar and Singh (2007), Lalramnghinglova (2003), Nisha and Sivadasan (2007), Prajapati *et al.* (2006), Prasad *et al.* (1996), Sharma (2003), Sharma and Rana (2000), Viswanathan (1997).

***Centratherum anthelminticum* (L.) Kuntze.**

Syn.: *Conyza anthelmintica* (L.) Willd.; *Serratula anthelmintica* Roxb.; *Vernonia anthelmintica* L.

Family: Asteraceae

Eng., H. and Sans. Names	:	*Eng.*: Purple Fleabane; *H.*: Banjira, Kalijira, Somraj, Somraji, Vapchi; *Sans.*: Agnibija, Bakuchee, Somraji, Vakushi.
Distribution	:	Wasteplaces, hedges and cultivated lands up to 1,500m throughout India.
Reproductive Cycle	:	January-May.
Part Used	:	Resin.
Folk Use	:	Alcoholic extract of resin powder given to children for expelling thread worms.
Reference	:	Chatterjee and Pakrashi (1997).

Cerasus vulgaris **Mill.**

Syn.: *Prunus cerasus* L.; *P. vulgaris* (Mill.) Schur.

Family: Rosaceae

Eng. and H. Names	:	*Eng.*: Dwarf Cherry, Red Cherry, Sour Cherry, Wilol Cherry; *H.*: Alubalu.
Distribution	:	Grows in Kashmir and Garhwal Himalayas; cultivated in N.W. Himalayas up to 2,700m.
Reproductive Cycle	:	April-June.
Part Used	:	Leaf.
Folk Use	:	Leaf infusion useful in convulsion of children.
References	:	Agarwal (2003), Ambasta (1986), Anonymous (1969), Bhattacharjee (2001).

Ceriscoides turgida **(Roxb.) Tirveng.**

Syn.: *Gardenia montana* Roxb.; *G. turgida* Roxb.; *Randia turgida* (Roxb.) Tirveng.

Family: Rubiaceae

H. and Sans. Names	:	*H.*: Ghurga, Ghurgia, Karamba, Khuriari, Khurrur, Mhaner, Thanella; *Sans.*: Bharangi, Bharngi, Mahapindi.
Distribution	:	Sub-Himalayan tract from Nepal-Jumna, ascending to 1,350m; also in Rajputana, Myanmar and C. and S. India.
Reproductive Cycle	:	April-June.
Part Used	:	Root.
Folk Use	:	Roots employed by Santals as a remedy for indigestion in children.
References	:	Anonymous (1956), Bhattacharjee (2001), Kirtikar and Basu (1984), Parrotta (2001), Watt (1972).

Ceropegia candelabrum **L.**

Syn.: *C. acuminata* Dalz. and Gibs.; *C. candelabrum* subsp. *tuberosa* (Roxb.) H. Huber.; *C. tuberosa* Roxb.

Family: Asclepiadaceae

Eng., H. and Sans. Names	:	*Eng.*: Glabrous Goglet Flower; *H.*: Patalatumbi;

		Sans.: Bhutumbi, Devi, Divyatumbi, Gartalabu, Nagatumbi.
Distribution	:	S. India and Maharashtra.
Reproductive Cycle	:	September-October.
Part Used	:	Tuber.
Folk Use	:	Tubers used in bowel complaints of the children.
References	:	Agarwal (2003), Jain (1991), Kirtikar and Basu (1984).

Chenopodium album **L.***

Syn.: *C. giganteum* Don.; *C. laciniatum* L.; *C. nepalense* Hort.; *C. purpurascens* Ham.; *C. viride* L.

Family: Chenopodiaceae

Eng., H. and Sans. Names	:	*Eng.*: Bacon Weed, Hen-fat, Rag Jag, Wild Spinach;
		H.: Bathua, Bethuasag, Cheel;
		Sans.: Agaralohita, Chilika, Vastukah, Vasuka.
Distribution	:	Throughout India, in areas up to 4,200m (wild as well as cultivated).
Reproductive Cycle	:	May-October.
Part Used	:	Leaf.
Folk Uses	:	Finely powdered leaves used as a dusting powder around the external genitalia in children. 2 tsp of leaf juice given to children on empty stomach in the morning for a week as vermicide.
References	:	Asolkar *et al.* (1992), Kirtikar and Basu (1984), Nath *et al.* (2007).

Christella dentata **(Forssk.) Holttum.**

Family: Thelypteridaceae

Eng. Names	:	*Eng.*: Downy Wood Fern, Oak Fern.
Distribution	:	Throughout hilly regions of India, ascending to 1,800m.
Reproductive Cycle	:	Spores throughout the year.
Part Used	:	Whole Plant.
Folk Use	:	Fresh juice of plant applied on skin rashes of infants.
Reference	:	Yasodamma *et al.* (2009).

Chrozophora plicata **(Vahl.) A. Juss.** ***ex*** **Spreng.**

Syn.: *C. burmanni* Spreng.; *C. obliquifolia* (Vis.) Baill.; *C. prostata* Dalz. and Gibson.

Family: Euphorbiaceae

Eng., H. and Sans. Names	:	*Eng.*: Giradol, Turnsole; *H.*: Neel Kanthi, Shahdevi; *Sans.*: Suryavarta.
Distribution	:	Occurs throughout India, from Punjab-Travancore and from Bengal to Pegu and Myanmar.
Reproductive Cycle	:	January-June.
Parts Used	:	Whole Plant, Root and Root Bark.
Folk Uses	:	Plant useful in ringworm. Ash of roots as well as root bark useful for curing cough in children.
References	:	Agarwal (2003), Anonymous (1950), Kirtikar and Basu (1984), Pal *et al.* (2000), Watt (1972).

Chrozophora rottleri **(Geis.) A. Juss.** ***ex*** **Spreng.**

Syn.: *Croton rottleri* Geis.

Family: Euphorbiaceae

Eng., H. and Sans. Names	:	*Eng.*: Rottler's Chrozophora; *H.*: Shadevi, Shanavalli; *Sans.*: Suryavarta.
Distribution	:	Wastelands, roadsides and low dry lands almost throughout India.
Reproductive Cycle	:	January-June.
Parts Used	:	Whole Plant and Root Bark.
Folk Uses	:	Plant good against ringworm. Root bark useful in cough.
References	:	Chatterjee and Pakrashi (1997), Pal *et al.* (2000).

Cicer arietinum **L.**

Syn.: *C. album* Hort.; *C. grossum* Salisb.; *C. nigrum* Hort.; *C. physodes* Rchb.

Family: Fabaceae

Eng., H. and Sans. Names	:	*Eng.*: Bengal Gram, Chick Pea, Chicken-Pea, Gram; *H.*: But, Chana, Chanaka, Chunna; *Sans.*: Balobhojya, Canaka, Chanaka, Jivana, Sugandha.

Distribution	:	Cultivated in subtropical zones of India; commonly cultivated in Nilgiris.
Reproductive Cycle	:	February-May.
Part Used	:	Seed.
Folk Uses	:	White gram useful in rickets, and given to children in protein defficiency.
Reference	:	Agarwal (2003).

Cinchona calisaya **Wedd.**

Syn.: *C. amygdalifolia* Wedd.; *C. australis* Wedd.

Family: Rubiaceae

Eng. and H. Names	:	*Eng.*: Bolivian Bark, Cinchona Bark, Peruvian Bark, Yellow Bark; *H.*: Kuenine.
Distribution	:	Throughout India, grows best in the Himalaya and Sikkim.
Part Used	:	Bark.
Folk Use	:	Bark used in smallpox and acute tonsillitis.
Reference	:	Anonymous (1976b).

Cinnamomum camphora **(L.) Nees and Ebrem.**

Syn.: *Camphora officinarum* Nees *ex* Wall.; *Laurus camphora* L.

Family: Lauraceae

Eng., H. and Sans. Names	:	*Eng.*: Camphor Cinnamomum, Camphor Tree, Official Camphor; *H.*: Kapur, Karpur; *Sans.*: Chandra, Kapura, Karpura.
Distribution	:	Naturalized and planted in some gardens in India, up to 1,200m in the N.W. Himalayas.
Reproductive Cycle	:	March-July.
Part Used	:	Whole Plant.
Folk Uses	:	Warmed camphor liminent rubbed over chest at night to give relief in cough of children. Also, used in smallpox and measles.
References	:	Dhiman (2003), Sood *et al.* (2012).

***Cinnamomum impressinervium* Meissn.**

Syn.: *C. cacharense* R. Parker.

Family: Lauraceae

Distribution	:	Sikkim Himalayas.
Reproductive Cycle	:	July-December.
Part Used	:	Seed.
Folk Uses	:	Seed powder with honey given to children in dysentery and cough.
Reference	:	Agarwal (2003).

***Cinnamomum iners* Reinw. *ex* Blume**

Syn.: *C. aromaticum* Zoll.; *C. curtisii* Lukman.; *C. rauwolfii* Blume.

Family: Lauraceae

Eng., H. and Sans. Names	:	*Eng.*: Wild Cinnamom; *H.*: Darchini, Jangli-Darchini; *Sans.*: Patra, Patraka, Tamalapatra, Tamalapatram, Tejpatra, Tilaka.
Distribution	:	W. Ghats, S. India.
Part Used	:	Seed.
Folk Uses	:	Bruised seeds mixed with honey or sugar given to children in dysentery, cough and fever.
References	:	Agarwal (2003), Ambasta (1986), Kirtikar and Basu (1984), Watt (1972).

***Cinnamomum tamala* Nees and Eberm.**

Syn.: *C. albiforum* Nees.; *C. cassia* Don.; *Laurus albiflora* Wall.; *L. cassia* Roxb.; *L. tamala* Ham.; *L. tazia* Ham.; *Persea tamala* Sperng.

Family: Lauraceae

Eng., H. and Sans. Names	:	*Eng.*: Cassia Cinnamon, Indian Cassia Lignea; *H.*: Kikoa, Silkanti, Tajkalam, Tejpatt; *Sans.*: Ankusha, Dala, Patraka, Roma, Surasa, Tamalapatram, Vasanavhaya.
Distribution	:	Tropical and sub-tropical Himalayas (900-2,500m) and in Khasi and Jaintia hills (900-1,200m).
Reproductive Cycle	:	July-October.
Part Used	:	Stem Bark.

Folk Use	:	A paste of stem bark applied externally around neck twice a day till relief to cure mumps.
Reference	:	Nath and Maiti (2003).

***Cinnamomum zeylanicum* Breyn.**

Syn.: *C. verum* Presl.

Family: Lauraceae

Eng., H. and Sans. Names	:	*Eng.*: Ceylon Cinnamon, Cinnamon; *H.*: Dalchini, Daruchini, Qualamidarchini; *Sans.*: Darusita, Gudatvacha, Hridya, Sinhala, Tvaka, Utkata.
Distribution	:	Cultivated in S. India and in the coastal regions of W. India.
Reproductive Cycle	:	December-July.
Part Used	:	Bark.
Folk Use	:	Dried inner bark used in infantile diarrhoea.
Reference	:	Karnick (1994a).

***Cirsium wallichii* DC.**

Syn.: *Cnicus wallichii* Hook.f.

Family: Asteraceae

Eng. and H. Names	:	*Eng.*: Wallich's Thistle; *H.*: Bungsee.
Distribution	:	Temperate Himalayas, (Marri-Bhutan), ascending to 4,000m, Nilgiri mountains, up to 2,000m.
Reproductive Cycle	:	February-July.
Part Used	:	Whole Plant.
Folk Use	:	Young plant parts given to children in cold.
Reference	:	Kumar *et al.* (2000).

***Cissampelos pareira* L. (Pl. 5B)**

Syn.: *C. caapeba* L.; *C. delicatula* Miers.; *C. elata* Miers.; *C. eriantha* Miers.; *C. nepalensis* Rhode.; *Menispermum orbiculatum* L.

Family: Menispermaceae

Eng., H. and Sans. Names	:	*Eng.*: False Pareira, Pareira Brava Root, Velvet-Leaf; *H.*: Akanadi, Pari, Patha; *Sans.*: Laghupatha, Patha.

Distribution : Distributed throughout the tropical and sub-tropical India, commonly found in shady areas of the forests of Garhwal between 700-1,800m.

Reproductive Cycle : August-November.

Parts Used : Whole Plant, Leaf and Root.

Folk Uses : Plant considered useful in diarrhoea, stomach pain and bowel complaints of children. Decoction of roots used as a wash for children by the Pedis having pimples over their body. Paste of root used against infantile diarrhoea by the Mundas of Chota Nagpur. Decoction of roots (half tsp) also given twice a day for 3-7 days to cure jaundice. Crushed leaves applied on the palate in case of heat in children.

References : Chauhan (1999), Jain *et al.* (2003), Kirtikar and Basu (1984), Pal *et al.* (2000).

Cissus quadrangularis **L.**

Syn.: *C. edulis* Dalz.; *C. quadrangula* L.;
***Vitis quadrangularis* (L.) Wall. *ex* Wight.**

Family: Vitaceae

Eng., H. and Sans. Names : *Eng.*: Adamant Creeper, Bone Setter, Edible-Stemmed Vine;

H.: Hadjod, Hadjora;

Sans.: Amara, Asthisrnkhala, Kulisha, Vajravalli.

Distribution : Throughout the hotter parts of India, from the foot of W. Himalayas in Kumaon-Sri Lanka and Malacca.

Reproductive Cycle : May-August.

Parts Used : Leaf and Stem.

Folk Uses : Extract of tender leaves and stem mixed with breast milk, administered 1/2 tsp once a day for 3 days to infants for fever by Nuka Doras and Porjas.

Reference : Rao and Henry (1996).

Citrullus colocynthis **(L.) Schrad.**

Syn.: *C. colocynthoides* Pangala.; *Colocynthis officinalis* Schard.;
***C. vulgaris* Schrad.; *Cucumis colocynthis* L.**

Family: Cucurbitaceae

Eng., H. and Sans. Names : *Eng.*: Bitter Apple, Bitter Cucumber, Colocynth;

H.: Badi Indrayan, Ghorumba, Indrayan, Makkal, Tumba;

		Sans.: Atmaraksha, Brihatphala, Brihavaruni, Chitrala, Chitraphala, Chitravalli, Devi, Dirghavalli, Katurasa, Kaya.
Distribution	:	Cultivated and wild, throughout India in warm areas.
Reproductive Cycle	:	May-October.
Part Used	:	Root.
Folk Uses	:	Roots used in visceromegaly in children; also its paste applied to enlarged abdomen of children.
References	:	Kirtikar and Basu (1984), Prajapati *et al.* (2006).

Citrus aurantifolia **(Christm) Swingle.***

Syn.: *C. acida* Pers.; *C. excelsa* Wester.; *C. medica* L. var. *acida* Hook.f.

Family: Rutaceae

Eng., H. and Sans. Names	:	*Eng.*: Acid Lime, Citron Lime, Sour Lime; *H.*: Baranimbu, Bijaura, Kaghzi Nimbu, Lebu, Limbu, Limun, Nebu, Nibu, Nimbu; *Sans.*: Amlajambir, Bijapura, Dantaghata, Jamhira, Jantumari, Vijapura.
Distribution	:	Tropical parts of India including valleys of Himalayas and Khasia hills; also cultivated in W. Bengal, Punjab, M.P., Maharashtra, Karnatka, A.P., Tamil Nadu and U.P.
Reproductive Cycle	:	July-October.
Part Used	:	Leaf.
Folk Uses	:	Preparation comprises of its crushed tender leaves, rhizome of *Curcuma longa* and leaves of *Cynodon dactylon* in equal amount by weight along with a little fresh lime water given for vomiting and indigestion of children.
Reference	:	Sarma *et al.* (2002).

Citrus aurantium **var.** ***bergamia*** **(Risso) Brandis.**

Family: Rutaceae

Eng., H. and Sans. Names	:	*Eng.*: Bergamotte Orange, The Bergamot; *H.*: Limu, Limun, Nibu; *Sans.*: Dantaharshana, Dantakarshana, Jadyari, Jambha, Jambhaka, Jambhala, Jambir, Nimbu-ka, Rewatavakra-Sodhi.
Distribution	:	Hot valleys along the foot of Himalayas, from Garhwal eastwards to Sikkim and Khasia mountains.

Reproductive Cycle : July-October.

Part Used : Fruit.

Folk Uses : Juice of fruit useful as a cooling drink in smallpox and measles.

References : Kirtikar and Basu (1984), Watt (1972).

Cleistanthus collinus (Roxb.) Benth. *ex* Hook.f.

Syn.: *Amanoa collina* (Roxb.) Baill.; *Andrachne orbiculata* Roth.; *Bridelia collina* (Roxb.) Hook. and Arn.

Family: Euphorbiaceae

Eng., H. and Sans. Names : *Eng.*: Dog's Mercury;

H.: Garari, Garrar;

Sans.: Indrayava, Kaudigam, Nandi.

Distribution : Dry hills in various parts of India, from Shimla-Bihar and southwards to C. India and Deccan Peninsula.

Reproductive Cycle : Flowering during October-December.

Part Used : Whole Plant.

Folk Use : Plant used in various skin diseases of children.

Reference : Pal *et al.* (2000).

Clematis gouriana Roxb. *ex* DC.

Syn.: *C. canna* Wall.; *C. martini* H. Lev.; *C. substipulata* Kuntze.

Family: Ranunculaceae

Eng., H. and Sans. Names : *Eng.*: Indian Traveller's Joy, Gourian Clematis;

H.: Belkangu, Churanhar;

Sans.: Morata.

Distribution : W. temperate Himalayas-Kumaon, at 2,000-3,650m.

Reproductive Cycle : November-February.

Part Used : Whole Plant.

Folk Use : Plant good for healing injured eyes.

Reference : Pal *et al.* (2000).

Clematis hexapetala Pall.

Syn.: *C. angustifolia* Jacq.

Family: Ranunculaceae.

Eng. Name : *Eng.*: Six-Petalled Clematis.

Distribution	:	On slopes and by roadsides in China, Korea, occurs in ranges between 2,000-2,350m.
Reproductive Cycle	:	June-October.
Part Used	:	Root.
Folk Use	:	Used in tonsillitis.
Reference	:	Prajapati *et al.* (2006).

Cleome gynandra L.

Syn.: *C. pentaphylla* L.; *Gynandropis gynandra* (L.) Briq.; *G. pentaphylla* (L.) DC.

Family: Capparaceae

Eng., H. and Sans. Names	:	*Eng.*: Spider Flower, Wild Mustard; *H.*: Chorota, Jangli Hurhur, Karalia, Setakala Arak; *Sans.*: Ajagandha, Bramhagarbha, Charavella, Hulkul, Putimayusika, Sugandhika, Suryavarta, Tilaparni, Ugragandha.
Distribution	:	Abundant throughout the warmer parts of India and in all tropical countries.
Reproductive Cycle	:	Throughout the year but profuse in Summer.
Part Used	:	Seed.
Folk Use	:	Seed decoction used in dysentery of children.
Reference	:	Painuli and Maheshwari (1996).

Cleome viscosa L.

Syn.: *C. icosandra* L.; *Polanisia isocandra* Wight and Arn.; *P. viscosa* DC.

Family: Capparaceae

Eng., H. and Sans. Names	:	*Eng.*: Dog Mustard, Sticky Cleome, Wild Mustard; *H.*: Hulpul, Hurhur, Hurhureh, Jangliharrar, Kanphytia; *Sans.*: Adityabhakta, Bhaskareshta, Pasugandha.
Distribution	:	Throughout India, upto 1,524m.
Reproductive Cycle	:	July-September.
Part Used	:	Seed.
Folk Uses	:	Seeds valued for anthelmintic and rubefacient properties, and given twice daily for 2 days followed on third day by a dose of castor oil; also used in infantile convulsions.
References	:	Asolkar *et al.* (1992), Kirtikar and Basu (1984).

Clerodendrum indicum (L.) Kuntze.

Syn.: *C. siphonanthus* R. Br.; *Siphonanthus indicus* L.

Family: Lamiaceae

Eng., H. and Sans. Names : *Eng.*: Tuber-Flower;

H.: Bharangi;

Sans.: Bhargi, Brahmuni, Yashtika, Yushtika.

Distribution : Warmer parts of India near hedges and roadsides.

Reproductive Cycle : May-October.

Part Used : Root.

Folk Uses : Medicated oil obtained by boiling 20g of its root with seed oil of *Guizotia abyssinica* massaged twice a day for one month in case of rickets. Paste of roots also recommended for external application in marasmus of children.

References : Tirkey (2004), Watt (1972).

Clerodendrum infortunatum L.

Syn.: *C. calycinum* Turcz.; *C. viscosum* Vent.

Family: Lamiaceae

Eng., H. and Sans. Names : *Eng.*: Glory Tree;

H.: Bhant, Bhat, Ghato, Thunera;

Sans.: Barhibarha, Barhichuda, Bhandira, Bhantaka, Bhanti, Harita, Kukkura, Shirnaroma, Vikacha, Vikinaroma.

Distribution : Common in wasteplaces throughout the greater parts of India.

Reproductive Cycle : January-April.

Parts Used : Leaf and Root.

Folk Uses :

Fresh juice of leaves employed as a vermifuge, bitter tonic and malarious fevers in children. Bath with decoction of leaves of this plant along with leaves of *Phyllanthus acidus,* Neem, *Glycosmis pentaphylla* water prescribed against measles. 5g of root paste in mother's milk given to children as an anthelmintic.

References : Brahman and Saxena (1990), Dey and Bahadur (1973), Kirtikar and Basu (1984), Nath and Maiti (2003), Watt (1972).

Clerodendrum multiflorum **(Burm.f.) O. Kuntze.**

Syn.: *C. phlomidis* L.f.; *Volkameria multiflora* Burm.f.

Family: Lamiaceae

Eng., H. and Sans. Names	:	*Eng.*: Wild Killer; *H.*: Arni, Pirun, Urni; *Sans.*: Agnimantha, Agnimanthah, Jatha, Jayarini, Tapana, Vijaya.
Distribution	:	Throughout India, chiefly in U.P., Bihar, W. Bengal and Deccan Peninsula.
Reproductive Cycle	:	April-June.
Part Used	:	Root.
Folk Use	:	Roots used as a bitter tonic for children during convalescence from measles.
References	:	Agarwal (2003), Anonymous (1976b), Chatterjee and Pakrashi (1997), Dhiman (2003), Kirtikar and Basu (1984), Parrotta (2001), Watt (1972).

Clitoria ternatea **L.**

Syn.: *C. albiflora* Mattei.; *C. bracteata* Poir.; *C. parviflora* Raf.; *C. pilosula* Benth.

Family: Fabaceae

Eng., H. and Sans. Names	:	*Eng.*: Butterfly Pea, Candiflower, Clitoria; *H.*: Aparajita, Aprajit, Kalizer, Kowa, Visnukranti; *Sans.*: Aparajita, Asphota, Nilaghiria.
Distribution	:	Common garden plant throughout India; wild in Assam and S. India.
Reproductive Cycle	:	July-November.
Parts Used	:	Whole Plant, Root Bark, Bark and Seed.
Folk Uses	:	Plant useful in cataract and as an anthelmintic. Root bark very frequently administered to children in croup and pulmonary affections; and also act as an emetic, nauseant, expectorant and laxative. One tsp of bark extract given to children on empty stomach for 3 days as an anthelmintic by Soligas. About 2 tsp seed paste of *Clitoria ternatea* and that of *Ricinus communis* in equal parts by weight given on empty stomach in early morning for three alternative days as a vermifuge to children over 2 years of age.

References : Borthakur (1993), Kshirsagar and Singh (2007), Pal *et al.* (2000), Watt (1972).

Coccinia grandis (L.) Voigt.

Syn.: *Bryonia grandis* L.; *C. Coccinia cordifolia* Cogn.; *C. indica* (W.A.) Naud.

Family: Cucurbitaceae

Eng., H. and Sans. Names : *Eng.*: Ivy Gourd;

H.: Bimb, Kanduri, Kanturi, Kunderi;

Sans.: Bimbi, Bimbika, Chhardini, Gouhi, Kamboja, Katuka, Oshthi, Tundi, Vimba.

Distribution : Cultivated as well as wild throughout Indian Subcontinent.

Reproductive Cycle : March-December.

Part Used : Whole Plant.

Folk Use : Plant used against measles.

Reference : Pal *et al.* (2000).

Cocculus hirsutus (L.) Diels.

Syn.: *C. sepium* Colebr.; *C. villosus* Lam.; *Menispermum hirsutum* L.; *M. villosum* Lam.

Family: Menispermaceae

Eng., H. and Sans. Names : *Eng.*: Broom Creeper, Ink Berry;

H.: Chireta, Dier, Hier, Jalayamani, Jamti-Ki-Bel;

Sans.: Chilihintah, Garudi, Mahahala, Patalagarudi, Somavalli, Vasantitikta.

Distribution : E. India, occurs from subtropical Himalayas-S. India.

Reproductive Cycle : July-March.

Parts Used : Whole Plant, Root and Leaf.

Folk Uses : Plant used for inflammation of eyes. Roots with Bonduc nuts administered as a cure for belly-ache in children. Paste of its roots mixed with seeds of *Caesalpina crista* prescribed for children in enteritis and bilious dyspepsia. Leaves beneficial in bed wetting of children.

References : Agarwal (2003), Ambasta (1986), Anonymous (1950), Chatterjee and Pakrashi (1997), Goel and Rajendran (2000), Jain (1991), Kirtikar and Basu (1984), Kothari *et al.* (2003), Pal *et al.* (2000), Watt (1972).

***Cocos nucifera* L.**

Syn.: *Calappa nucifera* (L.) Kuntze.; *Cocos indica* Royle.; *C. nana* Griff.

Family: Arecaceae

Eng., H. and Sans. Names	:	*Eng.*: Cocoa Nut, Cocoanut Palm, Coconut Palm, Coconut Tree; *H.*: Narel, Narial, Nariyal, Nariyalkaper; *Sans.*: Dakshinatriya, Garikera, Jataphala, Langoli, Mahaphala, Nilataru, Payodhara, Rasaphala, Shiraphala, Varaphala.
Distribution	:	Throughout tropical India; chiefly cultivated in Karnataka, Kerala, Tamil Nadu, also found wild in S. India upto 1,000m above sea level.
Reproductive Cycle	:	Throughout the year.
Part Used	:	Nut.
Folk Uses	:	Paste of endocarp ash of coconut mixed with half the dose of powdered tobacco leaves, a few ml of hubble bubble liquid ('hidak mahee') applied 2-3 times daily against scratching inflammation on skin. Immature nuts employed as an astringent in sore throats of children.
References	:	Dey and Bahadur (1973), Kirtikar and Basu (1984), Singh and Singh (2003a), Watt (1972).

***Codiaeum variegatum* (L.) Rumph. *ex* A. Juss.**

Syn.: *C. variegatum* (L.) Blume.; *Croton variegatus* L.; *Crozophyla variegata* L. Raf.

Family: Euphorbiaceae

Eng. Name	:	*Eng.*: Croton, Garden Croton, Joseph's coat.
Distribution	:	Commonly cultivated as an ornamental plant in the Indian gardens.
Reproductive Cycle	:	Flowering in Summer.
Part Used	:	Leaf.
Folk Use	:	Poultice of leaf paste in water applied on abdomen of children in urinary troubles.
References	:	Agarwal (2003), Ambasta (1986).

***Codiocarpus andamanicus* (Kurz) R.A. Howard.**

Syn.: *Apodytes andamanica* Kurz.

Family: Icacinaceae

Distribution : Andaman and Nicobar islands.

Part Used : Leaf.

Folk Use : Warmed paste of pounded leaves in coconut oil tied on enlarged scrotum of children by Nicobarese.

Reference : Dagar and Dagar (2000).

***Coffea benghalensis* Hyne *ex* Roem. and Schult.**

Syn.: *C. horsfieldiana* Miq.

Family: Rubiaceae

Eng. and H. Names : *Eng.*: Bengal Coffee;

H.: Coffee.

Distribution : Common in entire N.E. regions.

Reproductive Cycle : February-December.

Part Used : Leaf.

Folk Use : Infusion of leaves used as bath for infants to reduce fever.

References : Asolkar *et al.* (1992), Jain (1991), Kumar (2002).

***Coix lacryma-jobi* L.**

Syn.: *C. agrestis* Lour.; *C. lacrymal* L.

Family: Poaceae

Eng., H. and Sans. Names : *Eng.*: Christ's Tears, Job's Tears;

H.: Baru, Dabhir, Gurlu, Samkru;

Sans.: Gvedhukah, Jargadi, Kshudra, Kunta.

Distribution : Found in Meghalaya, Sikkim, ascending to 1,500m in hills.

Reproductive Cycle : September-October.

Parts Used : Whole Plant, Leaf and Fruit.

Folk Uses : Plant used for dysentery. Leaves useful in smallpox. Paste of its one fruit pounded with a tuber of *Cyperus rotundus* and a little amount of ground tooth of Swine fed to infants in aphthae, candidiasis, strong diarrhoea and feebleness.

References : Asolkar *et al.* (1992), Barua *et al.* (2000), Jain (1991), Kumar (2002), Pal *et al.* (2000).

***Colchicum luteum* Baker.**

Family: Colchicaceae

Eng., H. and Sans. Names : *Eng.*: Golden Collyrium;

H.: Hirantutiya;

Sans.: Hiranyatutha.

Distribution : W. Himalayas (Kashmir-Chamba): 1,000-3,000m.

Reproductive Cycle : Generally in autumn.

Part Used : Whole Plant.

Folk Use : Plant beneficial in poliomyelitis.

Reference : Pal *et al.* (2000).

***Coleus forskohlii* (Willd.) Briq.**

Family: Lamiaceae

Eng., H. and Sans. Names : *Eng.*: Chebulic Myrobalan, Kaffir Potato (Wild variety);

H.: Hara, Harad, Harara, Gurmal;

Sans.: Gandira.

Distribution : Himalayas from Garhwal-Nepal ascending to 2,500m, Parasnath hills in Bihar and hills of Deccan Peninsula.

Reproductive Cycle : September-October.

Parts Used : Root and Leaf.

Folk Uses : Roots given to children in constipation. Warmed leaves applied on chest of the babies and children suffering from cough and cold.

References : Chatterjee and Pakrashi (1997), Nadanakunjidam (2003).

***Combretum album* Pers.**

Syn.: *C. decandrum* Roxb.; *C. roxburghii* Spreng.; *Pentaptera roxburghii* Tul.

Family: Combretaceae

Eng. and H. Names : *Eng.*: Variable Combretum;

H.: Punk.

Distribution : Found in Andaman Islands, E. India and Peninsular India.

Part Used	:	Whole Plant.
Folk Use	:	Plant used against ringworm.
Reference	:	Pal *et al.* (2000).

Combretum indicum (L.) De Filipps.

Syn.: *Quisqualis glabra* Burm.f.; *Q. indica* L.; *Q. longiflora* Presl.

Family: Combretaceae

Eng., H. and Sans. Names	:	*Eng.*: Chinese Honey Suckle, Rangoon Creeper; *H.*: Rangoon, Rangoon-Ki-Bel; *Sans.*: Rangoon-Ki-Bel.
Distribution	:	Cultivated in gardens throughout India, upto 300m.
Reproductive Cycle	:	February-October.
Part Used	:	Seed.
Folk Uses	:	Ripe seeds used in rickets and as anthelmintics for children.
References	:	Agarwal (2003), Ambasta (1986), Anonymous (1969), Chatterjee and Pakrashi (1997), Parrotta (2001), Prajapati *et al.* (2006), Watt (1972).

Combretum nanum Buch.-Ham. *ex* D. Don.

Family: Combretaceae

Eng. and H. Names	:	*Eng.*: Dwarf Bush Willow; *H.*: Kaikayi.
Distribution	:	Himalayas Terai from Sikkim-Punjab.
Reproductive Cycle	:	March-August.
Part Used	:	Root.
Folk Use	:	Root paste applied on head of infants to expel intestinal worms.
Reference	:	Aminuddin and Girach (1993).

Commiphora mukul (Hook. *ex* Stocks) Engl.

Syn.: *Balsamodendron mukul* Hook. *ex* Stocks.; *B. roxbhurghii* Stocks.; *Commiphora agallocha* Engl.; *C. wightii* (Arnott) Bhandari.

Family: Burseraceae

Eng., H. and Sans. Names	:	*Eng.*: Indian Bdellium Tree, Indian Bedellium, Mukul Myrrh Tree; *H.*: Gugal, Guggul, Mukul; *Sans.*: Agaru, Durga, Guggulu, Koushikaha.

Distribution	:	Occurs throughout W. and S. India.
Reproductive Cycle	:	March-April.
Part Used	:	Bark Resin.
Folk Uses	:	Bark resin used in chronic tonsillitis and marasmus.
References	:	Agarwal (2003), Anonymous (1950), Chatterjee and Pakrashi (1997), Dhiman (2003), Khare (2004).

Commiphora myrrha **(Nees.) Engl.**

Syn.: *Balsamodendron myrrha* T. Nees.; *Commiphora cispidata* Choiv.; *C. coriacea* Engl.; *C. molmol* (Engl.) Engl.

Family: Burseraceae

Eng., H. and Sans. Names	:	*Eng.*: Common Myrrh Tree, Haroboi Myrrh, Myrrh; *H.*: Bol, Gugal, Hirabol; *Sans.*: Bohah, Guggla, Rasagandha.
Distribution	:	Cultivated in India; grown as a garden plant in temperate India.
Reproductive Cycle	:	January-October.
Part Used	:	Whole Plant.
Folk Use	:	Plant useful in tonsillar infections.
Reference	:	Khare (2004).

Coptis teeta **Wall.**

Family: Ranunculaceae

Eng., H. and Sans. Names	:	*Eng.*: Coptis, Gold Thread, Mouth Root, Vegetable Gold; *H.*: Mamira, Mamiran; *Sans.*: Mishmitikta, Pita, Supita.
Distribution	:	Temperate regions of Mishmi hills at the northern frontier of Assam.
Reproductive Cycle	:	Flowering during September-October and Fruiting later.
Part Used	:	Rhizome.
Folk Use	:	Decoction of dried rhizome given twice daily for 4-5 days in infective hepatitis.
References	:	Borthakur *et al.* (2004), Gogoi *et al.* (2003).

Corallocarpus epigaeus **(Rotter) C. B. Clarke.**

Syn.: *Aechmandra epigaea* Arn.; *Bryonia epigaea* Rottler.; *Corallocarpus tavetensis* Gilg.

Family: Cucurbitaceae

Eng., H. and Sans. Names	:	*Eng.*: Indian Bryonia; *H.*: Akasgaddah, Rakas-Gaddah; *Sans.*: Nahika, Sukanasa.
Distribution	:	Punjab, Rawalpindee, Sind, Gujarat, Deccan Peninsula, Belgaum, Sri Lanka.
Reproductive Cycle	:	Flowering in June-July.
Part Used	:	Tuber.
Folk Use	:	Tubers grounded along with turmeric and onion applied externally for treating smallpox.
Reference	:	Murugesan *et al.* (2011).

Corchorus capsularis **L.**

Syn.: *C. cordifolius* Salisb.

Family: Tiliaceae

Eng., H. and Sans. Names	:	*Eng.*: Jute, White Jute; *H.*: Narcha, Pat, Titapat; *Sans.*: Kalasaka.
Distribution	:	Throughout the hotter parts of India.
Reproductive Cycle	:	August-October.
Part Used	:	Whole Plant.
Folk Use	:	Plant useful against worms in children.
Reference	:	Kirtikar and Basu (1984).

Coriandrum sativum **L. (Pl. 5C)***

Syn.: *C. globosum* Salisb.; *C. majus* Gouan.; *Selinum coriandrum* Krause.

Family: Apiaceae

Eng., H. and Sans. Names	:	*Eng.*: Coriander; *H.*: Dhania, Dhanya; *Sans.*: Dhanaka, Dhanyaka, Kustumburi.
Distribution	:	Cultivated throughout India.
Reproductive Cycle	:	February-April.

Parts Used	:	Root, Leaf and Seed.
Folk Uses	:	Powdered leaves and roots macerated with alcohol used in eruptions of measles in children. Cold infusion of seeds with little sugar useful in chickenpox and colics of children.
References	:	Chatterjee and Pakrashi (1997), Jain (1991), Kirtikar and Basu (1984), Parrotta (2001), Watt (1972).

Coscinium fenestratum **(Goetgh.) Colebr.**

Syn.: *C. maingayi* Pierre.; *C. peltatum* Merr.; *Menispermum fenestratum* Gaertn.

Family: Menispermaceae

Eng., H. and Sans. Names	:	*Eng.*: Calumba Wood, Ceylon Calumba Root, Columbo Wood, False Columba, Tree Turmeric; *H.*: Jhar-i-huldi; *Sans.*: Daruharidra, Darvi, Pitadru.
Distribution	:	S. India, Sri Lanka, Sumatra; also found in W. Ghats.
Reproductive Cycle	:	Flowering season from November-December.
Part Used	:	Wood.
Folk Use	:	Used as a cure for tetanus.
Reference	:	Kirtikar and Basu (1984).

Costus speciosus **(Koen.) Sm.**

Syn.: *Banksia speciosa* Koenig.; *Costus speciosus var. nepalensis* (Rosc.).; *Hellenia grandiflora* Retz.

Family: Costaceae

Eng., H. and Sans. Names	:	*Eng.*: Arabian Costus, Spiral Ginger, Wild Ginger; *H.*: Kemuka, Kena, Keu, Kusht, Kust, Kut; *Sans.*: Bramhatirtha, Kashmeera, Kashmira, Kemuka, Kushtha, Pushkara.
Distribution	:	Native of India and Malaya; found in plains throughout India upto 1,200m.
Reproductive Cycle	:	June-December.
Parts Used	:	Rhizome, Stem and Leaf.
Folk Uses	:	Paste of rhizome applied on body for chickenpox. Juice of stem used as drops for ear pain. Poultice of leaves applied locally for curing mumps by Girijanas.

References : Jamir and Rao (1990), Kshirsagar and Singh (2000a, 2007), Kothari and Londhe (2000), Rao and Henry (1996).

Cotula anthemoides L.

Syn.: *Artemisia nilotica* L.; *Cotula chinensis* Kitam.; *C. microcephala* DC.

Family: Asteraceae

Eng. and H. Names : *Eng.*: Babuna, Hlapi-E-Nyenyane or Tuingras, Madras Carpet;

H.: Babuna.

Distribution : Gangetic plains, from Rajmahal and Sikkim westwards to Punjab.

Reproductive Cycle : Flowering during March-June.

Part Used : Whole Plant.

Folk Use : Plant useful in poliomyelitis.

Reference : Pal *et al.* (2000).

Crocus sativus L.

Syn.: *C. autumnalis* Sm.; *C. officinalis* (L.) Honck.; *C. orsinii* Parl.; *Geanthus autumnalis* Raf.

Family: Iridaceae

Eng., H. and Sans. Names : *Eng.*: Saffron, Saffron Crocus, Spanish Saffron;

H.: Kesar, Zafran;

Sans.: Agnishekhara, Aruna, Asra, Dhira, Dipaka, Harichandana, Jaguda, Keshara, Vara.

Distribution : Cultivated mainly in Kashmir valley upto 2,000m and in Chaubhattia in U.P.

Reproductive Cycle : October-December.

Parts Used : Whole Plant and Flower.

Folk Uses : Plant useful in catarrhal affections of children. Dried stigmas and tops of styles used for promoting eruptions in measles; also useful in tonsils.

References : Ambasta (1986), Chauhan (1999), Chauhan (2001), Kaul (1997), Khare (2004), Kirtikar and Basu (1984), Sharma (2003), Watt (1972).

***Crossandra infundibuliformis* (L.) Nees.**

Syn.: *C. axillaris* Nees.; *C. coccinea* Dalzell and Gibson.; *C. undulaefolia* Salisb.; *Justicia infundibuliformis* L.

Family: Acanthaceae

Eng., H. and Sans. Names	:	*Eng.*: Crossandra, Firecracker Flower, Orange Crossandra; *H.*: Aboli, Priyadarsha; *Sans.*: Alankrita Kesara.
Distribution	:	Deccan Peninsula and Sri Lanka; cultivated in N. India, the Malaya Peninsula and Islands.
Reproductive Cycle	:	Blooms throughout the year.
Part Used	:	Bark.
Folk Use	:	Grounded bark with turmeric applied over body for all types of skin diseases.
Reference	:	Satapathy (2008).

***Crotalaria juncea* L.**

Syn.: *C. benghalensis* Lam.; *C. fenestrata* Sims.; *C. porrecta* Wall.; *C. sericea* Wild.; *C. tenuifolia* Roxb.

Family: Fabaceae

Eng., H. and Sans. Names	:	*Eng.*: Brown Hemp, Indian Hemp, Jubbalpore Hemp, Madras Hemp, Sun Hemp; *H.*: Kharif, Sannai, Sunn; *Sans.*: San, Sana, Sunn.
Distribution	:	Plains from the Himalayas-Sri Lanka, Myanmar, Pegu.
Reproductive Cycle	:	Flowering during May-September.
Part Used	:	Root.
Folk Use	:	10-20ml root decoction given twice a day for 7-10 days to cure measles.
Reference	:	Sharma and Singh (1999).

***Crotalaria linifolia* L.f.**

Syn.: *C. caespitosa* Roxb.; *C. diffusa* Wight and Arn.; *C. tecta* B. Heyne.

Family: Fabaceae

Eng. Name	:	*Eng.*: Rattle Pod.
Distribution	:	Tropical region; Sikkim, 300-1,350m, W. Peninsula, Sri Lanka, Myanmar, Pegu.

Part Used	:	Root.
Folk Use	:	Paste of roots with 'neem' leaves applied on body for chickenpox by Savaras.
Reference	:	Rao and Henry (1996).

Crotalaria prostrata **Willd.**

Family: Fabaceae

Eng. and H. Names	:	*Eng.*: Prostrate Rattlepod; *H.*: Chhuchhuni.
Distribution	:	Drier parts of India; plains from the upper Ganges to Sri Lanka, ascending to 2,000m in the Chenab valley.
Reproductive Cycle	:	Flowering during October-November.
Part Used	:	Root.
Folk Use	:	Roots used in infantile diarrhoea by the Mundas.
References	:	Kirtikar and Basu (1984), Parrotta (2001).

Crotalaria spectabilis **Roth.**

Syn.: *C. lescenaultii* DC.; *C. sericea* Retz.

Family: Fabaceae

Eng., H. and Sans. Names	:	*Eng.*: Showy Rottlepad; *H.*: Jhunjhunia, Thunjhunia; *Sans.*: Ghantarava.
Distribution	:	Tropical parts of India, ascending upto 1,000m in Kumaon.
Reproductive Cycle	:	October-March.
Part Used	:	Twig.
Folk Use	:	Extract of twigs used to treat children suffering from intestinal worms.
References	:	Asolkar *et al.* (1992), Parrotta (2001).

Croton persimilis **Mull. Arg.**

Syn.: *C. oblongifolius* Roxb.; *Oxydectes oblongifolia* Kuntze.

Family: Euphorbiaceae

Eng., H. and Sans. Names	:	*Eng.*: Oblong-Leaved Croton; *H.*: Chucka; *Sans.*: Kastidanti.

Distribution	:	Central, W. and S. India, also eastwards to Bengal.
Reproductive Cycle	:	January-April.
Part Used	:	Whole Plant.
Folk Use	:	Used in blindness.
Reference	:	Pal *et al.* (2000).

***Croton tiglium* L.**

Syn.: *C. acutus* Thunb.; *C. birmanicus* Mull. Arg.; *C. camaza* Perr.; *Tiglium cumingii* Klotzsch.

Family: Euphorbiaceae

Eng., H. and Sans. Names	:	*Eng.*: Croton Oil Plant, The Purging Croton; *H.*: Danti, Jamalgota; *Sans.*: Jayapala, Kanakaphala.
Distribution	:	Distributed in Bengal, Assam and southwards to Malacca, Myanmar and Sri Lanka, naturalized or cultivated.
Reproductive Cycle	:	Flowering in summer and fruiting in winter.
Part Used	:	Whole Plant.
Folk Uses	:	Plant useful in inflammation of skeletal muscles of children. Oil diluted with 9 or 10 parts of mustard oil or olive oil considered to be very useful in infantile bronchitis.
References	:	Thakur *et al.* (1989), Watt (1972).

***Cryptocoryne spiralis* Fisch. *ex* Wydler.**

Syn.: *Ambrosina spiralis* (Retz.) Roxb.; *Arum spirale* Retz.

Family: Araceae

Eng. Names	:	*Eng.*: Indian Ipecacuanha, Spiralis, The Spiral Water Trumpet.
Distribution	:	From Khandesh-N. Kanara, Calicut, Coromandel Coast, Sri Lanka, Bengal.
Reproductive Cycle	:	October-June.
Part Used	:	Rhizome.
Folk Uses	:	Rhizome used in cough and vomiting in infants.
References	:	Ambasta (1986), Kirtikar and Basu (1984).

Cryptolepis buchananii **Schult.**

Syn.: *C. reticulata* Wall.; *Echites cuspidata* Heyne.; *E. reticulata* Roth.; *Nerium reticulatum* Roxb.

Family: Asclepiadaceae

Eng., H. and Sans. Names	:	*Eng.*: Indian Sarsaparilla, Wax-Leaved Climber; *H.*: Dudhi, Kalidudhi, Karanta; *Sans.*: Krishnasariva.
Distribution	:	Grows throughout India from Kashmir-Assam, Myanmar, Coromandel, Travancore, ascending Himalayas to 1,350m.
Reproductive Cycle	:	August-October.
Parts Used	:	Whole Plant and Leaf.
Folk Uses	:	Preparation from the plant given to children by Santals to cure rickets. Leaves also used for the same.
References	:	Agarwal (2003), Anonymous (1976b), Islam and Jha (2003), Jain (1991), Kirtikar and Basu (1984), Pande *et al.* (2006), Rana *et al.* (1996), Subramaniam (2000), Watt (1972).

Cucumis melo **L.**

Syn.: *Bryonia callosa* Wall.; *C. acidus* Jacq.; *C. alba* Nakai.; *C. momordica* Roxb.; *C. utilissimus* Roxb.

Family: Cucurbitaceae

Eng., H. and Sans. Names	:	*Eng.*: Melon, Musk Melon, Sweet Melon; *H.*: Kakni, Kakri, Kharbuja, Khurbuj; *Sans.*: Amritavha, Karkati, Kharvuja.
Distribution	:	Cultivated throughout India.
Reproductive Cycle	:	May-August.
Part Used	:	Seed.
Folk Use	:	Bruised seeds applied to abdomen to cure tympanites in children.
Reference	:	Watt (1972).

Cucurbita pepo **L.***

Syn.: *Cucumis pepo* (L.) Dumort.; *C. zapallo* Steud.; *Cucurbita aurantia* Willd.; *C. variegata* Steud.

Family: Cucurbitaceae

Eng., H. and Sans. Names	:	*Eng.*: Pompkin, Pumpkin, Vegatable Marrow;

		H.: Kadimah, Konrar, Kumra, Petha;
		Sans.: Kurkaru, Kushmanda.
Distribution	:	Cultivated throughout India; native of N. America.
Reproductive Cycle	:	April-June.
Part Used	:	Seed.
Folk Use	:	Seeds used against tapeworms in children.
Reference	:	Prajapati *et al.* (2006).

Curculigo orchioides **Gaertn.**

Syn.: *C. malabarica* Wight.; *Hypoxis orchioides* Kurz.

Family: Hypoxidaceae

Eng., H. and Sans. Names	:	*Eng.*: Black Musle, Black Musli, Common Curculigo, Golden Eye Grass;
		H.: Kali Musli, Muslikand, Siyah Musli;
		Sans.: Bhutali, Hema Pushpi, Kharjuri, Talemnlika.
Distribution	:	Sub-tropical areas of C. Himalayas and W. Ghats.
Reproductive Cycle	:	August-September.
Parts Used	:	Whole Plant and Rhizome.
Folk Uses	:	Plant used to cure white spots on the eyes and poliomyelitis. Rhizome useful in unconsciousness state of children.
References	:	Asolkar *et al.* (1992), Jain (1991), Pal *et al.* (2000).

Curcuma angustifolia **Roxb.**

Family: Zingiberaceae

Eng., H. and Sans. Names	:	*Eng.*: East Indian Arrowroot, Narrow-leaved Turmeric;
		H.: Tavakhira, Tikhur;
		Sans.: Gavayodbhava, Pishtika, Talakshira, Yavaja.
Distribution	:	Moist deciduous Sal and mixed forests of M.P., Chhatisgarh and Jharkhand and Bengal, Bombay, Chennai and some of lower Himalyan ranges.
Reproductive Cycle	:	May-November.
Parts Used	:	Leaf and Rhizome.
Folk Uses	:	Rhizomes used for smallpox by Nagas. Paste of rhizomes and leaves with water given twice daily to infants as stomachic.
References	:	Ahmed and Borthakur (2005), Nath and Bordoloi (1989).

Curcuma aromatica **Salisb.***

Syn.: *C. wenyujin* Y. H. Chen and C. Ling.

Family: Zingiberaceae

Eng., H. and Sans. Names	:	*Eng.*: Cochin Turmeric, Wild Turmeric, Yellow Zedoary; *H.*: Ban-haridra, Jangli Haldi; *Sans.*: Aranyaharidra, Sholi, Vanaharidra.
Distribution	:	Throughout India; cultivated in Mizoram for the aromatic rhizome in Jhum lands or gardens; Bengal, W. Peninsula.
Reproductive Cycle	:	May-September.
Part Used	:	Rhizome.
Folk Uses	:	Rhizome used for preventing eruptions of smallpox; also recommended for children as a substitute of arrowroot.
References	:	Kirtikar and Basu (1984), Lalramnghinglova (2003), Saikia and Nath (2003a), Tripathi and Goel (2001).

Curcuma caesia **Roxb.**

Syn.: *C. kuchoor* Royle.

Family: Zingiberaceae

Eng., H. and Sans. Names	:	*Eng.*: Black Zedoary; *H.*: Kali Haldhi, Narkachura; *Sans.*: Krishna Haridra, Nar-Kachura, Rajani.
Distribution	:	Cultivated to some extent in Bengal.
Reproductive Cycle	:	June-November.
Part Used	:	Rhizome.
Folk Use	:	Paste of rhizome applied on paralytic limbs of children.
Reference	:	Mudgal *et al.* (1999).

Curcuma longa **L.***

Syn.: *Amomum curcuma* Jacq.; *Curcuma domestica* Valeton.

Family: Zingiberaceae

Eng., H. and Sans. Names	:	*Eng.*: Indian Saffron, Long Turmeric, Turmeric; *H.*: Halad, Haldi; *Sans.*: Aneshta, Gauri, Haridra, Hemaragi, Jvarantika, Kshapa, Lakshmi, Uma, Varangi.

Distribution : Cultivated in humid climatic regions plains of India.

Reproductive Cycle : June-November.

Part Used : Rhizome.

Folk Use : Roasted turmeric used as an antidysentric for children. Powdered rhizome applied to heal umbilicus of new born child. Mixture of its 20g rhizome, 5g powder of 'ajwain' seeds and 20ml mustard oil pasted on the leaf of *Ricinus communis* tied for relief against chest pain in children. Rhizome powder used for curing scabs caused by smallpox, measles and chickenpox. Paste made up of its rhizome, 'motha' root, neem oil and ghee valued for skin diseases of infants. Rhizome with molasses given to children for intestinal worms in Manipur.

References : Asolkar *et al.* (1992), Khanna (2002a), Kirtikar and Basu (1984), Pal *et al.* (2000), Parrotta (2001), Saikia and Nath (2003a), Shivaprasad and Chandrashekar (2003), Sinha (1987), Thakur *et al.* (1989), Tripathi and Goel (2001), Vihari (1995).

***Curcuma montana* Roxb.**

Family: Zingiberaceae

Eng. Name : *Eng.*: Mountain Turmeric.

Distribution : Garo hills, Konkan and Anamalai hills.

Reproductive Cycle : June-November.

Part Used : Rhizome.

Folk Use : Rhizome paste applied for treating polio for a long time.

Reference : Singh *et al.* (2003).

***Curcuma zedoaria* (Christm.) Rosc.**

Syn.: *Amomum latifolium* Lam.; *A. zedoaria* Christm.; *Curcuma officinalis* Salisb.; *C. pallida* Lour.

Family: Zingiberaceae

Eng., H. and Sans. Names : *Eng.*: Cochin Turmeric, Round Zedoary, Zedoary;
H.: Kachura, Kalihaladi;
Sans.: Durlabha, Karchura, Mukhya, Sati.

Distribution : Wild in the E. Himalayas and moist deciduous forests of coastal tract of Kanara.

Reproductive Cycle : April-June.

Part Used : Rhizome.

Folk Uses	:	About 1g of rhizome rubbed in mother's milk given once daily for 3 consecutive days to infants having diarrhoea due to teething or exposure to cold. Rhizome chewed by mothers applied its paste together with their saliva to the head and body of children suffering from convulsions.
Reference	:	Kirtikar and Basu (1984).

Cuscuta reflexa **Roxb. (Pl. 5D)**

Syn.: *C. grandiflora* Wall.; *C. macrantha* G. Don.; *C. santapaui* Banerji and Das.; *C. verrucosa* Sweet.

Family: Convolvulaceae

Eng., H. and Sans. Names	:	*Eng.*: Giant Dodder;
		H.: Akashabela, Akashbel, Amarabela;
		Sans.: Akashvalli, Amaravallari, Amaravela, Khavalli.
Distribution	:	Common almost throughout India, grown in villages and roadsides.
Reproductive Cycle	:	October-January.
Part Used	:	Whole Plant.
Folk Use	:	Massage of the affected part with medicated oil obtained by boiling the plant in mustard oil good against polio. Poultice of plant and turmeric (1:1) massaged on the whole body of weak or thin children for a couple of weeks, starting from Saturday.
References	:	Hembrom (1991), Sahoo and Bahali (2003).

Cyamopsis tetragonoloba **(L.) Taub.**

Syn.: *C. psoraliodes* (L.) DC.; *Psoralla tetragonaloba* L.

Family: Fabaceae

Eng., H. and Sans. Names	:	*Eng.*: Cluster Bean;
		H.: Govar;
		Sans.: Bakuchi, Dridhabjja, Goraksaphalini, Gorani, Ksudrasimbi.
Distribution	:	Cultivated almost throughout India.
Reproductive Cycle	:	September-November.
Part Used	:	Pods.
Folk Use	:	Pods used as a tonic for children.
Reference	:	Agarwal (2003).

Cyclea arnottii Miers.

Syn.: *C. burmanni* Hook.f. and Thoms.; *C. peltata* Hook.f. and Thoms., non Diels.

Family: Menispermaceae

Eng., H. and Sans. Names	:	*Eng.*: Pataroot; *H.*: Parvel, Path, Patha, Pathi; *Sans.*: Patha.
Distribution	:	Throughout E. and W. Coasts of India, Assam, Tripura; rare in Mizoram, found in tropical evergreen forests.
Reproductive Cycle	:	August-January.
Part Used	:	Root.
Folk Uses	:	5ml decoction of root with sugar taken two times per day for fever and diarrhoea.
Reference	:	Lalramnghinglova (2003).

Cydonia oblonga Mill.

Syn.: *C. vulgaris* Pers.; *Pyrus cydonia* L.

Family: Rosaceae

Eng., H. and Sans. Names	:	*Eng.*: Common Quince, Quince; *H.*: Abi, Bihi; *Sans.*: Amritphala, Amrutaphala, Patala.
Distribution	:	Cultivated in N.W. India.
Reproductive Cycle	:	March-May.
Part Used	:	Fruit.
Folk Use	:	Fruits used against diarrhoea in children.
Reference	:	Prajapati *et al.* (2006).

Cymbopogon citratus (DC.) Stapf.

Syn.: *Andropogon citratus* DC.; *A. roxburghii* Nees *ex* Steud.

Family: Poaceae

Eng., H. and Sans. Names	:	*Eng.*: The Lemon Grass; *H.*: Gandha Trina; *Sans.*: Bhustrina, Chhatra, Guchhala.
Distribution	:	Known in the cultivated state, probably of Indian origin; now widely distributed over tropics of both hemispheres.

Reproductive Cycle : August-March.

Part Used : Leaf.

Folk Use : Lemon grass oil used as a stomachic for children.

Reference : Watt (1972).

Cymbopogon jwarancusa **(Jones) Schult.**

Syn.: *Andropogon jwarancusa* Jones.; *A. laniger* Duthie., non Desf.

Family: Poaceae

Eng., H. and Sans. Names : *Eng.*: Iwarancusa Grass, Russ Grass;
H.: Bur, Ghatyari, Khavi, Khoi, Lamjak, Panni, San;
Sans.: Amrinala, Laghu, Laya, Nalada.

Distribution : Wild in Himalayas from Kashmir – Assam up to 3,290m and in N.W. plains reaching down to Mumbai.

Reproductive Cycle : June-October.

Part Used : Whole Plant.

Folk Uses : Recommended as a valuable aromatic tonic in dyspepsia, and for mumps.

References : Jain (1991), Kapur and Nanda (1996), Kirtikar and Basu (1984).

Cymbopogon jwarancusa **subsp. *olivieri* (Boiss.) Soenarko**

Syn.: *Andropogon olivieri* (Boiss.) Bor.

Family: Poaceae

Eng. Names : *Eng.*: Herba Schcenanth of Pharmacists, The Juncus Odoratus.

Distribution : Native of lower Himalayan tract, extending through the plains of the N.W. regions and Punjab to Sind.

Reproductive Cycle : April-September.

Part Used : Whole Plant.

Folk Use : Plant used as a valuable aromatic tonic in dyspepsia.

Reference : Watt (1972).

Cymbopogon martinii **(Roxb.) Wats.**

Syn.: *Andropogon martinii* Roxb.; *A. pachnaodes* Trin.

Family: Poaceae

Eng., H. and Sans. Names : *Eng.*: Palmarosa, Rosha Grass, Rusa Grass;
H.: Gandhabel, Rusa Ghas;
Sans.: Dhyamakah, Rohisah.

Distribution	:	Throughout India, cultivated in dry localities.
Reproductive Cycle	:	August-December.
Part Used	:	Whole Plant.
Folk Use	:	Used in epileptic fits in children.
Reference	:	Prajapati *et al.* (2006).

Cymbopogon schoenanthus (L.) Spreng.

Syn.: *Andropogon circinnatus* Hochst. *ex* Steud.; *A. eriophorus* Willd.; *A. schoenanthus* L.

Family: Poaceae

Eng., H. and Sans. Names	:	*Eng.*: Camel Grass, Geranium Grass, Lemon grass; *H.*: Musel, Palakhari, Rohisha, Saundhiya, Rusaghas, Rousaghas; *Sans.*: Bhuti, Bhutika, Devajagdha, Dhyama, Katrina, Paura, Rohisha.
Distribution	:	Punjab, Sind, Baluchistan.
Part Used	:	Leaf.
Folk Use	:	Used against epileptic fits in children.
Reference	:	Kirtikar and Basu (1984).

Cynodon dactylon (L.) Pers. (Pl. 6A)*

Syn.: *C. maritimus* H. Bock.; *C. occidentalis* Willd.; *C. radiatus* Roth *ex* Roem. and Schult.; *Dactylon officinale* Vill.; *Digitaria dactylon* Scop.; *Panicum dactylon* L.; *Paspalum dactylon* DC.

Family: Poaceae

Eng., H. and Sans. Names	:	*Eng.*: Bahama Grass, Barmuda Grass, Conch Grass, Dhub Grass; *H.*: Dub, Dubra, Kabbar, Mangala, Nanda, Romghas, Shatmula, Vamini; *Sans.*: Amari, Bahuvirya, Durva, Niladurva.
Distribution	:	Wild, throughout India; also cultivated for decorative purposes.
Reproductive Cycle	:	September-October.
Part Used	:	Leaf.
Folk Use	:	10ml juice of leaves with little sugar given to children against vomiting.
Reference	:	Satapathy and Brahmam (2000).

Cynoglossum wallichii **G. Don.**

Syn.: *C. denticulatum* A. DC.; *C. edgeworthii* A. DC.; *C. glochidiatum* Wallich *ex* Benth.; *C. membranaceum* A. DC.

Family: Boraginaceae

Eng. Names	:	*Eng.*: Himalayan Hounds-Tongue, Prickly Hound's-Tongue.
Distribution	:	Tropical and temperate Himalaya from Kashmir-Sikkim (300-3,000m), C. India and Chota Nagpur-Bombay.
Reproductive Cycle	:	August-September.
Parts Used	:	Whole Plant and Root.
Folk Uses	:	Root as well as plant used for checking vomiting in infants.
References	:	Ambasta (1986), Anonymous (1950), Kapur (1996a), Kapur and Singh (1996), Kirtikar and Basu (1984), Sood *et al.* (2001), Srivastava *et al.* (2003c), Tiwari and Tiwari (1996), Viswanathan (2000).

Cyperus longus **L.**

Family: Cyperaceae

Eng. Names	:	*Eng.*: Cypress, Cypress Root, Galanga, Galingale, Sweet Galingale.
Distribution	:	Quetta, Mount Abu and westwards to the Atlantic.
Reproductive Cycle	:	Flowering during July-August.
Part Used	:	Tuber.
Folk Use	:	Julus prepare an enema from the tuber for children with stomach troubles.
Reference	:	Kirtikar and Basu (1984).

Cyperus rotundus **L.**

Syn.: *C. albidus* Herb.; *C. hexastachyos* Rottb.; *C. tenuiflorus* Royle.

Family: Cyperaceae

Eng., H. and Sans. Names	:	*Eng.*: Coco-Grass, Nut-Grass; *H.*: Motha, Mutha, Mustaka; *Sans.*: Abda, Arnoda, Gundra, Mustaka.
Distribution	:	Found in all the warm regions of India, ascending to 2,000m.
Reproductive Cycle	:	July-December.

Parts Used : Root and Rhizome.

Folk Uses : Its roots combined with other aromatics used for indigestion of children. Paste of one fruit of *Coix lacryma-jobi* pounded with a tuber of *Cyperus rotundus* and a grounded tooth of Swine fed to infants in aphthae, candidiasis, strong diarrhoea and feebleness. Paste of mixture of its rhizome along with plant of *Centella asiatica* given to children for fever and to check bedwetting. Massage with paste of equal amounts of fresh rhizomes of *Cyperus rotundus*, bulbs of *Allium cepa*, cloves of *Allium sativum*, fresh leaves of *Dactyloctenium aegyptium* and mustard oil over the chest, backbone, forhead, back of pinna, neck etc. gives relief against vomiting, diarrhoea, cutaneous diseases and worm infection in children.

References : Barua *et al.* (2000), Dhiman (2007), Saini (1996), Sharma *et al.* (2003a), Singh and Singh (2003a), Thomas and Britto (2003), Watt (1972).

***Cyphostemma auriculatum* (Roxb.) P. Singh and B. V. Shetty.**

Syn.: *Cayratia auriculata* (Roxb.) Gamble.; *Cissus auriculata* Roxb.

Family: Vitaceae

Eng., H. and Sans. Names : *Eng.*: Fox-Grape;

H.: Amalbel, Gidardrak, Ramchana;

Sans.: Gandirah.

Distribution : Throughout India on the hills, W. Coast, W. Ghats.

Reproductive Cycle : Flowering during June-October.

Part Used : Leaf.

Folk Use : Paste of leaves grounded with turmeric applied externally for chickenpox by Savaras.

Reference : Rao and Henry (1996).

***Dactylicapnos macrocapnos* (Prain) Hutch.**

Syn.: *Dicentra macrocapnos* Prain.

Family: Papaveraceae

Eng. Name : *Eng.*: Yellow Bleeding Heart.

Distribution : Distributed in Garhwal, Kumaon, W. and C. Nepal.

Reproductive Cycle : Flowering in April-October.

Part Used : Whole Plant.

Folk Use : 10ml decoction of 100g of plant boiled in 250ml water for 2 hours taken orally 3 times per day for two days against acute fever in infants.

Reference : Rajendran (2003).

Dactyloctenium aegyptium **(L.) Willd.**

Syn.: *Aegilops saccharina* Walter.; *Cenchrus aegyptius* (L.) P. Beauv.; *Eleusine mucronata* Stokes.; *E. prostata* Spreng.

Family: Poaceae

Eng. and H. Names : *Eng.*: Beach Wiregrass, Coast Button Grass, Comb Fringe Grass, Crowfoot Grass;

H.: Makra.

Distribution : Throughout the plains, ascending upto 1,900m in the hills.

Reproductive Cycle : May-October.

Parts Used : Leaf and Stem.

Folk Uses : Massage of medicated oil obtained by boiling equal amounts of fresh rhizomes of *Cyperus rotundus*, bulbs of *Allium cepa*, cloves of *Allium sativum*, fresh leaves of *Dactyloctenium aegyptium* in mustard oil done over chest, backbone, forhead, back of pinna, neck etc. for relief against vomiting and diarrhoea. Decoction of leaves used in smallpox. Culm decoction with roots of *Diplazium esculentum* in equal quantities given for asthma in children (20ml thrice daily for 2-3 months).

References : Sharma *et al.* (2003a), Singh and Singh (2003a), Nath *et al.* (2007).

Dactylorhiza hatagirea **(D. Don.) Soo.**

Syn.: *Orchis latifolia* auct.

Family: Orchidaceae

Eng., H. and Sans. Names : *Eng.*: Heart Orchid, Marsh Orchid, Salep;

H.: Salam, Salampanja, Salep;

Sans.: Munjatakah, Salampamisri.

Distribution : W. temperate Himalayas (Kashmir-Sikkim) up to 3,657m.

Reproductive Cycle : January – July.

Part Used : Whole Plant.

Folk Use : Plant used in poliomyelitis.

Reference : Pal *et al.* (2000).

***Dalbergia lanceolaria* subsp. *paniculata* (Roxb.) Thoth.**

Syn.: *D. hemsleyi* Prain.; *D. maymensis* Craib.; *D. nigrescens* Kurz.; *D. paniculata* Roxb.

Family: Fabaceae

H. Names	:	*H.*: Dhobein, Passi, Satpura.
Distribution	:	Throughout Western, C. and S. India and in Myanmar, common in dry deciduous forests upto about 450m.
Reproductive Cycle	:	May-October.
Part Used	:	Root.
Folk Use	:	Water extract of the root given to children to relieve diarrhoea in southern M.P. (Bastar).
Reference	:	Parrotta (2001).

***Dalbergia volubilis* Roxb.**

Family: Fabaceae

Eng., H. and Sans. Names	:	*Eng.*: Climbing Dalbergia; *H.*: Bankhara, Bhatia; *Sans.*: Sirisika.
Distribution	:	C. and E. Himalayas-Pegu and Sri Lanka.
Reproductive Cycle	:	Flowering during January-May.
Part Used	:	Stem Bark.
Folk Use	:	Warm filterate of its stem bark crushed with turmeric given, 1 tsp twice a day till cure, to infants as an anti-dysenteric.
Reference	:	Rao and Henry (1996).

Datura innoxia* Mill.

Syn.: *D. guayaquilensis* Kunth.; *D. meteloides* DC. *ex* Dunal.

Family: Solanaceae

Eng., H. and Sans. Names	:	*Eng.*: Thorn Apple; *H.*: Safed Datura; *Sans.*: Dhattura.
Distribution	:	Occasional, grown in Bangalore, Ahmedabad, Pune, Lucknow, Pilani and Jammu.
Reproductive Cycle	:	June-August.
Part Used	:	Fruit.

Folk Use	:	Paste of fruits pounded with lemon juice applied locally on mumps by Soligas.
References	:	Kshirsagar and Singh (2000a, 2007), Sood *et al.* (2012).

Datura metel L.

Syn.: *D. alba* Rumph. *ex* Nees.; *D. fastuosa* L.; *D. fastusoa* L. var. *alba* Bernh.; *D. humilis* Desf.; *D. laevis* Schkuhr.

Family: Solanaceae

Eng., H. and Sans. Names	:	*Eng.*: Apple of Peru, Black Datura, Devil's Trumpet; *H.*: Dhatura, Kaladhatura, Kanak, Sadadhatura; *Sans.*: Dhastura, Dhurta, Dhustoora, Unmatta.
Distribution	:	Temperate Himalayas from Kashmir-tropical parts of India.
Reproductive Cycle	:	April-June.
Parts Used	:	Leaf, Root and Seed.
Folk Uses	:	All these plant parts used in mumps.
References	:	Asolkar *et al.* (1992), Dhiman (2003), Jain (1991), Nayak *et al.* (2003a, 2003b), Watt (1972).

Decaspermum fructicosum J. R. Forst. and G. Forst.

Syn.: *Nelitris fruticosa* (J. R. Forst. G. Forst.) A. Gray.

Family: Myrtaceae

Eng. Name	:	*Eng.*: Christmas Bush.
Distribution	:	Occurs in Sikkim, Khasia hills, very common in dry forests, on ridges and hillside.
Part Used	:	Fruit.
Folk Use	:	Fruits used in loss of appetite in children.
Reference	:	Prajapati *et al.* (2006).

Dendrocnide harveyi (Seem.) Chew.

Syn.: *D. milnei* (Seem.) Chew.; *Laportea harveyi* Seem.; *L. milnei* Seem.; *Urticastrum milnei* (Seem.) Kuntz.

Family: Urticaceae

Distribution	:	Edge of forests and along streams from sea-level-mid-montane.
Reproductive Cycle	:	Throughout the year.
Part Used	:	Leaf.

Folk Use	:	Liquid squeezed from leaves valued for curing fits in children.
Reference	:	Prajapati *et al.* (2006).

Dendrophthoe falcata **(L.f.) Etting.**

Syn.: *Loranthus amplexifolius* Desr.; *L. falcatus* L.f.; *L. longiflorus* Desr.

Family: Loranthaceae

Eng., H. and Sans. Names	:	*Eng.*: Mistletoe; *H.*: Banda; *Sans.*: Vanda, Vrikshabhaksha, Vriksharuha.
Distribution	:	Parasite on trees throughout India.
Reproductive Cycle	:	December-April.
Parts Used	:	Whole Plant, Flower and Fruit.
Folk Uses	:	Paste of plant applied hot with 'ghee' on mumps. Powdered flowers (5g) given 2-3 times a day in smallpox and chickenpox. Sticky paste made of pounded fruits applied twice or thrice a day on mumps till cured.
References	:	Bhandary *et al.* (1996), Brahman and Saxena (1990).

Descurainia sophia **(L.) Webb.**

Syn.: *Sisymbrium sophia* L.; *Sophia chirurgurum* Lobel.

Family: Brassicaceae

Eng. and H. Names	:	*Eng.*: Flixweed, Fluxweed; *H.*: Khubkallana.
Distribution	:	Temperate Himalayas; Kashmir-Kumaon (1,524-4,270m); E. Himalayas and salt ranges.
Reproductive Cycle	:	July-August.
Parts Used	:	Whole Plant and Seed.
Folk Uses	:	Seeds as well as whole plant used against measles and chickenpox.
References	:	Jain (1991), Kala (2003), Kapur *et al.* (1996), Kaul (1997).

Desmodium gangeticum **(L.) DC.**

Syn.: *D. gangeticum* (L.) DC. var. *maculatum* (L.) DC.; *Hedysarum gangeticum* L.; *H. maculatum* L.

Family: Fabaceae

Eng., H. and Sans. Names	:	*Eng.*: Tick Tree;

		H.: Salpan, Salpani, Salun, Salwan, Sarivan;
		Sans.: Anshumati, Devi, Sala Parni, Shalaparni, Tanvi, Vidari.
Distribution	:	Himalayas (ascending to 1,700m on the Chenab)-Pegu and Sri Lanka.
Reproductive Cycle	:	Flowering during June-August and Fruiting in cold season.
Part Used	:	Whole Plant.
Folk Use	:	Plant kept below the pillow of children upto 1 year for inducing sleep.
Reference	:	Hembrom (1996).

Desmodium laxiflorum **DC.**

Family: Fabaceae

Eng. and Sans. Names	:	*Eng.*: Loose Flowered Desmodium;
		Sans.: Parsniparni.
Distribution	:	Himalayas, temperate and tropical regions from Gharhwal and Kumaon-Assam, ascending to 2,000m.
Reproductive Cycle	:	August-September.
Part Used	:	Root.
Folk Use	:	Roots used for curing smallpox.
Reference	:	Jain (1991).

Desmodium triflorum **(L.) DC.**

Syn.: *D. albiflorum* Cordem.; *D. bullamense* G. Don.; *Hedysarum stipulaceum* Burm.f.; *Meibomia triflora* (L.) Kuntze.

Family: Fabaceae

Eng., H. and Sans. Names	:	*Eng.*: Creeping Tickfoil;
		H.: Kundaliya;
		Sans.: Hamsapadi, Tripadi.
Distribution	:	Weed of open habitats in all plains districts; up to 900m in hills of Peninsular India and to 2,000m in the Himalayas.
Reproductive Cycle	:	August – November.
Parts Used	:	Whole Plant and Leaf.
Folk Uses	:	Juice of plant useful in cough. Leaves grounded with cow's milk given to children for diarrhoea due to indigestion and in convulsion.
References	:	Kirtikar and Basu (1984), Parrotta (2001).

Desmodium velutinum (Willd.) DC.

Syn.: *Hedysarum velutinum* Willd.

Family: Fabaceae

Eng., H. and Sans. Names	:	*Eng.*: Velvet-Leaf Desmodium, Villous leaf Desmodium; *H.*: Jagru, Lagavang, Lippa-Pank; *Sans.*: Prshniparni.
Distribution	:	Throughout most of India excepting driest regions, to an elevation of 1,200m.
Reproductive Cycle	:	August-November.
Part Used	:	Root.
Folk Use	:	Crushed roots used among the Kathodias of Rajasthan as a remedy for inability of infants to digest mother's milk.
Reference	:	Parrotta (2001).

Dichrostachys cinerea (L.) Wight and Arn.

Syn.: *Calliea cinerea* (L.) Roberty.; *Mimosa cinerea* L.; *M. glomerata* Forssk.; *M. nutans* Pers.

Family: Fabaceae

Eng., H. and Sans. Names	:	*Eng.*: Sickle Bush; *H.*: Kheri, Vartuli, Veerataru; *Sans.*: Bahuvaraka, Dirghamula, Mahakapitha, Viradru, Viravriksha.
Distribution	:	N.W. regions, W. Peninsula, Sri Lanka; found in dry regions throughout India.
Reproductive Cycle	:	Flowering during June-August.
Part Used	:	Root bark.
Folk Use	:	Extract of root bark crushed with cumin and sugar administered for chickenpox.
Reference	:	Rao and Henry (1996).

Dicliptera bupleuroides Nees.

Syn.: *D. buergeriana* Miq.; *D. cardiocarpa* Nees.; *D. crinite* Nees.; *Justicia canescens* Wall.

Family: Acanthaceae

Eng. Name	:	*Eng.*: Roxburgh's Foldwing.

Distribution	:	Assam and E. Bengal.
Reproductive Cycle	:	Flowering during November-June.
Part Used	:	Leaf.
Folk Use	:	Juice of leaves 1/2-1 tsp given to children suffering from fever and stomach troubles.
Reference	:	Chauhan (1999).

***Dillenia indica* L.**

Syn.: *D. elongata* Miq.; *D. speciosa* Thunb.

Family: Dilleniaceae

Eng., H. and Sans. Names	:	*Eng.*: Dillenia, Elephant apple;
		H.: Chalta, Chulta;
		Sans.: Bharija, Bhavya.
Distribution	:	Sub-Himalayan tract from Garhwal, Assam, Arunachal Pradesh, Manipur, Tripura, W. Bengal, Orissa, Bihar, C. and S. India.
Reproductive Cycle	:	June-August.
Part Used	:	Whole Plant.
Folk Use	:	Plant good against scabies in children.
Reference	:	Pal *et al.* (2000).

***Dioscorea hispida* Dennst.**

Syn.: *D. daemona* Roxb.; *D. hirsuta* Blume.; *D. lunata* Roth.; *D. mollissima* Blume.; *Helmia hirsuta* (Blume) Kunth.

Family: Dioscoreaceae

Eng. and Sans. Names	:	*Eng.*: Asiatic Bitter Yam, Intoxicating Yam;
		Sans.: Hastyaluka.
Distribution	:	Found growing upto 1,200m in the Himalayas.
Reproductive Cycle	:	April-September.
Part Used	:	Root.
Folk Use	:	Roasted root powder (5-10g) given two-three times a day with a cup of milk to children in ricketssiae and weakness.
Reference	:	Painuli and Maheshwari (1996).

Diospyros lanceifolia **Roxb.**

Syn.: *D. amoena* Wall. *ex* G. Don.; *D. grata* Wall. *ex* A. DC.; *D.hirsuta* Hiern.; *D. lucida* Wall. *ex* A. DC.; *D. sabtanense* Merr.

Family: Ebenaceae

Distribution : Assam, Meghalaya, Manipur and Bangladesh.

Reproductive Cycle : Fruiting in May-July.

Part Used : Leaf.

Folk Use : Leaf infusion prepared for bath of anaemic children.

Reference : Aminuddin and Girach (1991).

Diospyros malabarica **(Desr.) Kostel.**

Syn.: *D. embryopteris* Pers.; *D. peregrina* (Gaertn.) Gurke.; *Embryopteris glutenifera* Roxb.

Family: Ebenaceae

Eng., H. and Sans. Names : *Eng.*: Gaubpersimmon, Riber Ebony, Wild Mangosteen;

H.: Gab, Kalatendu, Makurkendi, Tendu;

Sans.: Anilsara, Rava, Ravana, Susara, Syandana, Tinduki, Tindula.

Distribution : Throughout India, Common in W. Bengal, S. India and sub-Himalayan tract; also cultivated in roadsides and villages.

Reproductive Cycle : April-November.

Parts Used : Flower and Fruit.

Folk Uses : Flowers and fruits given in hiccups of children.

Reference : Kirtikar and Basu (1984).

Diplazium esculentum **(Retz.) SW.***

Syn.: *D. ambiguum* (SW.) Hook.; *D. malabaricum* Spreng.; *D. pubescens* Link.; *Hemionitis esculenta* Retz.

Family: Woodsiaceae

Eng. Name : *Eng.*: Vegetable Fern.

Distribution : Almost throughout India, being common in the plains and on the hills upto 1,000m.

Reproductive Cycle : April-July.

Part Used : Fronds.

Folk Use : Young frond paste applied on skin infection of infants.

Reference : Mudgal (1999).

Diplocyclos palmatus **(L.) C. Jeffrey.**

Syn.: *Bryonia palmata* L.; *Bryonopsis laciniosa* (L.) Naud.

Family: Cucurbitaceae

Eng., H. and Sans. Names	:	*Eng.*: Lollipop Climber; *H.*: Bahupatra, Isvaralingi, Silvalingi; *Sans.*: Gargumaru, Lingini.
Distribution	:	On hedges and bushes throughout India upto 1,200m.
Reproductive Cycle	:	September-October.
Part Used	:	Whole Plant.
Folk Use	:	Plant used in smallpox.
Reference	:	Jain (1991).

Diplopenta odorata **Alef.**

Syn.: *Pavonia odorata* Willd.

Family: Malvaceae

H. and Sans. Names	:	*H.*: Sugandhabala; *Sans.*: Bala, Valakah.
Distribution	:	N.W. regions, Sindh, Banda, W. Peninsula, Myanmar and Sri Lanka.
Reproductive Cycle	:	All season.
Part Used	:	Leaf.
Folk Use	:	Two tsp of leaf juice along with black pepper administered orally twice a day on empty stomach for dysentery in babies.
References	:	Mohanty *et al.* (1996), Parrotta (2001).

Dipteracanthus prostratus **(Poir.) Nees.**

Family: Acanthaceae

Eng. Name	:	*Eng.*: Prostrate Wild Petunia.
Distribution	:	Throughout India, from Punjab and Assam-Sri Lanka.
Reproductive Cycle	:	July-December.
Part Used	:	Whole Plant.
Folk Use	:	Plant useful in dysentery.
Reference	:	Pal *et al.* (2000).

Dischidia benghalensis **Coleb.**

Syn.: *D. cuneifolia* Wall.; *D. spathulata* Blume.

Family: Asclepiadaceae

Eng. and H. Names	:	*Eng.*: Air Plant; *H.*: Hankha.
Distribution	:	Throughout E. India.
Reproductive Cycle	:	Flowers all year long.
Part Used	:	Leaf.
Folk Use	:	Leaf paste applied on mumps.
Reference	:	Sinha *et al.* (1996).

Drosera burmanni **Vahl.**

Syn.: *D. dietrichiana* Rchb.f.

Family: Droseraceae

Eng. and H. Names	:	*Eng.*: Burman's Sundew; *H.*: Mukhajali.
Distribution	:	Plains and hills of India upto 2,700m.
Reproductive Cycle	:	September-January.
Part Used	:	Whole Plant.
Folk Uses	:	Plant used in diarrhoea, dysentery and blood dysentery.
Reference	:	Pal *et al.* (2000).

Drymaria cordata **(L.) Willd. *ex* Schult.**

Syn.: *Alsine rotundifolia* Stokes.; *Bufonia rotundifolia* Buch.-Ham. *ex* Steud.; *Holosteum diandrum* SW.

Family: Caryophyllaceae

Eng. Names	:	*Eng.*: Drymaria, Heartleaf Drymary, Tropical Chickweed, West Indian Chickweed.
Distribution	:	Tropical and subtropical India and Sri Lanka, extending westwards to the Punjab and ascending the Himalayas to 2,350m in Sikkim.
Reproductive Cycle	:	October – November.
Parts Used	:	Whole Plant and Leaf.
Folk Uses	:	Paste of plant with a pinch of salt given daily to cure diarrhoea in children. Extract of about 25g leaves given 3-4 times a day to cure pneumonia in children of age

group between 6 months-5 years. Infusion of 10g leaves mixed with sugar (25g) administered orally once a day in tonsillitis. Leaves also used in smallpox.

References : Dutta and Nath (2000), Jain (1991), Srivastava *et al.* (2003b).

Dysphania ambrosioides **(L.) Mosyakin and Clemants.**

Syn.: *Atriplex ambrosioides* (L.) Crantz.; *Botrys anthelmintica* (L.) Nieuwl.

Family: Chenopodiaceae

Eng., H. and Sans. Names : *Eng.*: American Worm Wood, Maxican Tea;

H.: Khatua;

Sans.: Kshetravastuka, Swetachilli.

Distribution : Distributed in S. India, Maharashtra, Kashmir, W. Bengal.

Reproductive Cycle : June-September.

Part Used : Aerial Parts.

Folk Uses : Decoction of aerial parts good for cold and illness of small children.

Reference : Keller *et al.* (2011).

Echinacea angustifolia **DC.**

Syn.: *Brauneria angustifolia* (DC.) A. Heller.

Family: Asteraceae

Eng. Names : *Eng.*: Echinacea, Narrow-Leaf Echinacea, Narrow Leaf Purple Coneflower.

Distribution : Temperate Himalayas.

Reproductive Cycle : Flowering in late spring-mid summer.

Part Used : Root.

Folk Use : Dried roots used in tonsillitis.

Reference : Karnick (1994a).

Echinops cornigerus **DC.**

Syn.: *E. coriarius* C. B. Clarke.; *E. polygraphus* Czerniak.; *E. polygraphus* Tscherneva.

Family: Asteraceae

Eng. and H. Names : *Eng.*: Blue Globe Thistle, Globe Thistle;

H.: Kandara, Kanghi, Kantela.

Distribution : W. Himalayas (Kashmir-Garhwal); 2,000-3,000m.

Reproductive Cycle : July-November.

Part Used : Root.

Folk Use : Root extract given for promoting emergence of teeth in infants.

Reference : Sharma (2003).

***Eclipta prostrata* L.**

Syn.: *E. alba* (L.) Hassk.; *E. erecta* L.; *Verbesina prostrata* L.

Family: Asteraceae

Eng., H. and Sans. Names : *Eng.*: Trailing Eclipta;

H.: Bhangra, Mochkand, Shivet;

Sans.: Bhringara, Bhringaraja, Kesaraja, Keshya, Markara, Pararu, Patanga, Rangaka.

Distribution : Throughout India; ascending to 2,000m in Himalayas and other mountains.

Reproductive Cycle : March-June.

Parts Used : Whole Plant and Leaf.

Folk Uses : Powdered plant given for 1-5 weeks in curing jaundice of children. Juice of plant along with honey given to infants for catarrh, and with castor oil for expulsion of worms in children. Plant also useful in tetanus and eye complaints. One tsp of leaf juice along with honey given thrice a day as a remedy for catarrh in infants and for curing cough by Bettakurubas. 5-10ml macerated leaf juice given with lukewarm water once daily for 2-3 days to check dysentery in children.

References : Agarwal (2003), Anonymous (1952, 1976), Asolkar *et al.* (1992), Barua *et al.* (2000), Bondya *et al.* (2006), Chatterjee and Pakrashi (1997), Chauhan (1999), Dhiman (2007), Kirtikar and Basu (1984), Kshirsagar and Singh (2007), Pal *et al.* (2000), Parrotta (2001), Seetharam *et al.* (1999), Singh and Singh (2003a), Thakur *et al.* (1989), Watt (1972).

***Elaeocarpus sphaericus* (Gaertn.) K. Schum.**

Syn.: *E. ganitrus* Roxb. *ex* G. Don.; *Genitrus sphaerica* Gaertn.

Family: Elaeocarpaceae

Eng., H. and Sans. Names : *Eng.*: Bead Tree, Utrasum Bead Tree;

		H.: Rudraki, Rudraks, Rudraksha;
		Sans.: Rudraksha.
Distribution	:	Native to various parts of India, Nepal and Chittagong, Malacca, Concan, on the higher Ghats, Assam, Bengal, Maharashtra, parts of M.P. and Nepal.
Reproductive Cycle	:	June-December.
Part Used	:	Fruit.
Folk Use	:	Paste of beads applied for treating eruptions of smallpox.
Reference	:	Khare (2004).

***Elatostema platyphyllum* Wedd.**

Syn.: *E. ebracteatum* W. T. Wang.

Family: Urticaceae

Distribution	:	Tropical Himalayas; Kumaon, Sikkim 300-2,000m. Khasia mountains, 1,350-1,700m; Naga hills, on Kohima, 2,000m.
Reproductive Cycle	:	April-June.
Part Used	:	Leaf.
Folk Uses	:	3g paste of 15g fresh leaves grounded with 50ml cow milk given orally two times per day for 3 days to kill intestinal worms/ringworms of the children.
Reference	:	Rajendran (2003).

***Elephantopus scaber* L.**

Syn.: *E. carolinensis* G. Mey.; *E. sordidus* Salisb.

Family: Asteraceae

Eng., H. and Sans. Names	:	*Eng.*: Prickly-Leaved Elephant's Foot;
		H.: Gobhi, Samudulan;
		Sans.: Darvika, Gojihva, Hastipadi, Satamulika.
Distribution	:	Throughout India, from Punjab-Sri Lanka and Singapore.
Reproductive Cycle	:	September-January.
Parts Used	:	Whole Plant and Root.
Folk Uses	:	Plant useful against smallpox and scabies. Roots made into pills and given to children suffering from fever and debility of children. Root decoction given orally to cure

		amoebic dysentery. Root paste applied to cure pimples and tonsillitis in children.
References	:	Asolkar *et al.* (1992), Girach and Aminuddin (1995), Jain (1991), Jain and Patole (2001), Kirtikar and Basu (1984), Pal *et al.* (2000), Sur and Halder (2002).

Eleusine coracana **(L.) Gaertn.**

Syn.: *Cynodon coracanus* Raspail.; *Cynosurus coracanus* L.; *Eleusine pilosa* Gilli.; *E. reniformis* Divak.

Family: Poaceae

Eng., H. and Sans. Names	:	*Eng.*: African Millet, Finger Millet, Ragi;
		H.: Kado, Mandva, Maruva;
		Sans.: Bahupatrica, Guchha, Narttakah, Nrtyakundalah.
Distribution	:	Cultivated throughout S. India in areas up to 1,500m.
Reproductive Cycle	:	June-September.
Part Used	:	Grains.
Folk Uses	:	Decoction of freshly collected grains given orally to infants as remedy for diarrhoea and in measles and smallpox.
References	:	Asolkar *et al.* (1992), Rajan *et al.* (2003), Rawat and Chowdhury (1998).

Elsholtzia blanda **(Benth.) Benth.**

Syn.: *Aphanochilus blandus* Benth.; *Perilla elata* D. Don.

Family: Lamiaceae

Eng. and H. Names	:	*Eng.*: Lomba;
		H.: Jangli Tulsi.
Distribution	:	Myanmar, Java, Sumatra; Central and E. Himalayas from Nepal-Sikkim, Orissa and N.E. India.
Reproductive Cycle	:	September-January.
Parts Used	:	Aerial Parts and Inflorescence.
Folk Uses	:	Infusion of aerial parts given 10ml twice daily for children's diseases called 'nauhri' a combination of fever, cholera, skin diseases and inflammation. Inflorescence used in tonsillitis.
References	:	Lalramnghinglova (2003), Salam *et al.* (2011).

***Elsholtzia ciliata* (Thunb.) Hyland.**

Syn.: *E. cristata* Willd.; *E. formosana* Hayata.; *E. interrupta* Ohwi.; *E. minima* Nakai.

Family: Lamiaceae

Eng. Name : *Eng.*: Crested Late-Summer Mint.

Distribution : Extensively cultivated as spice and medicinal plant in India. Himalayas from Kashmir-Arunachal Pradesh, reaching upto 3,300m.

Reproductive Cycle : Flowering during August-October.

Part Used : Whole Plant.

Folk Use : Plant used in measles.

Reference : Prajapati *et al.* (2006).

***Elsholtzia communis* (Coll. and Hemsl.) Diels.**

Syn.: *Aphanochilus communis* (Collett and Hemsl.) Kudo.; *Dysophylla communis* Collett and Hemsl.

Family: Lamiaceae

Distribution : Found in Yunan (Mayanmar and Thailand).

Reproductive Cycle : October-December.

Parts Used : Whole Plant and Inflorescence.

Folk Use : Used for curing tonsils.

References : Khomdram *et al.* (2011), Salam *et al.* (2011).

***Elytraria acaulis* (L.f.) Lindau.**

Syn.: *E. crenata* Vahl.; *E. indica* Pers.; *Tubiflora acaulis* (L.f.) Kuntze.

Family: Acanthaceae

Eng. and H. Names : *Eng.*: Asian Scalystem;

H.: Dasmari, Patharchatta.

Distribution : Almost throughout India; found in Deccan Peninsula, extending northwards to the Gangetic plains and E. Himalayas.

Reproductive Cycle : August-December.

Parts Used : Whole Plant and Root.

Folk Uses : Infusion of whole plant given in cough of infants. Roots used in rickets; and given with mother's milk in marasmus and infantile diarrhoea. Root paste

		recommended for treatment of tonsillitis among Bhil, Dhanka, Dubada and Nayaka tribes of Khedbrahma region in N. Gujarat.
References	:	Agarwal (2003), Ambasta (1986), Anonymous (1952), Asolkar *et al.* (1992), Bhatt and Sabnis (1987), Jain (1991).

Embelia ribes Burm.f.

Syn.: *Antidesma grossularia* Raeusch.; *A. ribes* (Burm.f.) Raeusch.; *Embelia glandulifera* Wight.; *Samara ribes* Kurz.

Family: Myrsinaceae

Eng., H. and Sans. Names	:	*Eng.*: Embelia, Embelia Fruit; *H.*: Baberang, Baibrang, Bidang, Vidang, Wawrung; *Sans.*: Amogha, Bhasmaka, Bidanga, Chitra, Gahara, Ghosha, Kapali, Kevala, Mogha, Vara, Vella.
Distribution	:	Throughout India, from the C. Himalayas-Sri Lanka and Singapore, up to 1,700m. Common in the lower hills.
Reproductive Cycle	:	March-August.
Parts Used	:	Whole Plant and Fruit.
Folk Uses	:	Plant used against tape-worm, diarrhoea, dysentery, stomach pain and especially in blood dysentery, and poliomyelitis of children. Fruits used in flatulence and for spleen disorders in children.
References	:	Kirtikar and Basu (1984), Pal *et al.* (2000), Thakur *et al.* (1989), Watt (1972).

Embelia subcoriacea (C. B. Clarke) Mez.

Syn.: *E. nagushia* var. *subcoriacea* C. B. Clarke.

Family: Myrsinaceae

Distribution	:	Found in Assam and Meghalaya.
Reproductive Cycle	:	May-February.
Part Used	:	Leaf.
Folk Use	:	Decoction of leaves used for bathing as treatment of smallpox.
Reference	:	Lalramnghinghlova (1996).

Embelia tsjeriam-cottam **(Roem. and Schult.) A. DC.**

Syn.: *Ardisia basaal* Roem. and Schult.; *Dauceria acuta* Dennst.; *Embelia robusta* Roxb.; *E. villosa* Wall.

Family: Myrsinaceae

Eng., H. and Sans. Names	:	*Eng.*: Embelia; *H.*: Baberang, Bahabirang, Baibirang, Bhungi, Vindangah; *Sans.*: Vidanga, Vindangah.
Distribution	:	Throughout India, from Himalayas-Sri Lanka and Tenasserim very common.
Reproductive Cycle	:	August-January.
Parts Used	:	Bark and Leaf.
Folk Uses	:	Santal and Mahali tribes in Mayurbhanj district of Orissa use infusion of its bark and leaves as gargle for curing tonsillitis.
References	:	Jain (1991), Tribedi *et al.* (1982).

Emilia sonchifolia **(L.) DC. *ex* DC.**

Syn.: *Cacalia sonchifolia* Hort. *ex* L.; *Emilia purpurea* Cass.; *E. rigidula* DC.; *E. sinica* Miq.

Family: Asteraceae

Eng., H. and Sans. Names	:	*Eng.*: Red Tassle Flower, Sow Thistle Flower; *H.*: Hirankhuri; *Sans.*: Sasasruti, Sasasrutih.
Distribution	:	As a weed in cultivated fields and wasteplaces upto 1,500m throughout India.
Reproductive Cycle	:	April-June.
Parts Used	:	Whole Plant and Leaf.
Folk Uses	:	Pieces of plant hung around neck to cure fever in children. Plant also used in cataract and redness of eyes. Fresh cleaned whole plant chewed along with sugar and onion before sunrise given in tonsillitis. Decoction of leaves and stems given in infantile tympanitis and bowel troubles. 1 tsp of leaf juice given thrice a day till recovery for indigestion in children by Malekudiyas.
References	:	Agarwal (2003), Ambily *et al.* (2010), Anonymous (1952), Hembrom (1996), Kshirsagar and Singh (2007), Pal *et al.* (2000), Parrotta (2001), Prajapati *et al.* (2006).

Enicostemma littorale **Blume.**

Syn.: *Adenema nyssopifolia* G. Don.; *Cicendia nyssopifolia* Wight. and Arn.; *Enicostemma axillare* (Lam.) Raynal.

Family: Gentianaceae

Eng., H. and Sans. Names	:	*Eng.*: Indian gentian; *H.*: Chota Chirayita, Chota Kirayita, Katvinayi; *Sans.*: Mamajjakah, Nagajihva, Nahi.
Distribution	:	Throughout India from Punjab and Gangetic plains-Sri Lanka. More frequent near sea, not common in Bengal.
Reproductive Cycle	:	July-September.
Part Used	:	Leaf.
Folk Use	:	Leaves mixed with dried nutmeg given to young children for removing worms.
Reference	:	Thakor (2009).

Entada phaseoloides **Merril.**

Syn.: *E. pursaentha* DC.; *E. scandens* (L.) Benth.; *E. schefferi* Ridl.

Family: Fabaceae

Eng., H. and Sans. Names	:	*Eng.*: Climbing Entada, Elephant Creeper, Garber Bean, Nicker Bean, St. Thomas Bean; *H.*: Chian, Garbi, Gila; *Sans.*: Bidhanta.
Distribution	:	E. Himalayas and C. India, Nepal, ascending to 1,350m in Sikkim, W. Peninsula, Sri Lanka, Malacca.
Reproductive Cycle	:	November-May.
Parts Used	:	Whole Plant, Seed, Cotyledon and Kernel.
Folk Uses	:	Plant used for watering of eyes. Kernels and seeds used for mumps in Sikkim. Paste of cotyledons smeared over legs once a day for a week to increase joint strength in infants. Aqueous extract of powdered kernel used for washing hair of new born baby for removing stickiness and yellowish cutaneous affections and also to help in setting the soft veins to ensure proper growth and good health of the baby.
References	:	Ahmed and Borthakur (2005), Hajra and Chakraborty (1981), Jain (1991), Pal *et al.* (2000), Viswanathan *et al.* (2006).

Enydra fluctuans **DC.**

Family: Asteraceae

Eng., H. and Sans. Names	:	*Eng.*: Marsh Herb, Water Cress; *H.*: Haruch; *Sans.*: Achari, Bramhi, Chakrangi, Helanchi, Hilamochi, Mochi, Rochi, Vishaghni.
Distribution	:	Occurs in E. Bengal, Assam, Silhet, Malay Peninsula, Malaya.
Reproductive Cycle	:	December-March.
Part Used	:	Leaf.
Folk Use	:	Leaves used to cure smallpox.
Reference	:	Kirtikar and Basu (1984).

Ephedra gerardiana **Wallich. *ex* Stapf.**

Syn.: *E. vulgaris* Hook. A. Rich.

Family: Ephedraceae

Eng., H. and Sans. Names	:	*Eng.*: Ephedra; *H.*: Asmania, Khanda, Phok; *Sans.*: Soma.
Distribution	:	Occurs in drier regions of higher Himalayan ranges at about 4,000m; also cultivated.
Reproductive Cycle	:	March-September.
Part Used	:	Stem.
Folk Use	:	Stem used for controlling night wetting in children.
Reference	:	Jain (1994).

Equisetum arvense **L.**

Syn.: *E. calderi* B. Boivin.; *E. saxicola* Suksd.

Family: Equisetaceae

Eng., H. and Sans. Names	:	*Eng.*: Field-Horsetail; *H.*: Nari; *Sans.*: Ashwa - Puchha.
Distribution	:	Occurs in Himalayas, at high altitudes; found on stream banks, woods and along roadsides.
Part Used	:	Stem.
Folk Use	:	Dried sterile stem used in enuresis of children.
Reference	:	Karnick (1994a).

Equisetum giganteum **L.**

Syn.: *E. bolivianum* Gand.; *E. martii* Milde.; *E. ramosissimum* Desf.; *E. schaffneri* Milde.

Family: Equisetaceae

Eng. Names : *Eng.*: Giant Horsetail, Many Branched Horsetail.

Distribution : Found in Himalayas including Nepal.

Part Used : Root.

Folk Use : Root decoction given to infants in fever.

Reference : Jamir *et al.* (2011).

Eranthemum roseum **(Vahl.) R. Br.**

Syn.: *Daedalacanthus roseus* (Vahl.) T. Anders.; *Ecbolium roseum* (Nees) Kuntze.

Family: Acanthaceae

Eng., H. and Sans. Names : *Eng.*: Blue Eranthemum, Rosy Eranthemum;
H.: Dasmuli;
Sans.: Dhavan.

Distribution : W. and S. Deccan Peninsula from Bombay Ghats-Mangalore.

Reproductive Cycle : Flowering during November-April.

Part Used : Root.

Folk Use : Used in smallpox.

Reference : Jain (1991).

Erythrina stricta **Roxb.**

Syn.: *Corallodendron strictum* (Roxb.) Kuntze.; *Micropteryx stricta* (Roxb.) Duchass. and Walp.

Family: Fabaceae

Eng., H. and Sans. Names : *Eng.*: Coral Tree;
H.: Dhol Dak, Mandara, Mura;
Sans.: Mura, Paribhadra, Paribhadrah.

Distribution : Commonly met with in the deciduous forests of Konkan, N. Kanara, Malabar and Travancore; also found in Assam and Manipur, extending westwards to Nepal.

Reproductive Cycle : August-September.

Parts Used : Root and Root Bark.

Folk Uses : Root ash mixed with coconut oil applied locally to cure mumps. About 1 tsp juice of root bark given twice daily for 5 consecutive days for threadworm.

References : Borthakur (1993), Kshirsagar and Singh (2003, 2007).

Erythrina suberosa **Roxb.**

Syn.: *E. alba* Wight. and Arn.; *E. maxima* Wight. and Arn.; *Micropteryx suberosa* (Roxb.) Walp.

Family: Fabaceae

Eng., H. and Sans. Names : *Eng.*: Indian Coral-Tree;

H.: Dalap, Mandara, Pharad;

Sans.: Bahupushpa, Kantaki, Parijata.

Distribution : Himalayas-Sri Lanka, ascending to 1,350m in the northwest.

Reproductive Cycle : March-April.

Part Used : Whole Plant.

Folk Use : Plant used for curing mumps.

Reference : D'Cruz (2007).

Erythrina variegata **L.***

Syn.: *E. indica* Lam.; *E. variegata* L. var. *orientalis* L.

Family: Fabaceae

Eng., H. and Sans. Names : *Eng.*: East Indies Coral Tree, India Bean, India Coral-bean, Mochi Wood Tree;

H.: Mandara, Pamkara, Pangra, Pharhad;

Sans.: Mandar, Nimbataru, Parijata.

Distribution : Throughout India, grows wild in deciduous forests; cultivated in gardens as ornamental plant.

Reproductive Cycle : March-May.

Parts Used : Whole Plant, Leaf and Bark.

Folk Uses : Plant used as an anthelmintic, and for colic pain and stomach pain. Dry powder of leaves used for disinfecting skin of infants. Paste of basal portion of tender leaves of *Ananas comosus*, mature leaves of *Musa balbisiana*, tender leaves of *Erythrina variegata* and bark of *Pongamia pinnata* in equal quantities given for roundworm in children. Bark used as a medicine for eyes, especially for babies. Juice of bark and leaves rubbed on body to check itching and swellings due to excitement of owing bile in children by Bettakurubas.

References : Agarwal (2003), Gogoi and Borthakur (2001), Khare (2004), Kshirsagar and Singh (2007), Pal *et al.* (2000).

Eschscholzia californica **Cham.**

Syn.: *E. stricta* Greene.; *E. tenuisecta* Greene.; *E. tristis* Fedde.; *E. vernalis* Greene.; *E. yainacensis* Greene.

Family: Papaveraceae

Eng. Name : *Eng.*: Californian Poppy.

Distribution : Widely cultivated in Indian gardens.

Reproductive Cycle : Flowering in June-September.

Part Used : Aerial Parts.

Folk Uses : Aerial parts used in physical and psychological problems and for whooping cough in children.

References : Asolkar *et al.* (1992), Prajapati *et al.* (2006).

Eucalyptus globulus **Labill.***

Syn.: *E. gigantea* Dehn.; *E. glauca* DC.; *E. perfoliata* Desf.

Family: Myrtaceae

Eng., H. and Sans. Names : *Eng.*: Blue-Gum Eucalypt, Eucalyptus;
H.: Eucalyptus, Yukeliptas;
Sans.: Nilaniryasa, Tailaparnah.

Distribution : Found in sub-tropical regions in India.

Reproductive Cycle : August-September.

Part Used : Leaf.

Folk Uses : Infusion of leaves given internally in aphthous ulcerations in mouth and throat, and also for whooping cough in rachitic children.

References : Chatterjee and Pakrashi (1997), Parrotta (2001), Rajan *et al.* (2000).

Eucalyptus tereticornis **Sm.**

Syn.: *E. insignis* Naudin.; *E. populifolia* Desf.; *E. subulata* A. Cunn. *ex* Schauer.; *E. umbellatum* (Gaertn.) Domin.

Family: Myrtaceae

Eng. and H. Names : *Eng.*: Blue Gum, Eucalyptus;
H.: Safeda.

Distribution : Introduced into India, in Nilgiris, Maharashtra, U.P., Punjab and H.P. upto 1,200m.

Reproductive Cycle : March-April.

Part Used : Leaf.

Folk Use : Leaves used for whooping cough in rachitic children.

Reference : Rajan *et al.* (2000).

Euodia hortensis **J. R. Forst. and G.**

Family: Rutaceae

Eng. Names : *Eng.*: Island Musk, Pungent Shrub.

Distribution : Cultivated and naturalized in secondary forest, forest margins and thickets from sea-level to 500m.

Reproductive Cycle : April-August.

Part Used : Stem.

Folk Use : Liquid from stem used in treating children with convulsions.

Reference : Prajapati *et al.* (2006).

Eupatorium cannabinum **L.**

Syn.: *E. allaisii* Senneni.; *E. birmanicum* DC.; *E. cannabifolium* Salisb.; *E. decline* Edgew.; *E. trifidum* Vahl.

Family: Asteraceae

Eng. and H. Names : *Eng.*: Hemp Agrimony, Hemp Eupatorium, Water Hemp;

H.: Tongol Lati.

Distribution : Temperate Himalayas, abundant from 1,000-3,650m, Khasia mountains, 1,000-2,000m, Myanmar.

Reproductive Cycle : July-August.

Parts Used : Leaf and Twig.

Folk Uses : Crushed juice of leaves and twigs with or without macerated mass applied at the area of burns and scalds. Leaves kept with children suffering from smallpox probably to reduce virulence of infection.

References : Kumar (2002), Singh and Singh (2003a).

Euphorbia antiquorum L.

Syn.: *E. arborescens* Roxb.; *Tithymalus antiquorus* (L.) Moench.

Family: Euphorbiaceae

Eng., H. and Sans. Names : *Eng.*: Spurge Cactus, Triangular Spurge;
H.: Tidhara-Sehnr, Tindhara Sehund;
Sans.: Simhunda, Snuhi, Tridharaka, Vajrakantaka, Vajri.

Distribution : Throughout the hotter parts of India, common on dry hills of Bengal, Peninsula generally.

Reproductive Cycle : December-August.

Part Used : Root.

Folk Use : A plaster prepared from roots mixed with asafoetida applied externally to the stomach of a children suffering from worms.

References : Dhiman (2003), Kirtikar and Basu (1984), Watt (1972).

Euphorbia heyneana Spreng.

Syn.: *E. microphylla* B. Heyna. *ex* Roth.

Family: Euphorbiaceae

H. and Sans. Names : *H.*: Kahiree;
Sans.: Raajaadani.

Distribution : Common throughout the plains of India.

Reproductive Cycle : Round the year but luxuriant after rainy season.

Part Used : Whole Plant.

Folk Use : Paste of plant mixed with common salt, turmeric, garlic and pepper given orally with mother's milk for 3 days to treat dry cough and influenza in children.

Reference : Ram *et al.* (2002).

Euphorbia hirta L.*

Syn.: *E. bancana* Miq.; *E. capitata* Lam.; *E. globulifera* Kunth.; *E. nodiflora* Steud.

Family: Euphorbiaceae

Eng., H. and Sans. Names : *Eng.*: Asthma Weed;
H.: Lal Dudhi;
Sans.: Chara, Nagarjuni, Pusitoa.

Distribution : Common in wasteplaces, roadsides, gardens, ricefields, specially in warmer regions throughout India.

Reproductive Cycle : Maximum part of the year.

Parts Used : Whole Plant, Leaf and Latex.

Folk Uses : Plant used in children diseases, worms, cough, dysentery and colitis bronchial infections, asthma, warts, reddening of eyes and rickets. 1 tsp of fresh juice obtained by crushing the entire plant given, twice daily till cure against blood dysentery. Juice of leaves of the plant along with that of *Centella asiatica* given in dysentery. Latex applied to heal redness in eyes of children.

References : Anonymous (1976b), Chowdhury *et al.* (1996), Dey and Bahadur (1973), Ganesan and Kesavan (2003), Gogoi and Borthakur (2001), Jain (1991, 1994), Khan *et al.* (2003), Srivastava *et al.* (2003a).

Euphorbia hypericifolia L.

Syn.: *E. boliviana* Rusby.; *E. decumbens* W.; *E. papilligera* Boiss.

Family: Euphorbiaceae

Eng., H. and Sans. Names : *Eng.*: Euphorbia, Graceful Spurge;

H.: Dudhi, Dudhikalave, Hakshardana;

Sans.: Dugdhika.

Distribution : Common throughout hotter parts of India, occurring upto 1,350m on the Himalayas.

Reproductive Cycle : July-April.

Part Used : Whole Plant.

Folk Use : Given to children with milk in colic.

References : Kirtikar and Basu (1984), Watt (1972).

Euphorbia neriifolia L.

Syn.: *E. edulis* Lour.; *E. ligularia* Roxb. ex Buch.-Ham.; *Tithymalus edulis* (Lour.) H. Karst.

Family: Euphorbiaceae

Eng., H. and Sans. Names : *Eng.*: Common Milk Hedge;

H.: Patton-Ki-Send, Sehund, Sij, Thohar;

Sans.: Sehunda, Snuhi, Vajra, Vujri.

Distribution	:	Extensively cultivated; common in rocky places, mostly in Orissa and Deccan Peninsula.
Reproductive Cycle	:	February-April.
Parts Used	:	Leaf and Branches.
Folk Uses	:	One tsp extract prepared from warm leaf administered thrice a day for 3 days to treat cold and fever in children. Juice of roasted tender terminal portion of branches with molasses given for vomiting in bronchitis of children.
References	:	Dhiman (2003), Maliya (2004).

Euphorbia nivulia **Buch.-Ham.**

Syn.: *E. helicothele* Lem.

Family: Euphorbiaceae

Eng., H. and Sans. Names	:	*Eng.*: Dog's Tongue, Holy Milk Hedge, Leafy Milk Hedge; *H.*: Katathuhar, Sij, Thor; *Sans.*: Gandira, Guda, Patrasnuhi.
Distribution	:	Native of Himalayas, throughout drier regions in India.
Reproductive Cycle	:	Flowering in hot weather.
Parts Used	:	Leaf and Latex.
Folk Uses	:	Extract of 2-3 warmed leaves consumed once a day for 2 days to cure loose motion and stomach complaints in infants. Latex applied on mumps.
References	:	Kumar and Balakishnan (1996), Viswanathan *et al.* (2006).

Euphorbia plumerioides **Teijsm. *ex* Hassk.**

Syn.: *E. corynoclada* F. Muell.; *E. fidjiana* Boiss.

Family: Euphorbiaceae

Distribution	:	Open rocky habitats and cultivated from sea-level to 550m.
Part Used	:	Leaf.
Folk Uses	:	Filterate of leaves used to wean infants from their mothers and for protecting children from germs.
Reference	:	Prajapati *et al.* (2006).

Euphorbia serpens Kunth.

Syn.: *E. begoniifolia* Lehm.; *E. biramensis* Urb.; *E. inflexa* Urb. and Ekman.; *E. microclada* Urb.; *E. minutiflora* N.E. Br.

Family: Euphorbiaceae

Eng. and H. Names	:	*Eng.*: Matted Sandmat, Roundleaf Spurge; *H.*: Dudhi.
Distribution	:	Bengal, Bihar, Concan, Chennai, Tenasserim.
Reproductive Cycle	:	June-December.
Part Used	:	Whole Plant.
Folk Use	:	Decoction of plant beneficial in sickness 'Kambyryru Jere' of children.
Reference	:	Keller *et al.* (2011).

Euphorbia thymifolia L.

Syn.: *E. microphylla* Lam.; *E. rubicunda* Blume.; *E. rubrosperma* Lotsy.

Family: Euphorbiaceae

Eng., H. and Sans. Names	:	*Eng.*: Chicken Weed, Thyme-leaved Spurge; *H.*: Chotadudhi, Choti-dudhi; *Sans.*: Dugdhika, Laghududhika, Nagarjuni, Raktabindu.
Distribution	:	Throughout India in plains and lower hills, ascending to 2,800m in Kashmir.
Reproductive Cycle	:	August-December.
Parts Used	:	Whole Plant, Leaf and Seed.
Folk Uses	:	Juice of plant mixed with fresh goat's milk given to cure diarrhoea and blood dysentery in children. Plant also used for eye complaints and redness of eyes. Leaf extract given with mother's milk to infants suffering from dysentery. Dried leaves and seeds given to children in bowel complaints.
References	:	Ambasta (1986), Anonymous (1952), Chatterjee and Pakrashi (1997), Chowdhury *et al.* (1996), Kapur and Singh (1996), Khan *et al.* (2003), Kirtikar and Basu (1984), Pal *et al.* (2000), Parrotta (2001), Ranjan (2000), Singh (2000), Watt (1972).

Evolvulus alsinoides L.

Syn.: *E. angustifolius* Roxb.; *E. hirsutus* Lam.; *E. sericeus* Wall.

Family: Convolvulaceae

Eng., H. and Sans. Names	:	*Eng.*: Dwarf Morning Glory; *H.*: Shankpushpi, Shyamakranta; *Sans.*: Laghuvishnukranta, Nilapushpi, Shankpushpi, Tandikodebaha, Vishnukranta.
Distribution	:	Throughout India and Sri Lanka, ascending to 2,000m in Himalayas.
Reproductive Cycle	:	July-November.
Parts Used	:	Whole Plant and Root.
Folk Uses	:	Plant beneficial in teething of infants. Roots used by the Santals in intermittent fever of children.
References	:	Asolkar *et al.* (1992), Kirtikar and Basu (1984).

Fagonia cretica L.

Syn.: *F. arabica* L.; *F. bruguieri* DC.; *F. cretica* var. *arabica* T. Anders.; *F. mysorensis*Roth.

Family: Zygophyllaceae

Eng., H. and Sans. Names	:	*Eng.*: Khorasan Thorn; *H.*: Damahan, Hinguna, Ustarkhar, Usturgar; *Sans.*: Ajabhakshya, Ananto, Atmamuli, Dhanvi, Grahini, Kashaya, Rodini, Virupa.
Distribution	:	Grows in W. India, Rajasthan, Punjab, Gangetic plains and Deccan.
Reproductive Cycle	:	Flowering almost throughout the year.
Part Used	:	Whole Plant.
Folk Uses	:	Plant ash given to children suffering from anaemia and smallpox.
References	:	Agarwal (2003), Anonymous (1956), Asolkar *et al.* (1992), Khare (2004), Kirtikar and Basu (1984), Watt (1972).

Ferula asafoetida L.

Syn.: *F. foetida* Regel.

Family: Apiaceae

Eng., H. and Sans. Names	:	*Eng.*: Asafoetida; *H.*: Hing;

		Sans.: Balhika, Hingu, Shulanashana.
Distribution	:	Native of Afghanistan, Iran and Kashmir; W. Baluchistan, Chamba, Lahul and Spiti and Kinnaur in H.P.
Reproductive Cycle	:	March-April.
Parts Used	:	Gum-Resin from incision of tap root and Fruit.
Folk Uses	:	Gum-resin used in infantile convulsions. Fruits used in advanced stages of pneumonia and bronchitis of children.
References	:	Ambasta (1986), Chatterjee and Pakrashi (1997), Watt (1972).

***Ferula narthex* Boiss.**

Syn.: *Narthex asafoetida* Falr.

Family: Apiaceae

Eng., H. and Sans. Names	:	*Eng.*: Asafoetida, Devil's Dung, Persian Fennel-gaint;
		H.: Hing, Hingra;
		Sans.: Agudagandha, Balhika, Bhutari, Dipta, Grihini, Hinghka, Jatuka, Kesara.
Distribution	:	Open slopes, rocky substrate of temperate Himalayas, also occurs in Kashmir.
Reproductive Cycle	:	July-August.
Part Used	:	Gum.
Folk Uses	:	Gum produces remarkable effects in dyspnoea and advanced stages of pneumonia and bronchitis in children.
References	:	Bhattacharjee (2001), Jain (1994), Kirtikar and Basu (1984), Sharma (2003).

***Ficus benghalensis* L.**

Syn.: *F. banyan* Oken.; *F. cotoneifolia* Vahl.; *F. karet* Baill.; *Urostigma bengalensis* (L.) Gasp.

Family: Moraceae

Eng., H. and Sans. Names	:	*Eng.*: Banyan Tree;
		H.: Bar, Bargad, Bor;
		Sans.: Bahupada, Bhringi, Chira, Dhruva, Mana, Nyagrdhah, Rohina, Vatah, Yakshaves, Yamapriya.
Distribution	:	Throughout the plains and forest-tract of India, planted in avenues for shade upto 1,200m.

Reproductive Cycle : April-May.

Parts Used : Root and Latex.

Folk Uses : Juice from aerial roots administered orally to children in dysentery. Latex mixed with sugar given to children in cough, fever and dullness.

References : Asolkar *et al.* (1992), Mohanty *et al.* (1996), Parrotta (2001), Rajendran *et al.* (2003).

Ficus benjamina L.

Syn.: *F. comosa* Roxb.; *F. nuda* (Miq.) Miq.; *F. parvifolia* Oken.; *F. pendula* Link.

Family: Moraceae

Eng., H. and Sans. Names : *Eng.*: Weeping Fig;

H.: Chilkan, Kamrup, Pinwal, Zir;

Sans.: Kantalaka, Kini, Kuni, Mandirriksha, Tunna.

Distribution : Gardens throughout India; base of Himalayas, Khasia hills, Manipur, Bihar, M.P., Maharashtra, Tamil Nadu, Karnataka and Andamans.

Reproductive Cycle : April-September.

Part Used : Whole Plant.

Folk Use : Used in scabies.

Reference : Pal *et al.* (2000).

Ficus carica L.*

Syn.: *F. caprificus* Risso.; *F. colchica* Grossh.; *F. colombra* Gasp.; *F. regina* Miq.; *F. rugosa* Miq.

Family: Moraceae

Eng., H. and Sans. Names : *Eng.*: Common Fig, Fig Tree, The Fig;

H.: Anjir;

Sans.: Anjira, Kakodumbar, Manjula.

Distribution : Cultivated in N.W. and S. India.

Reproductive Cycle : September-January.

Part Used : Fruit.

Folk Uses : Pulp of fig mixed with vinegar and sugar very useful in bronchitic affections in children. Syrup of figs considered a mild laxative for children.

References : Khare (2004), Watt (1972).

Ficus hispida* L.f. (Pl. 6B)

Syn.: *Covellia assamica* Miq.; *C. daemomum* Miq.; *Ficus daemomum* Koem.

Family: Moraceae

Eng., H. and Sans. Names	:	*Eng.*: Creeping Fig; *H.*: Daduri, Dagurin, Gobla, Kagsha, Katgularia; *Sans.*: Ajaji, Kakodumbara, Kushthaghni, Phalguni, Rajika.
Distribution	:	Throughout India, from Punjab in the N.W.-Malacca and Sri Lanka.
Reproductive Cycle	:	June-September.
Parts Used	:	Leaf, Fruit and Latex.
Folk Uses	:	2-3 droplets of oil obtained by making a funnel of its leaves filled with cotton seeds and a burning cowdung cake kept over it applied 2-3 times daily to the affected area (earache and ear infection). Fruit paste applied on neck of children for throat inflammation. Latex along with mother's milk given to infants for sound sleep and to check diarrhoea.
References	:	Prusti (2007), Sharma and Singh (1999), Singh and Kumar (2000), Singh and Singh (2003a).

***Ficus racemosa* L.**

Syn.: *Covellia glomerata* Miq.; *Ficus chittagonga* Miq.; *F. glomerata* Roxb.; *F. mollis* Miq.

Family: Moraceae

Eng., H. and Sans. Names	:	*Eng.*: Cluster Fig, Country Fig, Gular Fig; *H.*: Gular, Lelka, Paroa, Umar; *Sans.*: Brahmavriksha, Haritaksha, Jantuphala, Udumbarah, Yojanaphala.
Distribution	:	Outer Himalayas, plains and low hills of India, from Rajputana and the Salt Range to the Khasia mountains, Myanmar, Deccan Peninsula and Sri Lanka.
Reproductive Cycle	:	April-July.
Parts Used	:	Whole Plant, Inflorescence, Galls and Latex.
Folk Uses	:	Plant used against mumps. Boiled fleshy part of hypanthodium given twice a week for a month to cure enuresis (bed-wetting) in children. Ripe fruits used in smallpox. Latex used in pulmonary disorders and mumps. Small blister-like galls on the leaves soaked in

milk mixed with honey given to prevent pitting in smallpox.

References : Biswas *et al.* (2009), Dhiman (2007), Jain (1991), Kirtikar and Basu (1984), Painuli and Maheshwari (1996), Pande *et al.* (2006), Pathak and Pathak (2007), Sood *et al.* (2012), Watt (1972).

***Ficus religiosa* L. (Pl. 6C)**

Syn.: *F. affinior* Griff.; *Urostigma affine* Miq.; *U. religiosum* Gasp.

Family: Moraceae

Eng., H. and Sans. Names : *Eng.*: Bo Tree, Peepal Tree, Sacred Fig Tree;

H.: Pipal, Pipar, Pipli;

Sans.: Asvatthah, Bodhidruma, Chaityadru, Devatma, Gajapatra, Kapitana, Mangalya, Nagabandhu, Pavitraka, Vriksharaja, Yajnika.

Distribution : Wild in the sub-Himalayan forests, Bengal and in C. India, planted elsewhere.

Reproductive Cycle : January-May.

Parts Used : Whole Plant, Leaf and Bark.

Folk Uses : Plant beneficial in smallpox. Two leaves of this plant attached on either side of chapati prepared from seeds of *Eleusine coracana* given for 2 days to children having stuttering problem. Bark rubbed with honey applied to aphthous sores of children. 1 tsp mixture of 10g bark ash mixed with 100 ml water given four times to treat throat infection in children.

References : Chatterjee and Pakrashi (1997), Chauhan *et al.* (2007), Dhiman (2003), D'Rozario *et al.* (1999), Farooq (2005), Khanna (2002a), Kshirsagar and Singh (2007).

***Flacourita indica* (Burm.f.) Merr.**

Syn.: *F. latifolia* Cooke.; *F. occidentasis* Blatter.; *F. ramontchi* L. Herit.; *F. sapida* Roxb.; *F. sepiaria* Roxb.

Family: Flacourtiaceae

Eng., H. and Sans. Names : *Eng.*: Governor's Plum, Madagascar Plum;

H.: Baichi, Bilangru, Kakein, Katar;

Sans.: Aghori, Gopakanta, Svadukantaka, Wadukantaka.

Distribution : Gregarious in sub-Himalayan tract and outer Himalayas.

Reproductive Cycle	:	March-June.
Parts Used	:	Whole Plant and Bark.
Folk Use	:	Paste of bark and whole plant applied on throat for tonsillitis.
References	:	Girach (1992), Sikarwar and Kaushik (1992), Sood *et al.* (2012), Watt (1972).

Flacourtia jangomas **(Lour.) Raeusch.**

Syn.: *F. cataphracta* Roxb. *ex* Willd.; *Stigmarota jangomas* Lour.

Family: Flacourtiaceae

Eng., H. and Sans. Names	:	*Eng.*: Puneala Palm; *H.*: Paniala, Talispatri; *Sans.*: Kalameshi, Khadiraprani, Talisha, Talishapatra.
Distribution	:	Commonly cultivated in India, Manipur, Assam, W. Bengal, Bihar, Orissa, A.P., E. Ghats.
Reproductive Cycle	:	July-December.
Parts Used	:	Leaf and Shoot.
Folk Use	:	Juice of young leaves and shoot useful in infantile fever.
Reference	:	Chatterjee and Pakrashi (1997).

Flagellaria indica **L.**

Syn.: *F. angustifolia* Wall.; *F. catenata* Lour. *ex* B. A. Gomes.; *F. minor* Blume.; *F. philippinensis* Elmer.

Family: Flagellariaceae

Eng. Names	:	*Eng.*: Bush Cane, False Rattan, Whip Vine.
Distribution	:	Coastal India.
Reproductive Cycle	:	Throughout the year.
Part Used	:	Fruit.
Folk Use	:	Used in smallpox.
Reference	:	Agarwal (2003).

Flemingia chappar **Benth.**

Syn.: *Moghania chappar* (Benth.) Kuntze.

Family: Fabaceae

H. Name	:	*H.*: Salpan.
Distribution	:	Found in E. Himalayas and Bihar.
Reproductive Cycle	:	January-March onwards.

Part Used : Whole Plant.

Folk Uses : Used in cataract and eye complaints.

Reference : Pal *et al.* (2000).

Flemingia strobilifera **(L.) W. T. Aiton.**

Syn.: *F. bracteata* (Roxb.) Wight.; *F. fruticulosa* Benth.

Family: Fabaceae

Eng. and H. Names : *Eng.*: Luck Plant, Wild Hops;
H.: Kanphuta, Kassunt.

Distribution : Distributed throughout India, Himalayas from Shimla and Kumaon, ascending – 2,700m to Assam, Khasia, Chittagong, Siam, Malacca and Sri Lanka.

Reproductive Cycle : August-September.

Part Used : Leaf.

Folk Use : Leaves used as vermifuge for children.

Reference : Agarwal (2003).

Flueggea virosa **(Roxb. *ex* Willd.) Royle.**

Syn.: *F. obovata* (Willd.) Wall.; *F. sinensis* Baill.; *Securinega virosa* (Roxb. *ex* Willd.) Baill.

Family: Euphorbiaceae

Eng., H. and Sans. Names : *Eng.*: White Berry- Bush;
H.: Bakarcha, Beri, Dalme, Patala, Rithoul;
Sans.: Apiyadruma, Dhusara, Nilishila.

Distribution : Throughout India, ascending upto 2,000m in hills and Andaman Islands.

Reproductive Cycle : May-August.

Parts Used : Root and Leaf.

Folk Uses : Root extract used for inducing sleep in children. Bath with decoction of leaves good for children suffering from smallpox, scabies and measles.

References : Agarwal (2003), Lalramnghinglova (1996, 2003).

Foeniculum vulgare **Mill. (Pl. 6D)***

Syn.: *F. capillaceum* Gilib.; *F. officinale* All.; *Ozodia foeniculacea* Wight.

Family: Apiaceae.

Eng., H. and Sans. Names : *Eng.*: Fennel, Finkel, Indian Sweet Fennel;

		H.: Barisounf, Saunf, Sonp, Sont;
		Sans.: Ahichhatra, Avakapshpi, Bhuripushpa, Gandhadhika, Karavi, Madhurica, Methica, Potika, Shaleya.
Distribution	:	Cultivated mostly as a garden or homeyard crop throughout India up to 2,000m.
Reproductive Cycle	:	January-May.
Parts Used	:	Whole Plant, Fruit and Seed.
Folk Uses	:	Plant syrup with honey given for catarrh of the upper respiratory tract in children. Infusion prepared with 8-12g of fruits in 500cc of boiling water employed as an enema, in colic and flatulence of infants. Paste of seeds applied on abdomen of children to cure intestinal troubles.
References	:	Anonymous (1956), Asolkar *et al.* (1992), Chatterjee and Pakrashi (1997), Karnick (1994a), Khare (2004), Kirtikar and Basu (1984).

Forsythia suspensa **(Thunb.) Vahl.**

Syn.: *F. fortunei* Lindl.; *Rangium suspensum* (Thunb.) Ohwi.; *Syringa suspensa* Thunb.

Family: Oleaceae

Eng. Names	:	*Eng.*: Lain Qiao, Weeping Forsythia.
Distribution	:	Thickets or grassy areas, on slopes or gullies, upto 2,200m.
Reproductive Cycle	:	March-July.
Part Used	:	Fruit.
Folk Uses	:	Fruits used in tonsillitis and mumps.
Reference	:	Prajapati *et al.* (2006).

Fragaria nubicola **(Hook.f.) Lindl. *ex* Lacaita***

Syn.: *F. vesca* var. *nubicola* Hook.f.

Family: Rosaceae

Eng. and H. Names	:	*Eng.*: Himalayan Strawberry;
		H.: Badri, Gand Kaphal, Kiplaya.
Distribution	:	Temperate Himalayas from Murree and Kashmir, 1,700-3,350m - Sikkim, 2,000-4,350m.
Reproductive Cycle	:	April-September.

Part Used : Fruit.

Folk Use : 4-5 tsp juice of fruit given thrice a day to small children for hastening recovery from weakness after long illness.

Reference : Arya *et al.* (1999).

Fraxinus ornus **L.**

Syn.: *F. montana* Salisb.; *Ornus europaea* Pers.; *O. ornus* (L.) H. Harst.

Family: Oleaceae

Eng. and H. Names : *Eng.*: Manna Ash, South European Flowering Ash;
H.: Shirkhist.

Distribution : S. Europe and W. Asia, extending from Spain through S. France and Germany, Italy, and Greece to Asia Minor and Syria.

Reproductive Cycle : Flowering in April-May.

Part Used : Whole plant.

Folk Use : Used as mild laxative for children.

Reference : Ambasta (1986).

Galium rotundifolium **L.**

Syn.: *G. decipiens* Ehrh.; *G. linckii* J. F. Gmel.; *G. rotundum* Thunb.; *Trichogalium rotundifolium* (L.) Fourr.

Family: Rubiaceae

Eng. Name : *Eng.*: Round- Leaved Bedstraw.

Distribution : Throughout Himalayas and Khasia mountains 1,350-3,350m, common in W. Peninsula and Nilgiri mountains.

Reproductive Cycle : Flowering in July-October.

Part Used : Whole Palnt.

Folk Use : Plant used in tonsillitis.

References : Jain (1991), Mukherjee *et al.* (1986), Pande *et al.* (2006).

Garcinia mangostana **L.**

Syn.: *Mangostana garcinia* Gaertn.

Family: Clusiaceae

Eng. and H. Names : *Eng.*: King's-fruit, Mangosteen, Purple mangosteen;
H.: Mangustan.

Distribution : Native and cultivated in the Malaya and S. Tenasserim, Karela, Nilgiri hills, mostly in S. India.

Reproductive Cycle	:	November-June.
Part Used	:	Fruit.
Folk Uses	:	Fruits used for chronic diarrhoea and dysentery in children.
References	:	Chatterjee and Pakrashi (1997), Kirtikar and Basu (1984).

Garcinia xanthochymus **Hook.f.** ***ex*** **T. Aderson.**

Syn.: *G. tinctoria* (DC.) Roxb. Dunn.; *Xanthochymus pictorius* Roxb.; *X. tinctorius* DC.

Family: Clusiaceae

Eng., H. and Sans. Names	:	*Eng.*: Egg Tree; *H.*: Dampel, Ota, Tamal, Tamul; *Sans.*: Avika, Bhavana, Bhavya, Lamphala, Tamala, Tapinjha.
Distribution	:	Distributed widely in the lower hill forests of E. Himalayas, Meghalaya, Sikkim, W. Bengal, Maharashtra, Gujarat, Tamil Nadu, Kerala, A.P., Andamans
Reproductive Cycle	:	Flowering during spring and fruiting during summer.
Part Used	:	Whole Plant.
Folk Uses	:	Used in colic pain, dysentery, diarrhoea and as an anthelmintic.
Reference	:	Pal *et al.* (2000).

Gardenia gummifera **L.f.**

Syn.: *G. arborea* Roxb.; *G. inermis* F. Dietr.; *Genipa arborea* (Roxb.) Baill.; *G. gummifera* (L.f.) Baill.

Family: Rubiaceae

Eng., H. and Sans. Names	:	*Eng.*: Gummy Cape Jasmine; *H.*: Dikamali, Dikmali, Mali; *Sans.*: Nadi-Hingu, Pindava.
Distribution	:	Grows from N. India-Deccan Peninsula, extending northwards-Bundelkhand and parts of Bihar.
Reproductive Cycle	:	March-August.
Parts Used	:	Whole Plant, Resin and Gum.
Folk Uses	:	Plant used in poliomyelitis. Resin/gum obtained from leaf buds and shoots given to children in nervous

disorders, in treatment of worms and diarrhoea during teething in children. Resin also rubbed on gums to allay irritation. Body of children suffering with pneumonia also massaged with a mixture of the gum and castor oil.

References : Agarwal (2003), Ambasta (1986), Anonymous (1956), Khare (2004), Kirtikar and Basu (1984), Kumar and Lal (2003), Pal *et al.* (2000), Upadhyay and Chauhan (2003), Watt (1972).

Gardenia resinifera Roth.

Syn.: *G. lucida* Roxb.; *Genipa resinifera* (Roth) Baill.

Family: Rubiaceae

Eng., H. and Sans. Names : *Eng.*: White Emetic Nut;

H.: Dekamari, Dikamali, Dikemali;

Sans.: Jantuka.

Distribution : Occurs naturally in deciduous forests of the western parts of Indian Peninsula and dry districts of Tamil Nadu as well as in southern Bangladesh and Myanmar; frequently cultivated in gardens.

Reproductive Cycle : March-November.

Part Used : Gum.

Folk Uses : Gum given to children to treat nervous disorders and diarrhoea and irritation due to teething.

Reference : Parrotta (2001).

Gardenia taitensis DC.

Syn.: *G. weissichii* H. St. John.

Family: Rubiaceae

Eng. Names : *Eng.*: Tahitian Gardenia, Tiare.

Distribution : Occurs on coastal limestone rock, occasionally in coastal forest near the shore.

Reproductive Cycle : Flowering in warmer months of the year.

Parts Used : Whole Plant and Flower.

Folk Uses : Plant used for treating inflammations in children. Flowers used for treating childhood diseases.

Reference : Prajapati *et al.* (2006).

Gelsomium elegans **(Gardner and Chapm.) Benth.**

Syn.: *G. sumatranum* (Blume) Boerl.; *Leptopteris sumatrana* Blume.; *Medicia elegans* Gardner and Champ.

Family: Loganiaceae

Eng. and H. Names : *Eng.*: Chinese Gelsemium;
H.: Gelsimium.

Distribution : Khasi, Lushai hills and Manipur.

Reproductive Cycle : Flowering: May-November; fruiting: July-February of next year.

Part Used : Rhizome.

Folk Use : Rhizome used in poliomyelitis.

Reference : Agarwal (2003).

Geranium walliachinum **D. Don. *ex* Sw.**

Family: Geraniaceae

Eng. and H. Names : *Eng.*: Shocking Blue, Wallich Cranesbill;
H.: Laljhari.

Distribution : Temperate regions of Uttrakhand between 2,200-3,400m.

Reproductive Cycle : February-April.

Part Used : Root.

Folk Use : Roots given in stomach disorders of infants.

References : Bhattacharjee (2001), Sharma (2003).

Givotia moluccana **(L.) Sreem.**

Syn.: *Croton moluccanus* L.; *Givotia rottleriformis* Griff. *ex* Wight.

Family: Euphorbiaceae

Eng. Name : *Eng.*: White Catamaran Tree.

Distribution : Deccan Peninsula, common in C. ranges of the Ghats from Dharwar and Bellary southwards.

Reproductive Cycle : April-August.

Part Used : Endosperm.

Folk Use : Extract of 10g powdered endosperm mixed with 100ml of cow's milk given orally to children once a day for 3 days to improve digestion.

Reference : Muthukumarsamy *et al.* (2003).

Glinus lotoides L.

Syn.: *M. glinus* A. Rich.; *Mollugo hirta* Thunb.

Family: Molluginaceae

Eng., H. and Sans. Names : *Eng.*: Lotus Sweetjuice;

H.: Gandibudi;

Sans.: Bhissata, Okharadi.

Distribution : Grows throughout India and Sri Lanka.

Reproductive Cycle : Almost throughout the year.

Part Used : Whole Plant.

Folk Use : Plant juice given to children in weakness.

References : Agarwal (2003), Kirtikar and Basu (1984).

Glochidion ellipticum Wight.

Syn.: *Bradleia wightiana* Wall.; *Diasperus assamicus* (Muell. Arg.) Kuntze.; *Glochidion assamicum* (Muell. Arg.) Hook.f.

Family: Euphorbiaceae

Distribution : W. Himalayas; Kumaon, Bengal; Dacca, foot hills and low valleys of the Himalayas from Sikkim eastwards-upper Assam, Chittagong, upper Myanmar.

Reproductive Cycle : October-March.

Part Used : Stem Bark.

Folk Use : Fresh bark juice given for stomachache in children and infants.

Reference : Mudgal (1999).

Gloriosa superba L.

Syn.: *G. simplex* Don.; *Methonica superba* Lamk.

Family: Liliaceae

Eng., H. and Sans. Names : *Eng.*: Climbing Lily, Malabar Glory Lily, Supper Lily, Tiger Lily;

H.: Karihari, Kuthari, Languli;

Sans.: Agnimukhi, Agnisikha, Ailni, Garbhaghatini, Kalikari, Langli, Vishalya.

Distribution : Native to tropics, occurs in humid hilly districts of India.

Reproductive Cycle : June-October.

Parts Used : Root and Tuber.

Folk Uses : Roots used in smallpox; its paste applied to abdomen after applying turmeric paste for 3 days for 'pirhi' disease (a boil-like thing developed inside the abdomen below rib-cage on left side in children). Tuber paste applied on body for chickenpox.

References : Brahmam and Saxena (1990), Pande *et al.* (2006), Rao and Henry (1996).

Glossocardia bidens **(Retz.) Veldkamp.**

Syn.: *Glossogyne bidens* (Retz.) Alston.; *G. pinnatifida* DC.; *Zinnia bidens* Retz.

Family: Asteraceae

Eng. and H. Names : *Eng.*: Cobbler's Tack;

H.: Bandhani.

Distribution : W. and N. India, extending eastwards to Bengal and southwards to Deccan, mostly in the plains.

Reproductive Cycle : Flowers in much of the month except in winter months.

Part Used : Whole Plant.

Folk Uses : Plant used against diarrhoea, dysentery and blood dysentery.

Reference : Pal *et al.* (2000).

Glycine max **(L.) Merr.**

Syn.: *Dolichos soja* L.; *Glycine angustifolia* Miq.; *G. gracilis* Skvortsov.; *Soja hispida* Moench.

Family: Fabaceae

Eng., H. and Sans. Names : *Eng.*: Soyabean;

H.: Bhat, Ram Kurthi;

Sans.: Garikulay.

Distribution : Himalayas tropical region, Kumaon-Sikkim, Khasia and Ava; often cultivated.

Reproductive Cycle : August-October.

Part Used : Leaf.

Folk Use : Macerated mass of fresh leaves kept on sting's (bee and wasp) spot to cure reaction in children.

Reference : Singh and Singh (2003a).

***Glycosmis pentaphylla* (Retz.) DC.**

Syn.: *Chionotria monogyna* Walp.; *C. rigida* Jack.; *Glycosmis arborea* (Roxb.) DC.; *Limonia pentaphylla* Retz.

Family: Rutaceae

Eng., H. and Sans. Names	:	*Eng.*: Tooth-Brush Plant; *H.*: Ban-Nimbu, Girgiti, Potali; *Sans.*: Ashvashakota, Vadadru, Vananimbuka.
Distribution	:	Throughout tropical and sub-tropical Himalayas, ascending to 2,350m in Sikkim from Satluj river in N.W. southwards to upper Assam, Travancor, Malacca and Sri Lanka.
Reproductive Cycle	:	September-June.
Parts Used	:	Whole Plant, Root, Root Bark and Leaf.
Folk Uses	:	Plant used in dyspepsia or indigestion, stomach pain, blood dysentery and as an anthelmintic. 1g of root powder given either with water or butter milk for deworming in children. Root bark extract mixed with jaggery given orally for cold and cough. Bath with decoction of leaves of this plant and those of 'Neem', *Phyllanthus acidus, Clerodendron viscosum* in water good against measles.
References	:	Nath and Maiti (2003), Pal *et al.* (2000), Rao and Henry (1996), Upadhyay and Chauhan (2000).

***Gmelina arborea* L.**

Syn.: *G. rheedii* Hook.; *Premna arborea* Roth.

Family: Verbenaceae

Eng., H. and Sans. Names	:	*Eng.*: Candahar Tree, Cashmere Tree, Coomb Teak, White Teak; *H.*: Gambhari, Kambar, Sewan, Shiwan; *Sans.*: Ashveta, Bhadraparni, Ghambhari, Kasmari, Suphala, Vidari.
Distribution	:	Throughout the greater parts of India and Andamans, upto 1,900m; also planted in gardens and avenues.
Reproductive Cycle	:	February-June.
Part Used	:	Whole Plant.
Folk Use	:	Plant used in smallpox.
References	:	Asolkar *et al.* (1992), Pande *et al.* (2006).

Gnaphalium uliginosum L.

Syn.: *G. tomentosum* Hoffm.

Family: Asteraceae

Eng. Name : *Eng.*: Marsh Cudweed.

Distribution : W. Himalayas; Kashmir 1,350-2,350m.

Reproductive Cycle : Flowering in July-October.

Part Used : Aerial Parts.

Folk Use : Used in tonsillitis.

Reference : Prajapati *et al.* (2006).

Gossypium arboreum L.

Syn.: *G. indicum* Lam.; *G. intermedium* Tod.; *G. neglectum* Tod.; *G. obustifolium* Roxb.; *G. sanguineum* Hazsl.

Family: Malvaceae

Eng., H. and Sans. Names : *Eng.*: Cotton, Tree Cotton;

H.: Kapas;

Sans.: Karpasamu.

Distribution : Found in plains of India, gardens, Kashmir; not generally cultivated.

Reproductive Cycle : July-September.

Part Used : Petals.

Folk Use : Extract of petals soaked in human or cow's milk used as a soothing and effective application for conjunctivitis of infants.

References : Kirtikar and Basu (1984), Sood *et al.* (2012), Watt (1972).

Gossypium hirsutum L.

Syn.: *G. birkinshawii* G. Watt.; *G. harrisii* G. Watt.; *G. maxicanum* Tod.; *G. palmeri* G. Watt.

Family: Malvaceae

Eng., H. and Sans. Names : *Eng.*: American Cotton, Mexican Cotton, Upland Cotton, Wild Cotton;

H.: Karpas, Karpasi;

Sans.: Chitradandakah, Guda, Karpasi, Karpasika.

Distribution : Cultivated and naturalized throughout tropics and sub-tropics.

Reproductive Cycle : Flowering in summer and autumn.

Part Used : Seed.

Folk Uses : Medicated oil obtained by keeping a burning piece of cow dung cake over cotton seeds half filled into rolled leaf of the plant, applied 2/3 times daily to the affected area (earache and ear infection).

Reference : Singh and Singh (2003a).

Grewia microcos **L.**

Syn.: *G. affinis* Lindl.; *G. ulmifolia* Roxb.; *Microcos paniculata* Roxb.; *M. syauntoniana* G. Don.

Family: Tiliaceae

Eng. and H. Names : *Eng.*: Microcos;

H.: Shiral.

Distribution : N.E. parts of India, W. Ghats and the Andamans.

Reproductive Cycle : March-December.

Part Used : Whole Plant.

Folk Use : Plant used in smallpox.

Reference : Kirtikar and Basu (1984).

Grewia rothii **DC.**

Syn.: *G. bicolor* Roth.

Family: Tiliaceae

Eng. Names : *Eng.*: Bastard Brandy Bush, False Brandy Bush, Two Coloured Grewia.

Distribution : E. Bengal, Assam, Sikkim and Chittangong, Bundelkund.

Reproductive Cycle : April-October.

Part Used : Flower.

Folk Use : Dried flowers pounded with honey given to young boys to check excessive night emission.

Reference : Panda and Das (2000).

Guaiacum officinale **L.**

Syn.: *G. bijugum* Stokes.; *G. breynii* Spreng.

Family: Zygophyllaceae

Eng. and Sans. Names : *Eng.*: Brazil Wood, Gum Guaiacum (resin), Lignum Vitae (Wood);

Sans.: Amrutdaru, Bridhamitra, Jeevdaru.

Distribution	:	Inhabitant of W. Indies, introduced into India. Also, it grows in the arid plains stretching form Florida Keys of USA to Venezuela.
Reproductive Cycle	:	March-October.
Part Used	:	Gum.
Folk Use	:	Used in tonsillitis.
References	:	Ambasta (1986), Anonymous (1956), Mitaliya and Bhatt (2003).

***Gymnema sylvestre* R. Br. *ex* Schult**

Syn.: *G. melicida* Edgew.; *Periploca sylvestrica* Willd.

Family: Asclepiadaceae

Eng., H. and Sans. Names	:	*Eng.*: Periploca of the Woods, Small Indian Ipecacuanha, Vine; *H.*: Chhotadudhilata, Gurmar, Merasimgi; *Sans.*: Ajaballi, Ajashringi, Bahalchakshu, Chakrashreni, Grihadruma, Karnika, Maurvi, Medhasingi, Nandivruksha, Putrashringi, Vatika, Vishani, Vrikshikali.
Distribution	:	Common in hills of Bihar, Orissa, M.P. and Deccan Peninsula.
Reproductive Cycle	:	July-December.
Part Used	:	Whole Plant.
Folk Uses	:	Decoction of plant helps to enhance memory, intellect and growth of young children.
Reference	:	Khare (2004).

***Gymnosporia montana* (Roth.) Benth**

Syn.: *Catha montana* (Roth.) G. Don.; *Celastrus montanus* Roth.; *C. obovatus* Hochst. *ex* Oliv.; *C. phyllacanthus* L. Her.

Family: Celastraceae

Eng., H. and Sans. Names	:	*Eng.*: Thorny Staff Tree; *H.*: Baikal, Kingani, Tondarsaijhad, Vingar; *Sans.*: Sudhavriksha, Vikankata.
Distribution	:	Occurs throughout the drier parts of C. southwestern and northwestern India.
Reproductive Cycle	:	Flowering in October.
Parts Used	:	Shoot and Leaf.

Folk Uses : Decoction of shoot used for colic, dysentery and diarrhoea among children. Leaves in milk used as vermifuge for children.

References : Agarwal (2003), Ambasta (1986), Anonymous (1956).

Gynocardia odorata R. Br.

Syn.: *Chaulmoogra odorata* Roxb.

Family: Flacourtiaceae

Eng., H. and Sans. Names : *Eng.*: Lukrabo Seeds, The Chaulmugra Oil;

H.: Chalmugra, Chavulmungri, Chhalmugra;

Sans.: Alaskapha, Kusthapa, Tuvarka.

Distribution : Sikkim and Khasia mountains eastwards to Chittagong, Rangoon and Tenasserim.

Reproductive Cycle : March-April.

Part Used : Seed.

Folk Use : Seed oil used in marasmus.

Reference : Watt (1972).

Haematoxylum campechianum L.

Syn.: *Cymbosepalum baronii* Baker.; *Haematoxylon campechianum* L.

Family: Fabaceae

Eng., H. and Sans. Names : *Eng.*: Campeachy Wood, Logwood Tree, Sappan Wood Tree;

H.: Bakam, Patang;

Sans.: Patang, Pattaranjaka.

Distribution : Native to tropical America; introduced in Indian gardens for its delicate foliage and fragrant flowers.

Part Used : Heart Wood.

Folk Use : Used in infantile diarrhoea.

References : Kirtikar and Basu (1984), Watt (1972).

Haldina cordifolia (Roxb.) Ridsdale

Syn.: *Adina cordifolia* (Roxb.) Hook.f. *ex* Brandis.; *Nauclea cordifolia* Roxb.

Family: Rubiaceae

Eng., H. and Sans. Names : *Eng.*: Saffron Teak, Yellow Teak;

H.: Haldu, Kadami, Ladiva;

Sans.: Dharakadambah, Girikadambah.

Distribution : Deciduous and semi-evergreen forests throughout India, especially in E. Ghats, Karnataka and Kerala.

Reproductive Cycle : June-December.

Part Used : Whole Plant.

Folk Uses : Plant used in stomach pain and fever.

Reference : Pal *et al.* (2000).

Hedera helix **L. (Pl. 7A)***

Family: Araliaceae

Eng. and H. Names : *Eng.*: Bindwood, English Ivy, Ivy, Lovestone;

H.: Lablab.

Distribution : Throughout Himalayas, 2,000-3,350m; Khasia hills, 1,300-2,000m.

Reproductive Cycle : June-November.

Parts Used : Shoot and Leaf.

Folk Uses : Young shoots considered good in rickets. Fine paste made from tender leaves applied on mumps.

References : Baburaj *et al.* (2000), Chhetri (1994).

Hedera nepalensis **K.Koch.**

Syn.: *Cissus wallichiana* Turcz.; *Hedera himalaica* Tobler.

Family: Araliaceae

Eng. and H. Names : *Eng.*: Barren Ivy, Ground Ivy, Himalayan Ivy, Nepal Ivy, Wood-bind;

H.: Bano, Kural, Mithiari.

Distribution : Throughout Himalayas: 2,000-3,500m and Khasia hills: 1,350-2,000m; grows in some places from W. Europe – Japan.

Reproductive Cycle : Flowering during September-October.

Part Used : Leaf.

Folk Use : Fine paste of leaves applied on mumps.

Reference : Chhetri (1994).

Hedychium coronarium **Koen.**

Syn.: *H. flavum* Roxb.; *H. gandasulium* Ham.; *H. maximum* Roscoe.; *Kaempferia hedychium* Lam.

Family: Zingiberaceae

Eng., H. and Sans. Names : *Eng.*: Butterfly Ginger, Common Ginger Lily, Garland Flower, White Ginger;

		H.: Dolan Champa;
		Sans.: Shati.
Distribution	:	Throughout India, ascending upto 2,000m.
Reproductive Cycle	:	June-October.
Parts Used	:	Rhizome and Leaf.
Folk Uses	:	Extract of rhizome prescribed in bronchitis; also its decoction gargled for relief from tonsillitis and bad breath. Leaf juice given for tonsillitis.
References	:	Sharma *et al.* (2003a), Singh *et al.* (2003), Sinha (1987).

Hedychium spicatum **Buch.-Ham.** ***ex*** **J. F. Smith**

Syn.: *H. acuminatum* Rosc.

Family: Zingiberaceae

Eng., H. and Sans. Names	:	*Eng.*: Spiked Ginger Lily, The Lesser Galangal of Ainslie;
		H.: Gandhapalashi, Kapurakachari;
		Sans.: Amlanisha, Gandhasshati, Karpurakachali, Sati.
Distribution	:	Himalayas, N.E. India, ascending upto 2,300m.
Reproductive Cycle	:	August-September.
Part Used	:	Rhizome.
Folk Use	:	Paste of rhizome along with turmeric powder used in smallpox.
References	:	Singh *et al.* (2003), Sinha (1987).

Hedyotis scandens **Roxb.**

Syn.: *H. polycarpa* R. Br. *ex* G. Don.; *H. subscandens* Griff.

Family: Rubiaceae

Distribution	:	Confined to Himalayas and Sub-Himalayan regions, 700-2,000m from Nepal-Sikkim, Khasia mountains and Silhet, 0-2,000m and Chittagong, abundant.
Reproductive Cycle	:	August-April.
Parts Used	:	Whole plant and Root.
Folk Uses	:	Plant used in diarrhoea, dyspepsia and indigestion. Decoction of its roots given to children for dysentery by Garos of Meghalaya.
References	:	Kumar *et al.* (1980), Pal *et al.* (2000).

Helicteres isora L.

Syn.: *H. chrysocalyx* Miq. *ex* Mast.; *H. grewiaefolia* DC.; *H. roxbhurghii* G. Don.; *Ixora versicolor* Hassk.

Family: Sterculiaceae

Eng., H. and Sans. Names : *Eng.*: East Indian Screw Tree;

H.: Bhendu, Jonkaphal, Kapasi, Maraphali, Marorphali, Marosi;

Sans.: Avartani, Mrigashinga.

Distribution : Common in deciduous and semi-evergreen forests throughout India; found wild in Shivalik tract for some distance into Punjab.

Reproductive Cycle : May-August.

Parts Used : Whole Plant, Fruit and Seed.

Folk Uses : Plant used in ringworm and other skin diseases. Extract of fruits given orally to children by Bhills of Rajasthan to relieve diarrhoea. Fruits beneficial in body pain, rickets, polio, antiseptic, griping of bowels, flatulence of children, and killing intestinal worms. Fruits tied as an amulet in the neck to treat diseases of malnutrition, fever, giddiness and tenesmus in stomach among children. Its decoction given 1 tsp once daily till cure to infants for stomach disorders. Paste of its seeds and raw turmeric with mustard oil applied on defective limbs of children. Medicinal oil obtained by boiling its seeds in mustard oil applied externally in stomach complaints and aching body of the children.

References : Agarwal (2003), Asolkar *et al.* (1992), Chauhan (1999), Dhiman (2003), Farooq (2005), Girach and Aminuddin (1995), Hembrom (1996), Hosagoudar and Henry (1996a), Jain (1991), Kirtikar and Basu (1984), Kshirsagar and Singh (2007), Pal *et al.* (2000), Pandey and Rout (2006), Parrotta (2001), Rao and Henry (1996), Sur and Halder (2002), Topno and Ghosh (2000), Watt (1972).

Heliotropium indicum L.

Syn.: *Eliopia riparia* Raf.; *E. serrata* Raf.; *Tiaridium anisophyllum* G. Don.; *T. indicum* (L.) Lehm.

Family: Boraginaceae

Eng., H. and Sans. Names : *Eng.*: Indian Turn Sole;

H.: Bithua, Chirghas, Hattajuri, Nagadanti, Nilkatte;

Sans.: Bhurundi, Duralabha, Hastisunda, Srihastini, Vrischikali.

Distribution : Throughout India.

Reproductive Cycle : April-October.

Part Used : Whole Plant.

Folk Use : Plant used for skin diseases and eye complaints of children.

Reference : Pal *et al.* (2000).

Hemidesmus indicus **(L.) R. Br. *ex* Schult.**

Syn.: *H. indicus* (L.) R. Br.; *Periploca indica* L.

Family: Asclepiadaceae

Eng., H. and Sans. Names : *Eng.*: Country Sarasaparilla, Indian Sarsaprilla;

H.: Anantmul, Magrabu, Salsa, Sariva, Shewat;

Sans.: Anantamulah, Sarba, Sariva.

Distribution : Upper Gangetic plains, eastwards to W. Bengal and from M.P.-S. India.

Reproductive Cycle : July-December.

Parts Used : Whole Plant, Root Bark and Root.

Folk Uses : Hot infusion of roots, root bark or whole plant with milk and sugar given to children for chronic cough and diarrhoea. Root juice given to children in tonsillitis. Root decoction given to check the attack of epilepsy and wasting diseases among children. Root paste eaten to cure dysentery and fever of babies; also its paste with common salt applied to treat tongue eruptions in children.

References : Asolkar *et al.* (1992), Chatterjee and Pakrashi (1997), Khare (2004), Kirtikar and Basu (1984), Kumar and Goel (2008), Panda and Das (2000), Rao and Henry (1996), Yasodamma *et al.* (2009), Watt (1972).

Hemionitis arifolia **(Burm.f.) T. Moore**

Syn.: *Asplenium arifolium* Burm.f.; *Hemionitis sagittata* Fee.; *H. toxotis* Trevis.

Family: Adiantaceae

Eng. Names : *Eng.*: Heart Fern, Strawberry Fern, Tongue Fern.

Distribution : Plains and mountains of S. India upto 1,200m, Bihar, W. Bengal and Orissa.

Reproductive Cycle : August-January.

Part Used	:	Root.
Folk Use	:	Root paste applied externally to treat polio.
Reference	:	Balasubramanian and Prasad (1996).

Heteropogon contortus **(L.) Beauv.** ***ex*** **Roem. and Schult.**

Syn.: *Andropogon contortus* L.

Family: Poaceae

Eng. and H. Names	:	*Eng.*: Bellary Grass, Spear Grass; *H.*: Kher, Kumeria, Parwa, Sura.
Distribution	:	Dry places throughout India, ascending upto 2,350m in Himalayas. Occurs in Punjab, U.P., Bihar, Bengal, Assam, Orissa, M.P., Bombay and Chennai.
Reproductive Cycle	:	August-January.
Part Used	:	Whole Plant.
Folk Uses	:	Plant good against stomach heat, dyspepsia and indigestion.
Reference	:	Pal *et al.* (2000).

Hibiscus rosa-sinensis **L.**

Family: Malvaceae

Eng., H. and Sans. Names	:	*Eng.*: Bengal Rose, China Rose, Chinese Hibiscus, Shoe Flower; *H.*: Gulhar, Jasum, Jasun, Odhul; *Sans.*: Arkapriya, Aruna, Harivallabha, Japa, Odhrapushpa, Pratika, Pushpi, Roga, Trisandhya.
Distribution	:	Cultivated in gardens throughout India.
Reproductive Cycle	:	February-October.
Part Used	:	Flower.
Folk Uses	:	Flowers used in infantile eczema; also applied on mumps in Sikkim.
References	:	Srivastava *et al.* (1987), Thomas and Britto (2003).

Holarrhena antidysenterica **(L.) Wall.**

Syn.: *Echites antidysentrica* Roxb.; *E. pubescens* Buch.-Ham.; *Holarrhena codaga* G. Don.; *H. malaccensis* Wight.; *H. pubescens* Wall.

Family: Apocynaceae

Eng., H. and Sans. Names	:	*Eng.*: Bitter Oleander, Conessi Bark, Easter Tree, Ivory Tree, Kurchi Bark, Tellicherry Bark;

		H.: Indrajau, Karchi, Karra, Kuar, Kurchi, Kureya;
		Sans.: Girimallika, Indrayava, Kalinga, Kutaja, Mallikapushpa.
Distribution	:	Tropical parts of India, abundant in the sub-Himalayan tracts; also planted on roadsides.
Reproductive Cycle	:	May-January.
Parts Used	:	Whole Plant, Root and Seed.
Folk Uses	:	Plant useful in stomach pain, dysentery, blood dysentery and used as dewormer. Fresh root juice given internally on empty stomach for deworming in children. Root paste applied in the anal region of children to expel worms. Seeds in doses of 1 to 2 grains useful in dysentery, diarrhoea of young infants and as a tonic in emaciation and general debility in doses of 4-5 grains.
References	:	Brahmam and Saxena (1990), Pal *et al.* (2000), Prusti (2007), Watt (1972).

Holoptelea integrifolia **(Roxb.) Planch**

Family: Ulmaceae

Eng., H. and Sans. Names	:	*Eng.*: Indian Elm, Kanju;
		H.: Chilbil, Kanju;
		Sans.: Cirabilvah, Putikaranhah.
Distribution	:	Outer lower ranges of the Himalayas, from Jammu-Oudh, ascending to 700m. From Banda and Bihar-Travancore and from Pegu-Martaban, Sri Lanka in hot drier regions.
Reproductive Cycle	:	April-May.
Parts Used	:	Whole Plant and Leaf.
Folk Uses	:	Plant used in ringworm. Poultice of young leaves applied for one hour on back-bone to cure rickets.
References	:	Pal *et al.* (2000), Singh and Prakash (1996).

Hordeum distichon **L.**

Syn.: *H. deficiens* Steud. *ex* A. Braun.; *H. irregulare* Aberg and Wiebe.; *H. nudum* (L.) Ard.

Family: Poaceae

Eng. Name	:	*Eng.*: Two-Rowed Barley.
Distribution	:	Grown in temperate region worldwide.
Reproductive Cycle	:	July-August.

Part Used : Grain.

Folk Uses : Barley given to babies for preventing curd formation in stomach and for minor infections or diarrhoea.

Reference : Prajapati *et al.* (2006).

Hordeum vulgare **L.***

Syn.: *H. sativum* Jessen.

Family: Poaceae

Eng., H. and Sans. Names : *Eng.*: Common Barley;

H.: Jau, Jawa, Jawakha;

Sans.: Akshata, Divya, Hayapriya, Yava.

Distribution : Widely cultivated throughout warm and dry regions of India.

Reproductive Cycle : Mainly in winters.

Part Used : Grains.

Folk Use : A light flake prepared for weak children as a nutritive tonic.

Reference : Agarwal (2003).

Houttuynia cordata **Thunb.**

Syn.: *H. emeiensis* Z. Y. Zhu and S. L. Zhang.; *H. foetida* Loudon.; *Polypara cordata* Kuntze.

Family: Saururaceae

Eng. Names : *Eng.*: Chameleon Plant, Fishwort, Heart Leaf, Lizard Tail, Tsi.

Distribution : Tropical Himalayas (Garhwal – Sikkim): 300-1,700m, Assam and Khasia mountains.

Reproductive Cycle : April-October.

Parts Used : Whole Plant and Leaf.

Folk Uses : Whole plant including leaves used in measles.

References : Agarwal (2003), Ambasta (1986), Anonymous (1959), Prajapati *et al.* (2006), Tiwari and Tiwari (1996).

Huperzia phlegmaria **(L.) Rothm.**

Syn.: *Lepidotis phlegmaria* (L.) P. Beauv.; *Lycopodium phlegmaria* L.

Family: Lycopodiaceae

Eng. Names : *Eng.*: Coarse Tassel Fern, Common Tassel Fern.

Distribution	:	W. Bengal, Assam, Sikkim, Arunachal Pradesh, Kerala and Andaman Nicobar Islands.
Part Used	:	Whole Plant.
Folk Use	:	Plant paste applied over the body before bath to prevent skin diseases in children.
Reference	:	Murugesan *et al.* (2011).

***Hybanthus enneaspermus* (L.) F. Muell.**

Syn.: *Ionidium enneaspermum* (L.) Vent.

Family: Violaceae

Eng., H. and Sans. Names	:	*Eng.*: Spade Flower, Pink Ladies Slipper;
		H.: Ratanpurush;
		Sans.: Amburuha, Charati, Padmacarini.
Distribution	:	Throughout the warmer parts of India from U.P. southwards to the Deccan Peninsula.
Reproductive Cycle	:	September-October.
Part Used	:	Root.
Folk Use	:	Santals employ roots in bowel complaints of children.
References	:	Agarwal (2003), Ambasta (1986), Chatterjee and Pakrashi (1997), Anonymous (1959), Kirtikar and Bsau (1984), Parrotta (2001), Ranjan (2000), Watt (1972).

***Hydrastis canadensis* L.**

Syn.: *Warnera canadensis* Mill.; *W. tinctoria* Rafin.

Family: Ranunculaceae

Eng. Names	:	*Eng.*: Eye Balm, Eye Root, Golden Seal, Indian Turmeric.
Distribution	:	Temperate Himalayas.
Reproductive Cycle	:	Flowering in spring.
Part Used	:	Rhizome.
Folk Use	:	Used in smallpox.
Reference	:	Prajapati *et al.* (2006).

***Hydrocotyle javanica* Thunb.**

Syn.: *H. polycephala* Wight. and Arn.

Family: Apiaceae

Eng. and H. Names	:	*Eng.*: Java Pennywort;
		H.: Mahagotukola.

Distribution : Himalayas from Kashmir-Bhutan: 700-2,650m, Khasia mountains, 700-2,000m. Mountains of Malabar and Sri Lanka, Myanmar, Malaya.

Reproductive Cycle : March.

Part Used : Leaf.

Folk Use : Two tsp of leaf juice given orally 2-3 times daily for 3-4 days to children in tonsillitis and stomachache (5ml twice daily).

References : Lalramnghinglova (2003), Srivastava *et al.* (2003b).

***Hydrocotyle sibthorpioides* Lam.**

Syn.: *H. rotundifolia* Roxb.

Family: Apiaceae

Eng., H. and Sans. Names : *Eng.*: Lawn Pennywort;

H.: Mandukaparni;

Sans.: Khulkhuri.

Distribution : Himalayas, Manipur and Meghalaya.

Reproductive Cycle : April-July.

Parts Used : Whole Plant and Leaf.

Folk Uses : Juice of whole plant given to growing children for improving memory. 1 tsp of leaf paste given twice a day to children passing green and watery stools or diarrhoea.

References : Nath *et al.* (2007), Panda (1996), Srivastava *et al.* (1987).

***Hygrophila auriculata* (K. Schum.) Heine**

Syn.: *Asteracantha longifolia* Nees.; *Barleria auriculata* Schumach.; *Hygrophila spinosa* T. Anders.

Family: Acanthaceae

Eng., H. and Sans. Names : *Eng.*: Asteracantha, Long-leaved Barleria;

H.: Gokshura, Juliakanta, Kamtakaliya, Talamkhana, Untkataru;

Sans.: Atichhatra, Ikshugandha, Iksurah, Kakilakshya, Pichhila, Vajra.

Distribution : Sub-Himalayan forests of northeastern U.P.

Reproductive Cycle : September-June.

Part Used : Leaf.

Folk Use	:	Paste of handful of leaves pounded together with black pepper eaten twice daily to treat infantile diarrhoea.
Reference	:	Girach *et al.* (1997).

Hyptis suaveolens **(L.) Poit.**

Syn.: *Ballota suaveolens* L.; *Bysteropogo graveolens* Blume.

Family: Lamiaceae

Eng., H. and Sans. Names	:	*Eng.*: Bush Tea, Horehound, Hyptis, Mint Weed, Wild Spikenard; *H.*: Nili Tulasi, Wilayati Tulasi; *Sans.*: Bhustana.
Distribution	:	Deccan Peninsula, Cachar, Nicobar Islands, found in N.E. and Peninsular India.
Reproductive Cycle	:	March-September.
Part Used	:	Leaf.
Folk Uses	:	3-4 drops of juice mixed with water used for taking bath for treating cough and cold by Apatani people.
Reference	:	Rawat and Chowdhury (1998).

Ichnocarpus frutescens **R. Br.**

Syn.: *Apocynum frutescens* L.; *Echites frutescens* Roxb.; *Ichnocarpus dascalyx* Miq.; *I. radicans* Wall.

Family: Apocynaceae

Eng., H. and Sans. Names	:	*Eng.*: Black Pepper; *H.*: Dhudhi, Kalidudhi, Siamalata; *Sans.*: Palindi, Paravalli, Saria, Syamalata, Ulpalasariba.
Distribution	:	Occurs throughout India, W. Himalayas; Sirmore-Nepal, 300-650m. Upper Gangetic plains, Delhi-Bengal, Assam, Silhet, Chittagong, Myanmar, the Deccan Peninsula and Sri Lanka.
Reproductive Cycle	:	September-April.
Parts Used	:	Whole Plant and Stem.
Folk Uses	:	Plant used in measles. Stem tied as an amulet on the waist of children having prolapsed anus.
References	:	Chatterjee and Pakrashi (1997), Farooq (2005), Girach and Aminuddin (1995).

Ilex paraguariensis **A. St. Hilaire**

Syn.: *I. domestica* Reiss.; *I. mate* St.-Hil.; *I. paraguaiensis* Lamb.

Family: Aquifoliaceae

Eng. Names	:	*Eng.*: Mate, Paraguay Tea, Yerba Mate.
Distribution	:	Cultivated in some gardens in N. India.
Reproductive Cycle	:	Flowering in October-May.
Part Used	:	Bud.
Folk Use	:	Macerated buds used for treating measles.
Reference	:	Keller *et al.* (2010).

Imperata cylindrica **Beauv.**

Syn.: *I. arundinacea* Cyrill.; *Lagarus cylindrica* L.

Family: Poaceae

Eng., H. and Sans. Names	:	*Eng.*: Blady Grass, Cogongrass, Thatch Grass; *H.*: Dabh, Khans, Sirhu, Uloo; *Sans.*: Darbha, Yagnika.
Distribution	:	Plains and hills of Punjab, Bengal, N.E. India; common throughout Mizoram.
Reproductive Cycle	:	April-May.
Part Used	:	Stolon.
Folk Use	:	One tsp juice of its stolon and black pepper given on empty stomach for expelling worms.
Reference	:	Srivastava *et al.* (2003b).

Incarvillea emodi **(Royle *ex* Lindl) Chatterjee.**

Syn.: *Amphicome emodi* Royle *ex* Lindl.

Family: Bignoniaceae

Eng. Name	:	*Eng.*: Himalayan Incarvillea.
Distribution	:	W. Himalayas, 700-3,000m, from Kashmir-Nepal.
Reproductive Cycle	:	Flowering in March-April.
Part Used	:	Whole Plant.
Folk Use	:	Plant used in fever.
Reference	:	Pal *et al.* (2000).

Indigofera cassioides **DC. (Pl. 7B)**

Syn.: *I. arborea* Roxb.; *I. elliptica* Roxb.; *I. pulchella* Roxb.; *I. violacea* Roxb.

Family: Fabaceae

Eng. and H. Names	:	*Eng.*: Cassia Indigo; *H.*: Hakna, Nil, Sakena.
Distribution	:	Throughout Himalayas and hills of India proper, ascending to 150m in the Kumaon.
Reproductive Cycle	:	July-November.
Part Used	:	Fruit.
Folk Use	:	A wreath made of one fruit in white thread given for wearing to children by tribals as a charm against fever.
Reference	:	Singh and Narain (2008).

Indigofera cordifolia **Roth.**

Family: Fabaceae

Eng. and H. Names	:	*Eng.*: Heart Leaf Indigo; *H.*: Bekara, Gokhru.
Distribution	:	Plains throughout India proper, ascending to 1,350m in the Chenab valley.
Reproductive Cycle	:	August-December.
Part Used	:	Leaf.
Folk Use	:	Extract of leaves mixed with *Terminalia* fruits taken orally for loose motion (2 tsp twice daily).
References	:	Singh (2002, 2003).

Indigofera linifolia **(L.f.) Retz.**

Syn.: *Anila linifolia* (L.f.) Kuntze.; *Heysarum linifolium* L.f.; *Indigofera albicans* Span.; *I. roxburghii* Tausch.

Family: Fabaceae

Eng. and H. Names	:	*Eng.*: Narrow-Leaf Indigo; *H.*: Torbi, Torki.
Distribution	:	Common in Indian plains, ascending Himalayas.
Reproductive Cycle	:	July-October.
Part Used	:	Whole Plant.
Folk Uses	:	Plant used in stomach heat, blood dysentery, colic pain, and eye complaints.
Reference	:	Pal *et al.* (2000).

Indigofera tinctoria **L.**

Syn.: *I. bergii* Vatke.; *I. indica* Lam.; *I. houer* Forssk.; *I. tinctoria* Blanco.

Family: Fabaceae

Eng., H. and Sans. Names	:	*Eng.*: Common Indigo, Indian Indigo; *H.*: Gouli, Lil, Nil; *Sans.*: Aklika, Dronika, Gramya, Nilini, Rangapatri, Vijaya.
Distribution	:	Widely cultivated in many parts of India.
Reproductive Cycle	:	August-May.
Parts Used	:	Whole Plant and Root.
Folk Uses	:	Plant used in chronic bronchitis, asthma, stomach pain, blood dysentery and colic pain of children. Paste made from root smeared on tongue of newborn babies to cure vomiting.
References	:	Ansarali and Sivadasan (2009), Kirtikar and Basu (1984), Pal *et al.* (2000).

Inula helenium **L.**

Syn.: *Aster helenium* (L.) Scop.; *A. officinalis* All.;
***Helenium grandiflorum* Gilib.**

Family: Asteraceae

Eng., H. and Sans. Names	:	*Eng.*: Elecampane; *H.*: Kaashmira, Kushtha-Bheda; *Sans.*: Kaashmira, Kushtha-Bheda, Pauskaram.
Distribution	:	Temperate Himalayas, N.W. India, Kashmir 1,700-2,350m.
Reproductive Cycle	:	July-August.
Parts Used	:	Rhizome and Root.
Folk Use	:	Dried root and rhizome used for irritating cough in children.
Reference	:	Karnick (1994a).

Ipomoea coptica **(L.) Roth *ex* Roem. and Schult.**

Syn.: *Convolvulus thonningii* Schumach and Thonn.

Family: Convolvulaceae

Eng. Names	:	*Eng.*: Alamo Vine, Egyptian Morning Glory.
Distribution	:	W. India and Deccan and Sri Lanka.

Reproductive Cycle	:	March-November.
Part Used	:	Whole plant.
Folk Use	:	Herb used for chest complaint.
References	:	Ambasta (1986), Kirtikar and Basu (1984).

Ipomoea mauritiana Jacq.

Syn.: *Batatas paniculata* (L.) Choisy.; *Convolvulus paniculatus* L.; *Ipomoea digitata* Baker and Rendle.; *I. paniculata* (L.) R. Br.

Family: Convolvulaceae

Eng., H. and Sans. Names	:	*Eng.*: Giant Potato; *H.*: Bhuyikohada, Bidar Kand, Bilaikand; *Sans.*: Bhumikushmanda, Kasiravidari, Vidari.
Distribution	:	Deciduous and evergreen forests and coastal tracts throughout India.
Reproductive Cycle	:	July-December.
Parts Used	:	Leaf and Tuberous Root.
Folk Uses	:	Paste of leaves mixed with castor seeds applied on head to reduce high fever in children. Tuberous roots used for emaciation in children.
References	:	Anonymous (1959), Chatterjee and Pakrashi (1997), Prajapati *et al.* (2006), Yasodamma *et al.* (2009), Watt (1972).

Ipomoea turbinata Lag.

Syn.: *Calonyction muricatum* (L.) G. Don.; *Convolvulus muricatus* L.

Family: Convolvulaceae

Eng. and H. Names	:	*Eng.*: Lavender Moonvine, Purple Moonflower; *H.*: Michai.
Distribution	:	Himalayas, 300-1,700m frequent from Kangra-Sikkim, Deccan hills.
Reproductive Cycle	:	Flowering in October-January.
Part Used	:	Whole Plant.
Folk Use	:	Plant used for curing fever in children.
Reference	:	Pal *et al.* (2000)

Iris domestica **(L.) Goldblatt and Mabb.**

Syn.: *Belamcanda chinensis* (L.) DC.; *B. flabellata* Grey.; *Epidendrum domesticum* L.

Family: Iridaceae

Eng. Names	:	*Eng.*: Blackberry Lily, Leopard Flower, Leopard Lily.
Distribution	:	Cultivated as an ornamental plant all over India.
Reproductive Cycle	:	July-November.
Part Used	:	Rhizome.
Folk Use	:	Rhizome used in tonsillitis.
References	:	Agarwal (2003), Ambasta (1986), Anonymous (1948), Prajapati *et al.* (2006).

Iris germanica **L. (Pl. 7C)**

Syn.: *I. alba* Savi.; *I. amoena* DC.; *I. nepalensis* Wall. *ex* Lindl.; *I. sulcata* Wall.

Family: Iridaceae

Eng., H. and Sans. Names	:	*Eng.*: Bearded Iris, German Iris, Orris, Rhizomatous Iris; *H.*: Keore-Ke-Mul; *Sans.*: Padma-Pushkara.
Distribution	:	Cultivated in Kashmir.
Reproductive Cycle	:	June-July.
Part Used	:	Rhizome.
Folk Use	:	Used in diarrhoea in infants.
Reference	:	Karnick (1994a).

Ixora coccinea **L.**

Syn.: *Pavetta coccinea* (L.) Blume.

Family: Rubiaceae

Eng., H. and Sans. Names	:	*Eng.*: Flame of the Woods, Jungle Flame Ixora, Sacred Ixora; *H.*: Rajana, Rangan, Rookmini; *Sans.*: Bandhujivaka, Bandhuka, Paranti, Raktata.
Distribution	:	Occurs naturally in forests along the W. coast of India; commonly planted in gardens throughout the country.
Reproductive Cycle	:	In rainy season.
Parts Used	:	Whole Plant and Leaf.

Folk Uses : Plant used in ringworm. Leaves used as an external wash to treat infantile skin ailments.

References : Pal *et al.* (2000), Parrotta (2001).

Jasminum arborescens Roxb.

Syn.: *J. glabellum* Wall.; *J. latifolium* Roxb.; *J. punctatum* Wall.; *J. roxburghianum* Wall.

Family: Oleaceae

Eng., H. and Sans. Names : *Eng.*: Tree Jasmine, Wild Jasmine;

H.: Bela, Chameli, Navamallika, Saptala;

Sans.: Navamallica, Saptola.

Distribution : Upper Gangetic plains, ascending to 1,000m, on the Himalayas, Bengal, C. and S. India, hills of Ganjam and *Viz.*agapatam.

Reproductive Cycle : March-September.

Part Used : Leaf.

Folk Use : Juice of its half leaf and 4 leaves of *Sesbania grandiflora* mixed with two grains each of black pepper and dried borax given in honey for obstruction of bronchial tubes by viscid phlegm.

References : Kirtikar and Basu (1984), Watt (1972).

Jasminum grandiflorum L. (Pl. 7D)

Syn.: *J. officinale* (L.) forma var. *grandiflorum* (L.) Kobuski.

Family: Oleaceae

Eng., H. and Sans. Names : *Eng.*: Catalonion Jasmine, Common Jasmine, Spanish Jasmine;

H.: Chameli, Jati, Juhi;

Sans.: Bahugandha, Chetaki, Janesha, Jati, Malati, Mallika, Manodna, Punyagandha, Sukemari, Surupa, Veshika.

Distribution : Kashmir; 1,000-3,000m, frequent.

Reproductive Cycle : May-July.

Parts Used : Flower and Berries.

Folk Use : Both used in tetanus.

References : Baburaj *et al.* (2000), Rajan *et al.* (2000).

Jasminum multiflorum **(Burm.f.) Andr.**

Syn.: *J. hirsutum* Willd.; *J. pubescenes* Willd.; *Nyctanthe multiflora* Burm.f.; *N. pubescenes* Retz.

Family: Oleaceae

Eng., H. and Sans. Names : *Eng.*: Common Jasmine, Downy Jasmine, Musk Jasmine;

H.: Chameli, Kundo, Kundphul;

Sans.: Atimukha, Damana, Kundah, Mahamoda, Makaranda, Monohara, Monomoda, Muktapushpa, Tarapushpa, Vanahasa.

Distribution : India, Himalayas-Sri Lanka and Tenasserim; Common.

Reproductive Cycle : November-April.

Part Used : Leaf.

Folk Use : Leaf paste and coconut oil applied on anus of children infected with pin worms.

Reference : Dagar (2003).

Jasminum rottlerianum **Wall. *ex* DC.**

Syn.: *J. reticulatum* Wall. and G. Don.

Family: Oleaceae

Sans. Name : *Sans.*: Vanamalliga.

Distribution : Commonly found in W. Ghats upto 1,700m from Konkan southwards-Kerala.

Reproductive Cycle : June-August.

Part Used : Leaf.

Folk Use : Leaves reported to be employed in preparation for eczema in children.

Reference : Anonymous (1959).

Jatropha curcas **L.**

Syn.: *Castiglionia lobata* Ruiz and Pav.; *Curcas adansonii* Endl.; *C. indica* A. Rich.; *Jatropha acerifolia* Salisb.; *Manihot curcas* (L.) Crantz.; *Ricinus americanus* Mill.

Family: Euphorbiaceae

Eng., H. and Sans. Names : *Eng.*: Physic Nut, Poison Nut, Purging Nut;

H.: Chandarjot, Jungli Errand, Pahari Errand, Ratanjot, Safed Arand;

Sans.: Akhuparnika, Dravanti, Kananaerand, Nayagrodhi, Paravatay Eranda.

Distribution	:	Throughout India, common near villages, cultivated and naturalized, throughout the tropics.
Reproductive Cycle	:	May-October.
Parts Used	:	Leaf, Stem Bark, Latex, Seed.
Folk Uses	:	Leaf juice applied for inflammation of tongue in babies. Stem bark grounded with breast milk or cow's milk administered orally to children against fever. Latex and dried seeds used for suppressing measles.
References	:	Anonymous (1959), Hosagoudar and Henry (1996b), Rajan *et al.* (2000), Rao and Henry (1996).

Jatropha glandulifera **Roxb.**

Syn.: *Adenoropium roxburghii* Kostel.

Family: Euphorbiaceae

Eng., H. and Sans. Names	:	*Eng.*: Purple Jatropha;
		H.: Ban Jayapal, Janglie-Randi, Undarbibi;
		Sans.: Aranyajayapala, Nilkumba, Vanajayapala.
Distribution	:	Deccan and Karnataka.
Reproductive Cycle	:	Throughout the year.
Part Used	:	Root.
Folk Use	:	Root powder in water given to children for abdominal enlargements.
References	:	Agarwal (2003), Ambasta (1986), Anonymous (1959), Kirtikar and Basu (1984), Watt (1972).

Jatropha gossypifolia **L.**

Syn.: *J. elegans* Kl.; *Manihot gossypiifolia* (L.) Crantz.

Family: Euphorbiaceae

Eng., H. and Sans. Names	:	*Eng.*: Bellyache Bush, Tua-Tua;
		H.: Bherenda, Verenda;
		Sans.: Raktaeramada.
Distribution	:	Grows almost all over India, often cultivated for its gregarious flowers. Often naturalized as a common weed in open degraded habitats such as roadsides throughout most of the plain districts of India.
Reproductive Cycle	:	February-May.
Part Used	:	Leaf.

Folk Use	:	Leaf juice used for sores on children's tongue.
References	:	Agarwal (2003), Anonymous (1959), Kirtikar and Basu (1984), Parrotta (2001).

Jatropha multifida **L.**

Syn.: *J. Janipha* Blanco.; *Manihot multifida* (L.) Crantz.

Family: Euphorbiaceae

Eng., H. and Sans. Names	:	*Eng.*: Chicaquil, Coral Plant, French Physic Nut, Tortora; *H.*: Bhadradanti; *Sans.*: Bhadradanti, Jayavaha, Vishabhadra.
Distribution	:	Occurs in S. India.
Reproductive Cycle	:	Coral plant blooms on and off all year long, and especially during hot weather.
Part Used	:	Whole Plant.
Folk Use	:	Plant used in eye irritation.
Reference	:	Pal *et al.* (2000).

Juglans regia **L. (Pl. 8A)***

Syn.: *J. arguta* Wall.; *J. regia* L. var. *kumaonica* DC.

Family: Juglandaceae

Eng., H. and Sans. Names	:	*Eng.*: Common Walnut, European Walnut, Persian Walnut; *H.*: Akhor, Akhrot, Khor; *Sans.*: Akhoda, Akschoda, Aksotah.
Distribution	:	Throughout Himalayas and hills of Assam; commonly found at Chewan, Utterkashi between 1,500-2,000m.
Reproductive Cycle	:	February-April.
Parts Used	:	Fruit and Oil Cake.
Folk Uses	:	Walnut with 20 raisins given daily for 2 weeks to check bed wetting in children in the night. Oil cake also given to children as a general tonic.
References	:	Pande and Joshi (2001), Verma and Chauhan (2006).

Justicia adhatoda **L. (Pl. 8B)***

Syn.: *Adhatoda zeylanica* Medik.

Family: Acanthaceae

Eng., H. and Sans. Names	:	*Eng.*: Malabar Nut;

		H.: Adusa, Arusha, Banasaa;
		Sans.: Bhisangmata, Raktappittaghni, Vasaka.
Distribution	:	Throughout plains of India and in the sub-Himalayan tracts upto 1,450m.
Reproductive Cycle	:	December-May.
Parts Used	:	Whole Plant, Shoot and Leaf.
Folk Uses	:	Plant useful in malarial fever, leucoderma, intestinal worm, cold and cough. One tsp of roasted shoots mixed with honey given once a day till relief for cough. Decoction of leaves taken with drinking water to prevent affliction in smallpox. Leaf extract used in tonsillitis. Lodhas use leaf juice, common salt and honey as a cure for cough and cold in children.
References	:	Borthakur (1993), Chauhan (1999), D'Rozario *et al.* (1999), Kumar (2002), Pal (2000), Sharma (2003).

Justicia gendarussa **Burm.f.**

Syn.: *Gendarussa vulgaris* Nees.

Family: Acanthaceae

Eng., H. and Sans. Names	:	*Eng.*: Warer Willow;
		H.: Nilinargandhi, Udisambhalu, Vilayatri Bakes;
		Sans.: Krishanirgunda, Nilanirgundi, Vataghni.
Distribution	:	Throughout India.
Reproductive Cycle	:	December-April.
Part Used	:	Leaf.
Folk Use	:	Leaf juice given for colic.
References	:	Agarwal (2003), Anonymous (1959), Chatterjee and Pakrashi (1997), Kumar (2002), Subramaniam (2000).

Kaempferia galanga **L.**

Syn.: *Alpinia sessilis* Kon.

Family: Zingiberaceae

Eng., H. and Sans. Names	:	*Eng.*: Black Thorn, East India Galingale;
		H.: Chandramula, Kacora, Sugandhavacha;
		Sans.: Chandramulika, Sugandhavacha.
Distribution	:	Throughout N.E. region.

Reproductive Cycle : Flowering during August-September.

Part Used : Rhizome.

Folk Use : Decoction of rhizome used for tonsillitis.

Reference : Kumar (2002).

Kaempferia rotunda **L.**

Syn.: *K. longa* Jacq.; *K. versicolor* Salisb.

Family: Zingiberaceae

Eng., H. and Sans. Names : *Eng.*: Resurrection Lily, Round Rooted Galangal;

H.: Bhuichampa, Champa;

Sans.: Bhuchampaca, Bhumichampa.

Distribution : Throughout India; from Himalayas-Sri Lanka and Malaya Peninsula.

Reproductive Cycle : April-June.

Part Used : Rhizome.

Folk Uses : Powdered rhizome with croton oil in small doses given in mumps; also its paste applied for the same.

References : Ambasta (1986), Aswal (2001), Islam (1996), Kirtikar and Basu (1984), Lalramnghinglova (1996), Rahman and Yusuf (1996), Tripathi and Goel (2001), Watt (1972).

Kalanchoe lanciniata **(L.) DC.**

Syn.: *Cotyledon lanciniata* L.; *Kalanchoe schweinfurthii* Penzig.; *Vereia laciniata* (L.) Willd.

Family: Crassulaceae

Eng., H. and Sans. Names : *Eng.*: Laciniate Kalanchoe;

H.: Hamsagar;

Sans.: Hamasagara.

Distribution : Throughout S. India, tropical regions of Deccan Peninsula, Bengal, Patna, Dacca, Myanmar, Malacca.

Reproductive Cycle : During late summer the yellowish-orange flowers appear.

Part Used : Leaf.

Folk Use : Leaves used in smallpox.

References : Agarwal (2003), Asolkar *et al.* (1992).

Kallstroemia pubescens **(G.Don.) Dandy**

Syn.: *K. caribaea* Rydb.; *K. glabrata* Rydb.; *K. minor* Hook.f.; *Tribulus glabrata* Rydb.

Family: Zygophyllaceae

Eng. Name : *Eng.*: Caribbean Caltrop.

Distribution : Found in W. Bengal.

Part Used : Whole Plant.

Folk Use : Plant used in stomach heat, dyspepsia, indigestion and as an anthelmintic.

Reference : Pal *et al.* (2000).

Kickxia ramosissima **Janchen.**

Family: Scrophulariaceae

Eng. Names : *Eng.*: Branched Cancerwort Myanmar, Branched Fluellen.

Distribution : Throughout India.

Reproductive Cycle : February-June.

Part Used : Whole Plant.

Folk Use : Plant used to cure scabies.

Reference : Pal *et al.* (2000).

Lactuca serriola **L.**

Syn.: *L. scariola* L.

Family: Asteraceae

Eng., H. and Sans. Names : *Eng.*: Garden Lettuce, Prickly Lettuce;

H.: Kahu, Khas, Salad;

Sans.: Vanya kahu.

Distribution : W. Himalayas, 2,000-4,000m.

Reproductive Cycle : August-september.

Part Used : Leaf.

Folk Uses : Dried leaves used to calm restlessness and excitability in children.

Reference : Karnick (1994a).

Lactuca virosa **L.**

Syn.: *L. lactucarii* Lamotte.; *L. patersonii* Menezes.

Family: Asteraceae

Eng. Names : *Eng.*: Acrid Lettuce, Wild Lettuce.

Distribution : Distributed in N.W. Himalayas; found between 1,800-3,300m.

Reproductive Cycle : Flowers from July until September.

Part Used : Whole Plant.

Folk Uses : Wild lettuce given to encourage a sound night sleep or calm overactivity or over stimulation in children.

Reference : Prajapati *et al.* (2006).

Lagenaria siceraria **(Molina) Standl.***

Syn.: *Cucumis lagenaria* (L.) Dumort.; *Cucurbita lagenaria* L.; *Lagenaria vulgaris* Seringe.

Family: Cucurbitaceae

Eng., H. and Sans. Names : *Eng.*: Bitter Bottle Gourd, Bitter Calbash Gourd, Bottle-Gourd, White Pumpkin;

H.: Alkaddu, Golkaddu, Kadvilauki, Lau, Laue, Lauka, Titalauki, Tumri;

Sans.: Alabu, Katutumbi, Tiktalabu.

Distribution : Cultivated throughout India.

Reproductive Cycle : Throughout the year but mostly during winters.

Part Used : Leaf.

Folk Use : Juice of leaves given in diarrhoea.

Reference : Watt (1972).

Lannea coromandelica **(Houtt.) Merr.**

Syn.: *Calesium grande* Kuntze.; *Haberlia grandis* Dennst.; *Lannea grandis* (Dennst.) Engl.; *Odina wodier* Roxb.

Family: Anacardiaceae

Eng., H. and Sans. Names : *Eng.*: Jhingam, Wodier;

H.: Jingan, Kaimil, Kimul, Mohin, Moyen;

Sans.: Ajasringi, Jhingini, Jingini, Manjari, Parvati, Pramodini.

Distribution : Throughout the hotter parts of India, from extreme N.W. and along the foot of the Himalayas (ascending to

1,350m) to Assam, Myanmar, Tenasserim, the Andaman Islands and Sri Lanka.

Reproductive Cycle : February-June.

Part Used : Whole Plant.

Folk Use : Plant used in poliomyelitis.

Reference : Pal *et al.* (2000)

Lantana camara* auct. (Pl. 8C)

Syn.: *L. aculeata* L.; *L. camara* L. var. *aculeata* (L.) Moldenke.; *L. myxa* L.

Family: Verbenaceae

Eng., H. and Sans. Names : *Eng.*: Lantana Weed, Wild Sage;

H.: Caturang, Desi Lanatan, Ghaneri;

Sans.: Caturangi, Vanacchedi.

Distribution : Native of tropical America, now naturalized in many parts of India in areas receiving from 75 to over 500cm annual rainfall on a variety of soil types as a troublesome prickly weed.

Reproductive Cycle : June-October.

Parts Used : Whole Plant, Leaf, Flower and Bark.

Folk Uses : Plant used in mumps. Extract of leaves administered orally in fever, tuberculosis and tetanus. Infusion of flowers given as pectoral for children. Bark used in tetanus.

References : Ambasta (1986), Anonymous (1962), Hosagoudar and Henry (1996b), Nayak *et al.* (2003a-c), Pande *et al.* (2006), Parrotta (2001), Prajapati *et al.* (2006), Samant *et al.* (2001a), Seetharam *et al.* (1999), Sood *et al.* (2012), Subramaniam (2000).

***Lantana indica* Roxb.**

Syn.: *L. collina* Decne.; *L. latifolia* Tausch.

Family: Verbenaceae

Eng. Name : *Eng.*: Wild Sage.

Distribution : Wastelands and near cultivated fields throughout India, also on the riverbanks of Bengal.

Reproductive Cycle : During the greater parts of the year.

Part Used : Leaf.

Folk Use : Leaves used in smallpox. Also, 10ml juice of its leaves given once daily for 21 days in chickenpox whereas juice administered orally only once to children in fits.

References : Farooq (2005), Hosagoudar and Henry (1996c), Pande *et al.* (2006).

Lasianthus longicaudus **Hk.f.**

Syn.: *L. japonicus* subsp. *longicaudus* (Hook.f.) C. Y. Wu. and H. Zhu.; *Litosanthes longicauda* (Hook.f.) Deb and M.G. Gangop.; *Nonatelia longicauda* (Hook.f.) Kuntze.

Family: Rubiaceae

Distribution : Mishmi hills (upper Assam), Khasia mountains, 1,700-2,350m.

Part Used : Leaf.

Folk Use : Leaves grounded with ginger given 2 tsp twice daily to children for relief from cough.

Reference : Ahmed and Borthakur (2005).

Launaea nudicaulis **Hook.**

Family: Asteraceae

Eng. Name : *Eng.*: Bold-Leaf Launaeae.

Distribution : More or less throughout the plains of India, from Bengal and Bihar-Punjab, ascending W. Himalayas-2,700m in the Kumaon, Sind and Deccan.

Reproductive Cycle : September-March.

Part Used : Leaf.

Folk Use : Leaves applied as poultice on head of children to check fever.

References : Bhalla *et al.* (1996), Kirtikar and Basu (1984).

Launaea procumbens **(Roxb.) Ramayya and Rajagopal.**

Syn.: *L. fallax* (Jaub. and Spach) Kuntze.; *Microrhynchus fallax* Jaub. and Spach.

Family: Asteraceae

Eng. and H. Names : *Eng.*: Creeping Launaea;

H.: Jangi Gobi, Van-Gobi.

Distribution : Throughout India, occasional in wasteplaces.

Reproductive Cycle : March-November.

Part Used : Leaf.

Folk Use	:	Paste of leaves applied externally to forehead of children during fever.
References	:	Farooq (2005), Ray (2003), Srivastava *et al.* (2003a).

Launaea sarmentosa **(Willd.) Alston**

Syn.: *L. pinnatifida* Cass. in Ann.; *Prenanthes sarmentosa* Willd.

Family: Asteraceae

Eng. and H. Names	:	*Eng.*: Beach Launaea; *H.*: Bankau, Pathri.
Distribution	:	Sandy coasts of India and Sri Lanka.
Reproductive Cycle	:	July-September.
Part Used	:	Whole Plant.
Folk Use	:	Plant used as soporific for children.
References	:	Kirtikar and Basu (1984), Nayak *et al.* (2003c), Watt (1972).

Lawsonia inermis **Lam.**

Syn.: *L. alba* Lam.; *L.purpurea* Lamk.; *L. spinosa* L.

Family: Lythraceae

Eng., H. and Sans. Names	:	*Eng.*: Cypress Shrub, Egyptian Priven, Henna; *H.*: Henna, Mehndi; *Sans.*: Madayantika, Medhini, Nakharanjani, Ranjaka, Sakachara.
Distribution	:	Cultivated throughout India, perhaps wild in W. region.
Reproductive Cycle	:	April-July.
Parts Used	:	Root and Leaf.
Folk Uses	:	Pounded roots, 'neem' leaves, coconut root and dried ginger given to infants to check diarrhoea. Leaves applied to soles of the feet in smallpox.
References	:	Dhiman (2003), Parrotta (2001), Watt (1972).

Lens culinaris **Medik.**

Syn.: *Ervum lens* L.; *Lens esculenta* Moench.; *L. lens* Huth.

Family: Fabaceae

Eng., H. and Sans. Names	:	*Eng.*: Lentil; *H.*: Masser, Masur, Masuri; *Sans.*: Gurubija, Masurah, Ragadoli, Sura.

Distribution	:	Throughout the colder parts of India, cultivated, indigenous in S.E. Europe and temperate W. Asia.
Reproductive Cycle	:	Flowering in March-May.
Parts Used	:	Whole Plant and Seed.
Folk Uses	:	Decoction of plant used for facilitating eruptions in smallpox. Paste of seeds used as an application to ulcers after smallpox.
References	:	Jain (1991), Kirtikar and Basu (1984), Watt (1972).

Leonotis nepetifolia **(L.) R. Br.**

Syn.: *Leonurus nepetifolius* (L.) Mill.; *Phlomis nepetifolia* Roxb.

Family: Lamiaceae

Eng., H. and Sans. Names	:	*Eng.*: Annual Lion's Ear, Christmas Candlestick, Lion's Ear; *H.*: Bara Guma, Hejurchei, Lal Guma; *Sans.*: Granthi, Granthika, Granthiparna, Granthiparni.
Distribution	:	Throughout hotter India, typically found near villages on roadsides and wasteplaces.
Reproductive Cycle	:	Flowering during November-December.
Parts Used	:	Whole Plant, Leaf, Flower, Stalk and Seed.
Folk Uses	:	Plant used for scabies in children. Bath with decoction of leaves and stalk good for strengthening weak children. Paste of flowers or its seeds fried in 'koronj' oil rubbed on itch and on head sores of small children.
References	:	Kirtikar and Basu (1984), Pal *et al.* (2000), Parrotta (2001), Sikarwar (2002).

Lepidagathis cristata **Willd.**

Syn.: *Acanthus cristatus* Rottler *ex* Nees.; *Selago acaulis* Nees.

Family: Acanthaceae

Eng. and H. Names	:	*Eng.*: Crested Lepidagathis; *H.*: Bukhar Jadi.
Distribution	:	Found in dry, rocky habitats from Rajasthan and Haryana southwards to Tamil Nadu.
Reproductive Cycle	:	September-April.
Part Used	:	Whole Plant.
Folk Use	:	Plant decoction (2 tsp) given twice a day for 7 days for measles.
References	:	Parrotta (2001), Upadhye *et al.* (1994).

Lepidagathis hamiltoniana Wall.

Family: Acanthaceae

H. Name	:	*H.*: Agni Khair.
Distribution	:	Bihar, Chota Nagpur, 300-1,000m.
Reproductive Cycle	:	September-March.
Parts Used	:	Flower and Seed.
Folk Use	:	Whorl of flower or seeds pounded and fried in 'koronj' oil, rubbed on itch on head sores of small children.
Reference	:	Kirtikar and Basu (1984).

Lepidium didymum L. (Pl. 8D)

Syn.: *Coronopus didymus* (L.) Sm.; *Senebiera didyma* (L.) Pers.; *S. pinnatifida* DC.

Family: Brassicaceae

Eng. and H. Names	:	*Eng.*: Bitter Cress, Lesser Swinecress, Wart Cress; *H.*: Jangli Hala.
Distribution	:	Naturalized in India and found throughout Indian plains and foot hills in N.S. India.
Reproductive Cycle	:	July-November.
Part Used	:	Whole Plant.
Folk Use	:	Lukewarm plant cytoplasm mixed with oil of *Sesamum indicum* used as drops to cure obstinate pus in the ear passage.
Reference	:	Gopal and Chandra (2003).

Lepionurus sylvestris Blume.

Syn.: *L. oblongifolius* (Griff.) Mast.; *Leptonium oblongifolium* Griff.

Family: Opiliaceae

Distribution	:	E. Bengal, Assam, Khasia mountains ascending to 700m, Abor hills, Tenasserim at Mergue, Malacca.
Reproductive Cycle	:	February-August.
Part Used	:	Root.
Folk Use	:	Roots used as poultice for headache in children.
References	:	Ambasta (1986), Lalramnghinglova (2003).

Leucas aspera (Willd.) Link.

Syn.: *L. dimidiata* Benth.; *Phlomis aspera* Willd.; *P. esculenta* Roxb.; *P. obliqua* Buch.-Ham. *ex* Hook.f.; *P. plukenetii* Roth.

Family: Lamiaceae

Eng., H. and Sans. Names : *Eng.*: Thumbe;
H.: Chota-italkusa, Chotalalkura;
Sans.: Drona Pushpi.

Distribution : Wasteland and roadsides upto 900m, Himalayas up to 2,000m from Shimla-Bhutan.

Reproductive Cycle : February-November.

Parts Used : Whole Plant, Stem, Leaf and Flower.

Folk Uses : Infusion of plant with honey given to children in cough and cold. A wreath made of piece of stem given for wearing to children by tribals against skin diseases. Infusion of tender shoots (25g) mixed with a little amount of fruit powder (1g) of *Capsicum annum* prescribed for gargling in tonsillitis twice a day. Vegetable of its leaves and tender twigs cooked with Lata fish given to children for promoting rapid growth. Mixture of its leaf juice mixed with lime applied locally on tonsils and mumps by Bettakurubas; also the juice recommended as an eyedrop in encephalopathy due to worm infestation in children. Leaf paste along with jaggery applied on forehead thrice a day for 5 days for treatment of rickets. Flower juice with water sprinkled to smallpox. Flowers eaten raw to cure tonsillitis. Also, flowers mixed with one tsp of honey given to children in cold and cough (3 days –thrice daily).

References : Agarwal (2003), Dutta and Nath (2000), Hembrom (1996), Khare (2004), Khomdram *et al.* (2011), Kshirsagar and Singh (2007), Mandal and Basu (1996), Nath *et al.* (2007), Parrotta (2001), Prajapati *et al.* (2006), Rajendran *et al.* (2003), Singh and Narain (2008), Singh and Prakash (1996), Singh *et al.* (1996), Subramaniam (2000), Watt (1972).

Leucas helianthemifolia Desf.

Syn.: *L. ternifolia* Desf.

Family: Lamiaceae

Distribution : S. Deccan, Pulney, Nilgiri and Anamallay hills, 1,700-2,350m.

Parts Used : Leaf and Flower.

Folk Use : Leaf and flower paste applied on body of infants followed by a bath after half-hour daily for a week to cure skin diseases.

Reference : Viswanathan *et al.* (2006).

***Limonia acidissima* L.**

Syn.: *Feronia elephantum* Correa.; *Limonia elephantum* (Correa) Panigrahi.

Family: Rutaceae

Eng., H. and Sans. Names : *Eng.*: Elephant Apple, Wood Apple;

H.: Beli, Katbel, Kathbel, Kavitha, Vilatti;

Sans.: Kapittha, Kapitthah.

Distribution : W. and S. India, Punjab, N.W. Himalayas, Shimla, Kumaon, Bihar, Bengal, Assam, Myanmar.

Reproductive Cycle : Flowering in March-May.

Parts Used : Leaf, Fruit and Root Bark.

Folk Uses : Juice of leaves given to children suffering from stomach troubles. Extract of root bark/fruit pulp crushed with black pepper administered for whooping cough in infants. Dried fruits considered beneficial in resisting contagion of smallpox.

References : Chatterjee and Pakrashi (1997), Kirtikar and Basu (1984), Parrotta (2001), Rao and Henry (1996), Watt (1972).

***Litchi chinensis* Sonn.**

Syn.: *Nephelium litchi* Camb.

Family: Sapindaceae

Eng., H. and Sans. Names : *Eng.*: Litchi, Lychee;

H.: Lichi;

Sans.: Raktaphala.

Distribution : Widely cultivated throughout India; chiefly in warmers parts like N. Bihar, Muzaffarpur, Saharanpur Dehradun, Haridwar, Muzaffarnagar, Gorakhpur and W. Bengal.

Reproductive Cycle : February-June.

Parts Used : Flower and Fruit.

Folk Uses : Unripe flowers and green fruits prescribed to children in smallpox.

References : Dhiman (2003), Kirtikar and Basu (1984).

Lithospermum officinale **L.**

Syn.: *Magarospermum officinale* Dene.

Family: Boraginaceae

Eng. Names : *Eng.*: Common Gromwell, European Gromwell, Gromwell.

Distribution : Found in Kashmir and Kumaon: 1,500-2,700m.

Reproductive Cycle : June-August.

Part Used : Root.

Folk Uses : Decoction of roots given in smallpox, measles and itch.

References : Ambasta (1986), Anonymous (1959).

Lobelia angulata **G. Forst.**

Syn.: *L. arenaria* (Hook.f.) Heenan and de Lange.; *L. littoralis* R. Cunn. *ex* A. Cunn.

Family: Campanulaceae

Eng. Names : *Eng.*: Creeping Pratia, Panakenake.

Distribution : Sikkim, N.E. India, scattered in Mizoram, on waysides and clearings in tropical secondary open forests.

Reproductive Cycle : April-August.

Part Used : Fruit.

Folk Use : 2 or 3 fresh fruits eaten against tonsillitis.

Reference : Lalramnghinglova (2003).

Lodoicea maldivica **(J. F. Gmel.) Pers.**

Syn.: *Cocos maldivicia* J. F. Gmel.; *Lodoicea sechellarum* Labill.

Family: Arecaceae

Eng., H. and Sans. Names : *Eng.*: Double Coconut Palm, Love Nut, Sea Coconut Palm, Seychelles Nut;

H.: Dariya-i-nariyal, Darya-ka-nariyal;

Sans.: Abdhinarikelah, Agnijarah, Arkaragah.

Distribution : A native of Maldives, cultivated in coastal parts of India.

Part Used : Whole Plant.

Folk Use	:	Plant mixed with root of nux-vomica given to children for colic.
References	:	Dhiman (2003), Kirtikar and Basu (1984), Watt (1972).

Lonicera japonica Thunb.

Syn.: *Caprifolium chinense* S. Watson *ex* Loudon.; *C. roseum* Lam.

Family: Caprifoliaceae

Eng. Names	:	*Eng.*: Hall's Honeysuckle, Japanese Honeysuckle.
Distribution	:	Grows wild in mountains, cultivated as an ornamental plant. Temperate Himalayas from Nepal-Bhutan, 2,000-3,500m, Khasia mountains, 1,350-2,000m.
Reproductive Cycle	:	May-June.
Part Used	:	Stem.
Folk Use	:	Leaf bearing stem used in measles.
Reference	:	Prajapati *et al.* (2006).

Ludwigia octovalvis (Jacq.) P. H. Raven

Syn.: *Jussiaea angustifolia* Lam.; *J. pubescens* L.; *J. suffruticosa* L.

Family: Onagraceae

Eng., H. and Sans. Names	:	*Eng.*: False Primrose, Jamaica Loostrife, Wild Clove, Willow Primrose; *H.*: Ban Long; *Sans.*: Bhu Lavangah.
Distribution	:	Throughout India excepting W. desert region and in Sri Lanka.
Reproductive Cycle	:	January-December.
Part Used	:	Root.
Folk Use	:	100g root and 50g 'gur' crushed with a cup of water given once daily for 3 days to cure infantile diarrhoea.
Reference	:	Tirkey (2004).

Luffa acutangula (L.) Roxb.

Syn.: *Cucumis acutangulus* L.; *Luffa acutangula* var. *amara* (Roxb.) Cl.; *L. kleinii* W. and A.; *L. umbellate* (Klein *ex* Willd.) Roem.

Family: Cucurbitaceae

Eng., H. and Sans. Names	:	*Eng.*: Ribbed Gourd, Ridged Gourd; *H.*: Jinga, Kali Tori, Tori; *Sans.*: Dhamargavah, Koshataki, Svadukosataki.

Distribution	:	Throughout greater parts of India, N.W. India; Sikkim, Assam, plains of E. Bengal, Sri Lanka.
Reproductive Cycle	:	August-December.
Parts Used	:	Whole Plant and Leaf.
Folk Uses	:	Plant used in tonsillitis. Juice of fresh leaves used as eyedrops to cure granular conjunctivitis in children.
References	:	Agarwal (2003), Anonymous (1962), Katewa *et al.* (2001), Kirtikar and Basu (1984), Parrotta (2001), Sood *et al.* (2012).

Luffa cylindrica **M. Roem.**

Syn.: *Bryonia cheirophylla* Wall.; *L. aegyptica* Mill.; *L. pentandra* Roxb.; *Momordica cylindrica* L.

Family: Cucurbitaceae

Eng., H. and Sans. Names	:	*Eng.*: Sponge Gourd, Vegetable Sponge, Wash Sponge; *H.*: Ghiya Tori, Nenua, Purula; *Sans.*: Dhirghapatolika, Hastiparna, Mahakosataki, Rajakoshatoki, Sapitaka.
Distribution	:	Throughout India, wild in wastelands especially along coastal areas; also cultivated.
Reproductive Cycle	:	August-December.
Part Used	:	Whole Plant.
Folk Uses	:	Used in diarrhoea, dyspepsia and indigestion.
Reference	:	Pal *et al.* (2000).

Lycopodium clavatum **L.**

Syn.: *L. inflexum* Sw.

Family: Lycopodiaceae

Eng. and H. Names	:	*Eng.*: Common Club Moss, Groundpine, Staghorn Clubmoss, Wolf's Foot Clubmoss; *H.*: Bendarli.
Distribution	:	N.E. Himalayas.
Reproductive Cycle	:	July-September.
Part Used	:	Spores.
Folk Use	:	Used as a local dessicative to prevent chapping of children's skin.
Reference	:	Bhattacharjee (2001).

Lygodium flexuosum **(L.) Sw.**

Syn.: *Hydroglossum flexuosum* (L.) Willd.; *Ophioglossum flexuosum* L.; *Ramondia flexuosa* (L.) Mirb.

Family: Schizaeaceae

Eng., H. and Sans. Names : *Eng.*: Maiden Hair Creeper;
H.: Mahadebjata;
Sans.: Valliparna.

Distribution : Throughout India, plains-hills ascending upto 1,500m.

Reproductive Cycle : February-December.

Part Used : Rhizome.

Folk Use : Rhizome paste applied all over the body for convulsion in children.

Reference : Prusti (2007).

Macrotyloma uniflorum **(Lam.) Verdc.**

Syn.: *Dolichos biflorus* Sensu auct.; *Dolichos uniflorus* Lam.

Family: Fabaceae

Eng., H. and Sans. Names : *Eng.*: Horsegram, Kool Tee, Madras Gram;
H.: Gahat, Kulthi;
Sans.: Kulaththa, Svetabija, Tamabija.

Distribution : Himalayas-Sri Lanka and Myanmar ascending to 1,000m in Sikkim; sometimes cultivated.

Reproductive Cycle : August-September.

Part Used : Whole Plant.

Folk Use : Plant used in smallpox.

References : Chauhan (1999), Sharma (2003).

Madhuca indica **J. F. Gmel.**

Syn.: *Bassia latifolia* Roxb.; *B. longifolia* Koen.; *Madhuca latifolia* Macbr.; *M. longifolia* (Koen.) Macbr.; *M. longifolia* (Koen.) Macbr. var. *latifolia* (Roxb.) Chev.

Family: Sapotaceae

Eng., H. and Sans. Names : *Eng.*: Butter Tree, Illipe, Mahua, Mowra;
H.: Mahua, Mauwa, Mohwa;
Sans.: Dalaphala, Garudapu, Lodhrapushpa, Madhava, Madhu, Madhuka, Priyala, Vanaprastha.

Distribution : Forests of Karnataka and Kerala, M.P. and Rajasthan; occurs in plains and lower hills of India up to 1,200m; common throughout W. Bengal, Bihar, M.P., Orissa, Punjab and U.P. including plains of Uttarakhand.

Reproductive Cycle : April-July.

Parts Used : Leaf. Flower, Seed and Bark.

Folk Uses : Leaves used in fast breathing. Flowers useful in strangury, verminosis, haemoptysis, hepatopathy and gastropathy. Seeds worn as amulet by children affected with tetanus and rickets. Expressed seed oil applied to arms of children for allaying extreme itchiness caused by the presence of intestinal worms. Bark used in tonsillitis.

References : Agarwal (2003), D'Rozario *et al.* (1999), Khanna (2002b), Khare (2004), Prajapati *et al.* (2006), Watt (1972).

Mallotus philippensis **(Lam.) Muell.-Arg.***

Syn.: *Croton distans* Wall.; *C. montanus* Willd.; *C. philippinensis* (Lam.) Muell.-Arg.; *Rattlera tinctoria* Roxb.

Family: Euphorbiaceae

Eng., H. and Sans. Names : *Eng.*: Kamala Tree, Monkey Face Tree;

H.: Kamila, Rohni, Rula;

Sans.: Kampillaka, Lohitanga, Nadivasa, Pikaksha, Rechanaka.

Distribution : Distributed rather sparsely in the sub-Himalayan tract, M.P. and Peninsular India.

Reproductive Cycle : August-March.

Part Used : Whole Plant.

Folk Uses : Plant used in stomach heat and as an anthelmintic.

Reference : Pal *et al.* (2000).

Malus pumila **Mill.***

Syn.: *M. communis* Poir.; *M. dasyphylla* Borkh.; *Pyrus malus* L.; *P. pumila* (Mill.) Steud.

Family: Rosaceae

Eng., H. and Sans. Names : *Eng.*: Apple;

H.: Seb, Sev;

Sans.: Simbitika.

Distribution : Grown widely in W. and C. Himalayas, Kashmir, H.P.; wild in N.W. Himalayas, ascending to 2,000m.

Reproductive Cycle : May-October.

Part Used : Fruit.

Folk Uses : Dried apple powder used for controlling infantile diarrhoea. Apple juice, syrup and vinegar reduce curd tension of milk used in infant feeding.

References : Agarwal (2003), Anonymous (1962).

***Mangifera indica* L. (Pl. 9A)**

Syn.: *M. domestica* Gaertn.

Family: Anacardiaceae

Eng., H. and Sans. Names : *Eng.*: Cuckoo's Joy, Mango, Mango Tree;

H.: Aam, Am, Amb;

Sans.: Amra, Chuta, Rasala, Sripriya, Vasantaduta.

Distribution : Tropical Himalayas from Kumaon-Bhutan hills and valleys of Bihar, Khasia mountains, Myanmar, oudh, W. Peninsula from Kandesh southwards.

Reproductive Cycle : March-April.

Parts Used : Leaf, Inflorescence, Fruit, Bark and Latex.

Folk Uses : Leaves used for septic wounds of children especially after vaccination; 'Dasehri' variety useful in rickets. Inflorescence used in infantile diarrhoea in Sikkim. Fruit stalk of mango and latex exudates of petiole smeared over bite or sting (bee or wasp) area to cure the disorder in children. Bark grounded to paste with water given twice a day to check diarrhoea in children.

References : Agarwal (2003), Hosagoudar and Henry (1996a), Kothari and Rao (2000), Mohanty *et al.* (1996), Singh and Singh (2003a), Srivastava *et al.* (1987).

***Manilkara kauki* (L.) Dubard**

Syn.: *Achras octodecemfida* Stokes.; *Imbricaria malabarica* Poir.; *Mimusops kauki* L.; *M. manilkara* G. Don.

Family: Sapotaceae

Eng., H. and Sans. Names : *Eng.*: The Obtuse Leaved Mimusops;

H.: Khirni;

Sans.: Khirni, Talavrinta.

Distribution : Myanmar at Amherst; Malacca; planted in Indian gardens.

Reproductive Cycle : April-May.

Parts Used : Root and Bark.

Folk Use : Bark and root grounded with water and honey given in infantile diarrhoea.

References : Ambasta (1986), Anonymous (1962), Kirtikar and Basu (1984), Watt (1972).

Marsilea minuta L.

Syn.: *M. quadrifolia* L.

Family: Marsileaceae

Eng., H. and Sans. Names : *Eng.*: Caupatiya;

H.: Chaupatia, Sunsuniya;

Sans.: Catuspatri, Sunisannak.

Distribution : Grows as a weed in marshy places, edges of canals and ponds throughout India.

Reproductive Cycle : November-March.

Part Used : Sporocarps.

Folk Use : Crushed sporocarps applied on throat with the help of a finger to cure throat inflammation in children.

Reference : Khanna (2002a).

Martynia annua L.

Syn.: *Carpoceras angulata* A. Rich.; *Martynia angulosa* Lam.; *M. diandra* Gloxin.

Family: Martyniaceae

Eng., H. and Sans. Names : *Eng.*: Baghnokhi, Snake's Head;

H.: Bichu, Hathajori;

Sans.: Kaknasika, Vysalnkaha.

Distribution : Found sporadically in wasteplaces roadsides, hedges in several, in parts of India having hot weather.

Reproductive Cycle : August-October.

Parts Used : Root, Leaf and Fruit.

Folk Uses : Used in tonsillitis.

References : Bhatt *et al.* (2002), Jain (1991), Mehrotra *et al.* (1995).

Matricaria chamomilla L.

Syn.: *M. recutita* L.; *M. suaveolens* L.

Family: Asteraceae

Eng., H. and Sans. Names	:	*Eng.*: German Chamomile, Wild Chamomile; *H.*: Babuna, Bauna, Sanamoti; *Sans.*: Babuna, Babunaj.
Distribution	:	An aromatic herb grown in Punjab, H.P. and upper Gangetic plains.
Reproductive Cycle	:	August-October.
Part Used	:	Flower.
Folk Uses	:	Flowers used in dentition troubles, stomach disorders, earache, neuralgic pains and convulsions in children.
References	:	Agarwal (2003), Ambasta (1986), Anonymous (1962), Karnick (1994a), Kirtikar and Basu (1984), Thakur *et al.* (1989).

Melanthera biflora (L.) Willd.

Syn.: *Wedelia biflora* (L.) DC.; *Wollastonia biflora* (L.) DC.

Family: Asteraceae

Eng. Names	:	*Eng.*: Beach Sunflower, Sea Daisy, Wedelia.
Distribution	:	Common in littoral and coastal areas, forest margins, margins of mangroves, roadsides, wasteplaces from sea-level to 500m.
Part Used	:	Leaf.
Folk Use	:	Juice of leaves used in tetanus.
Reference	:	Prajapati *et al.* (2006).

Melastoma malabathricum L.

Syn.: *M. affine* D. Don.; *M. esquirolii* H. Lev.; *M. normale* D. Don.; *M. polyanthum* Blume.

Family: Melastomataceae

Eng., H. and Sans. Names	:	*Eng.*: Indian Rhododerdron; *H.*: Phutki; *Sans.*: Tinisah.
Distribution	:	Throughout India, very abundant, from sea-level upto 2,000m.
Reproductive Cycle	:	October-March.

Parts Used	:	Whole Plant, Leaf and Root.
Folk Uses	:	Plant used in diarrhoea, stomach pain, dysentery and dyspepsia or indigestion. Powdered leaves applied over healing smallpox to prevent scar formation. Roots along with leaves used in smallpox.
References	:	Jain (1991), Pal *et al.* (2000), Sharma (2004).

Melilotus indicus **(L.) All.**

Syn.: *M. indica* (L.) All.; *M. parviflora* Desf.; *Sertula indica* (L.) Kuntze.; *Trifolium indicum* L.

Family: Fabaceae

Eng., H. and Sans. Names	:	*Eng.*: Annual Yellow Sweet Clover, Small Flowered Melilot; *H.*: Banmethi; *Sans.*: Banamethika, Vanamethika.
Distribution	:	Grows throughout India, ascending to 1,650m; common in N. India.
Reproductive Cycle	:	December-June.
Parts Used	:	Whole Plant and Seed.
Folk Uses	:	Whole plant along with seeds used in bowel complaints and infantile diarrhoea.
References	:	Agarwal (2003), Ambasta (1986), Anonymous (1962), Kirtikar and Basu (1984), Sharma (2004), Singh *et al.* (2007), Watt (1972).

Melocalamus compactiflorus **(Kurz.) Benth.**

Syn.: *Dinochloa compactiflora* (Kurz.) Mc Clure.; *Melocalamus fimbriatus* Hsueh and C.M. Hill.; *Pseuydotachyum compactiflorum* Kurz.

Family: Poaceae

Eng. and H. Names	:	*Eng.*: Climbing Pear Bamboo, Congest-Flowered Dinochloa; *H.*: Sairil.
Distribution	:	Bangladesh, Myanmar, E. Bengal-Bhutan, Assam and Tripura frequent in Mizoram in tropical wet evergreen forests and sub-tropical dense forests.
Part Used	:	Stem.
Folk Uses	:	Sap oozing out of cut stem given to children for curing influenza, cough and chest complaints.
References	:	Lalramnghinglova (1996, 2003).

***Melochia corchorifolia* L.**

Syn.: *Hibiscus donii* Walp.; *Lochemia corchorifolia* Arn.; *Melochia supina* L.; *M. pauciflora* Wall.

Family: Sterculiaceae

Eng. and H. Names	:	*Eng.*: Chocolate Weed, Red Weed, Wire Bush; *H.*: Bundahia, Bundava.
Distribution	:	Hotter parts of India and Kumaon 1,350m- Sikkim, Malacca and Sri Lanka.
Reproductive Cycle	:	Flowering Summer-Autumn.
Part Used	:	Leaf.
Folk Use	:	Leaf juice (1 tsp twice daily for 3 days) prescribed for children suffering from diarrhoea.
Reference	:	Saren *et al.* (2000).

***Mentha arvensis* L.**

Syn.: *M. parietariifolia* Becker *ex* Steud.

Family: Lamiaceae

Eng., H. and Sans. Names	:	*Eng.*: Corn Mint, Field Mint, Japanese Mint, The Marsh Mint; *H.*: Podina, Pudinah; *Sans.*: Putiha.
Distribution	:	Found in W. Himalayas, in Kashmir: 1,700-3,350m.
Reproductive Cycle	:	August-December.
Parts Used	:	Plant and Leaf.
Folk Uses	:	Cold infusion of the plant good carminative for infants. Juice of leaves given to check vomiting in children.
References	:	Chauhan (1999), Watt (1972).

***Mentha canadensis* L.**

Syn.: *M. haplocalyx* Briq.; *M. terebinthinacea* Willd. *ex* Steud.

Family: Lamiaceae

Eng. Names	:	*Eng.*: American Corn Mint, Bohe, Canadian Mint.
Distribution	:	Native to temperate regions of the northern hemisphere; widely cultivated in China.
Reproductive Cycle	:	Flowering in April-July.
Parts Used	:	Whole Plant and Aerial Roots.

Folk Uses	:	Decoction of the plant recommended for checking vomiting in children. Aerial parts used in measles.
References	:	Bhattacharjee (2001), Prajapati *et al.* (2006).

Mentha piperita L. (Pl. 9B)

Family: Lamiaceae

Eng., H. and Sans. Names	:	*Eng.*: Brandy Mint, Pepper Mint; *H.*: Gamathi Pudina, Paparaminta; *Sans.*: Pootiha, Pudina, Rochani.
Distribution	:	Native of Europe; naturalized in many parts of India; commercially cultivated in Tarai region.
Reproductive Cycle	:	July-September.
Parts Used	:	Whole Plant and Leaf.
Folk Uses	:	Plant used in infantile colic. Leaf powder mixed with honey used as a cholagogue.
References	:	Baburaj *et al.* (2000), Sikarwar (2002).

Mentha spicata L.

Family: Lamiaceae

Eng., H. and Sans. Names	:	*Eng.*: Garden Mint, Lamb Mint, Spearmint; *H.*: Pahari Pudina, Pudina; *Sans.*: Putiha, Rocani.
Distribution	:	Cultivated in Indian gardens.
Reproductive Cycle	:	July-September.
Part Used	:	Leaf.
Folk Use	:	Sweet infusion of leaves used in infantile stomach disorders.
References	:	Agarwal (2003), Anonymous (1962), Watt (1972).

Mesua ferrea L.

Syn.: *M. nagassarium* (Burm.f.) Kosterm.; *M. pedunculata* Wight.; *M. speciosa* Chois.

Family: Clusiaceae

Eng., H. and Sans. Names	:	*Eng.*: Ceylon Iron-wood, Iron-wood of Assam, Iron-wood Tree, Mesua; *H.*: Nagakesar, Naghar; *Sans.*: Champeya, Hema, Kanchana, Nagakesarah, Nagapushpah, Nogaya, Pinjara, Rukma.

Distribution	:	Mountains of E. Bengal, E. Himalayas and E. and W. Peninsula, Andaman islands; cultivated elsewhere in India.
Reproductive Cycle	:	February-October.
Parts Used	:	Whole Plant and Leaf.
Folk Uses	:	Plant used in poliomyelitis. Young leaf juice of the plant used against measles.
References	:	Nath and Maiti (2003), Pal *et al.* (2000).

***Meyna laxiflora* Robyns**

Family: Rubiaceae

Eng., H. and Sans. Names	:	*Eng.*: Muyna; *H.*: Muyna; *Sans.*: Pindituka.
Distribution	:	In hedges and wasteplaces in W. Bengal, Bihar and Orissa.
Reproductive Cycle	:	March-June.
Part Used	:	Leaf.
Folk Use	:	Paste of leaves applied on mumps by Jenukurubas.
References	:	Kshirsagar and Singh (2000b, 2007).

***Meyna spinosa* Roxb. *ex* Link.**

Syn.: *Pyrostria spinosa* (Roxb. *ex* Link) Miq.; *Vangueria pyrostria* Boerl.; *V. spinosa* (Roxb. *ex* Link) Roxb.

Family: Rubiaceae

H. and Sans. Names	:	*H.*: Moina, Mudana, Muyuna; *Sans.*: Pindi, Pinditaka, Pindu.
Distribution	:	N. Bengal-Canara, Pegu, Tenasserim and Myanmar.
Reproductive Cycle	:	March-June.
Part Used	:	Seed.
Folk Uses	:	Aliquot of one year old cotyledon (about 1 or 2) crushed with water and honey (1-2 tsp) given to children (2.5-5ml twice daily for 3-4 days) against vomiting and diarrhoea.
Reference	:	Singh and Singh (2003a).

Microglossa pyrifolia **(Lam.) Kuntze.**

Syn.: *Conyza pyrifolia* Lam.; *C. volubilis* Wall.; *Microglossa petiolaris* DC.; *M. volubilis* DC.

Family: Asteraceae

Distribution : Grows in E. India, Khasia mountains, Assam, Naga hills, Cachar, Myanmar and from Pegu-Penang.

Reproductive Cycle : Throughout the year.

Part Used : Whole Plant.

Folk Use : Plant extract given as enema to infants in fever.

References : Agarwal (2003), Anonymous (1962).

Micromelum minutum **Wight and Arn.**

Syn.: *Glycosmis subvelutina* F. Muell.; *Micromelum pubescens* Bl.

Family: Rutaceae

Eng. Name : *Eng.*: Lime Berry.

Distribution : Tropical Himalayas, Andaman and Assam, common in thickets, dense, dry or open forest, distributed areas and rocky coasts.

Reproductive Cycle : April-November.

Parts Used : Whole Plant and Leaf.

Folk Uses : Plant rubbed on skin to check skin irritation in children. Infusion of leaves used for treating toothache and teething problems in babies.

References : Agarwal (2003), Prajapati *et al.* (2006).

Mikania micrantha **Kunth.**

Syn.: *Eupatorium denticulatum* Vahl.; *Mikania alata* (G. Mey.) DC.; *M. sinuata* Rusby.

Family: Asteraceae

Eng. and H. Names : *Eng.*: American Rope, Chinese Creeper, Heart Shaped Mikania, Mile-a-Minute;

H.: Titaiya Baur.

Distribution : E. Assam, in the Duphla hills.

Reproductive Cycle : November-December (Flowering) and September-February (Fruiting).

Part Used : Whole Plant.

Folk Uses	:	Plant beneficial in itching, wound, diarrhoea, bowel complaints, dysentery, blood dysentery and as an anthelmintic.
Reference	:	Pal *et al.* (2000).

***Mimosa pudica* L.**

Syn.: *M. hispidula* Kunth.

Family: Fabaceae

Eng., H. and Sans. Names	:	*Eng.*: Humble Plant, Sensitive Plant, Touch-me-not; *H.*: Lajalu, Lajjavati, Lajwanti, Sharmpate; *Sans.*: Anjalikaraka, Anjalikarika, Gandamalika, Lajja, Lajjalu, Lajjika, Mahabhita, Mamaskari, Raktapadi.
Distribution	:	Spread throughout hotter parts of India; common in wasteplaces where the climate is moist and warm.
Reproductive Cycle	:	June-December.
Parts Used	:	Root and Leaf.
Folk Uses	:	Roots tied around the neck of children to cure cough. Leaves beneficial in smallpox.
References	:	Kirtikar and Basu (1984), Nayak *et al.* (2003a), Sen and Behera (2007), Parrotta (2001).

***Mimusops elngi* L.**

Syn.: *Imbricaria perroudii* Montrouz.; *Mimusops lucida* Poir.; *M. parvifolia* R. Br.

Family: Sapotaceae

Eng., H. and Sans. Names	:	*Eng.*: Bakul Tree, Bullet Wood, Indian Medlar Tree; *H.*: Bakul, Bolsari, Maulser, Maulsiri; *Sans.*: Angaka, Bakulah, Chirapushpa, Dhanvi, Karuka, Madhupanjara, Mukula, Sharadika, Surabhi, Tailanga, Vakula.
Distribution	:	Throughout S. Deccan and Malay Peninsula; common in N. India, frequently cultivated.
Reproductive Cycle	:	March-June.
Parts Used	:	Whole Plant, Flower and Seed.
Folk Uses	:	Plant used in smallpox. Half cup extract of 20-25 fresh flowers soaked overnight in water given in early morning for 15 consecutive days in sore throat accompanied with ringing cough. Seeds used for preparing suppositories for constipation in children.

References : Bhattacharjee (2001), Borthakur (1993), Chatterjee and Pakrashi (1997), Dhiman (2003), Kirtikar and Basu (1984), Pande *et al.* (2006), Parrotta (2001), Prajapati *et al.* (2006), Sharma (2003), Singh *et al.* (2007), Watt (1972).

Momordica charantia L. (Pl. 9C)*

Syn.: *Cucumis africanus* Bat.; *Momordica humilis* Wall.; *M. muricata* DC.; *M. senegalensis* Lamk.

Family: Cucurbitaceae

Eng., H. and Sans. Names : *Eng.*: Balsam Apple, Bitter Gourd, Carilla Fruit;

H.: Karela, Kareli, Karola;

Sans.: Ambuvallika, Karavella, Karvellak, Sushavi, Visakantaki.

Distribution : Found throughout India, upto 1,500m.

Reproductive Cycle : Throughout the year but chiefly during April-November.

Parts Used : Leaf and Fruit.

Folk Uses : Juice of leaves acts as a mild purgative; also its juice mixed with powder of *Curcuma longa* prescribed in measles with fever and smallpox. Fruits commonly prescribed as an anthelmintic. Fruit juice with honey given to prevent pox.

References : Agarwal (2003), Dey and Bahadur (1973), Khare (2004), Kirtikar and Basu (1984), Watt (1972).

Momordica cochinchinesis Spreng

Syn.: *M. macrophylla* Gage.; *M. meloniflora* Hand.-Mazz.; *M. mixa* Roxb.

Family: Cucurbitaceae

Eng., H. and Sans. Names : *Eng.*: Giant Spine Gourd;

H.: Gangerua, Gulkakra;

Sans.: Karkataka.

Distribution : Cultivated everywhere for its edible fruit, Bengal-Tenasserim; common.

Reproductive Cycle : June-October.

Part Used : Seed.

Folk Uses : Seed memberanes rich in â-carotene and used for treating infantile rickets, xerophthalmia and nyctalopia.

Reference : Prajapati *et al.* (2006).

Monochoria hastata (L.) Solms.

Syn.: *Carigola hastata* (L.) Raf.; *Pontederia hastata* L.

Family: Pontederiaceae

Eng., H. and Sans. Names	:	*Eng.*: Arrow-Leaf Monochoria, Arrow Leaf Pondweed, Hastate-Leaf Pondweed, Monochoria; *H.*: Launkia; *Sans.*: Nilotpalam.
Distribution	:	Common throughout India and Sri Lanka.
Reproductive Cycle	:	August-March.
Part Used	:	Whole Plant.
Folk Use	:	Plant considered good for reducing secretion of saliva in children.
References	:	Jain (1991), Kumar and Kumari (2007).

Monochoria vaginalis Presl.

Syn.: *M. plantaginea* (Roxb.) Solms.; *Pontederia vaginalis* Burm.f.

Family: Pontederiaceae

Eng., H. and Sans. Names	:	*Eng.*: Heart-Leaved False Pickerel Weed, Marshy Betelvine, Oval-Leaved Pondweed; *H.*: Neelolpal; *Sans.*: Indivarah.
Distribution	:	Ponds and marches throughout India, upto 1,500m.
Reproductive Cycle	:	Flowering during summer and winter season.
Part Used	:	Whole Plant.
Folk Use	:	Plant used in stomach heat.
Reference	:	Pal *et al.* (2000).

Morinda citrifolia L.

Syn.: *M. aspera* Wight and Arn.; *M. bracteata* Roxb.; *Samama citrifolia* (L.) Kuntze.

Family: Rubiaceae

Eng., H. and Sans. Names	:	*Eng.*: Indian Mulberry, The Togari Wood of Madras; *H.*: Ach, Ak, Al, Borraal; *Sans.*: Acchaka, Achchuka, Ashyuka.
Distribution	:	Cultivated sometimes wild throughout the hotter parts of India; indigenous in the Darjeeling Terai and outer

hills and on the Andamans and along the Konkan coast.

Reproductive Cycle : May-July.

Parts Used : Leaf and Stem Bark.

Folk Uses : Decoction of charred leaves mixed with a little mustard useful in infantile diarrhoea. Stem bark and leaf both used in smallpox.

References : Chatterjee and Pakrashi (1997), Kirtikar and Basu (1984), Watt (1972).

Morinda pubescens **Sm.**

Family: Rubiaceae

Eng., H. and Sans. Names : *Eng.*: Morinda Tree, Togari Wood of Madaras;

H.: Acc, Ach, Al;

Sans.: Achchhuka, Paphanah.

Distribution : Bengal, C. and S. India.

Reproductive Cycle : April-November.

Part Used : Leaf.

Folk Uses : Decoction of depetiolated leaves administered for stomach pain and dysentery in children.

Reference : Ganesan and Kesavan (2003).

Moringa oleifera **Lam.**

Syn.: *Hyperanthera decandra* Willd.; *Moringa pterygosperma* Gaertn.; *M. zeylanica* Pers.

Family: Moringaceae

Eng., H. and Sans. Names : *Eng.*: Drumstick Tree, Horse Radish Tree;

H.: Mungna, Sahinjna, Seyla, Shajna;

Sans.: Akshiba, Chalusha, Jalaproya, Sainjna, Shigru, Shobhanjana, Tilashigru.

Distribution : Indigenous to N.W. India, frequently cultivated in hedges and homeyards; plentiful on alluvial land, in or near sandy beds of rivers and streams.

Reproductive Cycle : January-May.

Parts Used : Whole Plant, Flower and Fruit.

Folk Uses : Plant used as antidote to measles and pox. Flowers taken as vegetable as a preventive measure against smallpox. Fruits used in tetanus.

References : Banerjee (2000b), D'Rozario *et al.* (1999), Saini (1996).

Morus rubra **L.**

Family: Moraceae

Eng. Names : *Eng.*: American Mulberry, Red Mulberry.

Distribution : Cultivated in many parts of India for edible fruits.

Reproductive Cycle : August-October.

Part Used : Bark.

Folk Use : A tea made from inner bark used as laxative for children.

Reference : Bhattacharjee (2001).

Mucuna pruriens **Baker, non DC.**

Syn.: *Dolichos pruriens* L.; *Mucuna prurita* (L.) Hook.; *Stizolobium pruritum* (Wight.) Piper.

Family: Fabaceae

Eng., H. and Sans. Names : *Eng.*: Common Cowitch, Cowhage, Velvet Bean;

H.: Goncha, Kaunch, Kiwach;

Sans.: Atamagupta, Badari, Chanda, Kapikacchu, Romalu.

Distribution : Throughout India and in Andaman and Nicobar islands.

Reproductive Cycle : October-May.

Parts Used : Whole Plant and Root.

Folk Uses : Plant used in poliomyelitis. 5ml expressed juice of stem and leaves from wilted plant taken on empty stomach for 3 days against respiratory problems in children. 2g powder of mixture of its roots and that of *Tragia involucrate* in equal proportions given once a day for 3 days to children suffering from fever and body pains.

References : Ambily *et al.* (2010), Pal *et al.* (2000), Satapathy (2008).

Murraya koenigii **(L.) Spreng (Pl. 9D)***

Syn.: *Bergera koenigii* L.

Family: Rutaceae

Eng., H. and Sans. Names : *Eng.*: Curry Leaf Tree;

H.: Barsanga, Gandhela, Kathnim, Kurry Patta, Mitha Neem;

Sans.: Krishna-nemba, Raman, Surabhininiba.

Distribution	:	Along the foot of the Himalayas from Garhwal-Sikkim, ascending to 1,700m, Bengal, Pegu and southwards to Travancor and Sri Lanka; often cultivated.
Reproductive Cycle	:	May-August.
Part Used	:	Leaf.
Folk Uses	:	Leaves chewed by children to check mouthsores and vomiting.
Reference	:	Dhiman (2007).

Murraya paniculata (L.) Jack

Syn.: *Chalcas paniculata* (L.) Millsp.; *Murraya exotica* L.; *M. sumatrana* Roxb.

Family: Rutaceae

Eng., H. and Sans. Names	:	*Eng.*: Orange Jasmine, The China Box Tree; *H.*: Bibsar, Juti, Kamini, Marchula; *Sans.*: Ekangi, Surabhininiba.
Distribution	:	Commonly found in shady ravine slopes upto 1,500m in Garhwal Himalayas.
Reproductive Cycle	:	March-September.
Parts Used	:	Leaf, Twig and Flower.
Folk Uses	:	Bath with decoction of its leaves and twigs given for curing stomach troubles. 1g paste made from equal quantity of leaves and flowers mixed with mother's milk given once a day for 2 days to infants suffering from swelling of small intestine.
References	:	Brahmam and Saxena (1990), Parrotta (2001), Viswanathan *et al.* (2006).

Musa balbisiana Colla.

Syn.: *M. paradisiaca* subsp. *seminifera* (Lour.) Baker.

Family: Musaceae

Distribution	:	Cultivated throughout India.
Reproductive Cycle	:	Throughout the year.
Parts Used	:	Leaf and Fruit.
Folk Uses	:	Paste of mature leaves of *Musa balbisiana*, tender leaves of *Erythrina variegate*, basal portion of tender leaves of *Ananas comosus*, and bark of *Pongamia pinnata* in equal quantities given against roundworms in children. Food

prepared by boiling half of an unriped banana with rootstock of *Sida rhombifolia* and roots of a *Leucas indica* in 250ml of milk until the volume reduced to one fourth of its original, fed to infants as stimulant to increase vigour.

References : Barua *et al.* (2000), Gogoi and Borthakur (2001).

Mussaenda frondosa L.

Syn.: *Gardenia frondosa* (L.) Lam.; *Musa belilla* Buch.-Ham.; *M. dovinia* Buch.- Ham.; *M. fruticosa* L.

Family: Rubiaceae

Eng., H. and Sans. Names : *Eng.*: White Lady;

H.: Bebina, Bedina;

Sans.: Rajatarih, Sriparnah.

Distribution : Tropical Himalayas from W. and E., Andamans and Deccan.

Reproductive Cycle : July-October.

Parts Used : Shoot, Leaf, Calyx and Fruit.

Folk Uses : Infusion of dried shoot or fruit given to children to check cough and cold and catarrh. 2g leaf paste mixed with milk consumed 2 times a day for four days as tonic for infants. 10ml extract of leafy calyx-lobes grounded with cow's milk administered orally to check vomiting in children.

References : Agarwal (2003), Ambasta (1986), Anonymous (1962), Chatterjee and Pakrashi (1997), Dhiman (2003), Kumar (2002), Parrotta (2001), Rajendran and Henry (1994), Viswanathan *et al.* (2006).

Mussaenda raiateensis J.W. More.

Family: Rubiaceae

Distribution : Common in forest clearings secondary forests, open ridges from sea level-mid montane.

Parts Used : Leaf and Bark.

Folk Uses : Infusion of leaves and bark applied as poultice for improving complexion of skin in children. Infusion of bark given to undernourished infants.

Reference : Prajapati *et al.* (2006).

Myrica esculenta* Ham.

Syn.: *M. nagi* Sensu. Hook.f.

Family: Myricaceae

Eng., H. and Sans. Names : *Eng.*: Box Berry, Box Myrtle;

H.: Kaephal, Kaiphal;

Sans.: Aranya, Katphala, Laghukashmarya, Nasanu, Rohini.

Distribution : Widly distributed in India, sub-tropical Himalayas from Ravi eastwards-Assam as well as in Lushar, Khasia, Jaintia and Naga hills between 900-2,100m.

Reproductive Cycle : August-December.

Part Used : Whole Plant.

Folk Use : Plant used in poliomyelitis.

Reference : Pal *et al.* (2000).

***Myristica fragrans* Houtt.**

Syn.: *M. aromatica* Lamk.; *M. moschata* Thunb.; *M. officinalis* Mart.

Family: Myristicaceae

Eng., H. and Sans. Names : *Eng.*: Nutmeg Tree;

H.: Jaiphal, Javitri, Jayapatri, Jotri;

Sans.: Jaiphala, Jajiphalam, Javitri.

Distribution : Cultivated in S. India, W. Bengal and Assam.

Reproductive Cycle : August-December.

Parts Used : Whole Plant, Fruit and Seed (nutmeg).

Folk Uses : Plant used in diarrhoea, stomach pain, dysentery and poliomyelitis. Fruit wreath worn by children in dentition problems. Seeds used to cure ailments in children during cold.

References : Pal *et al.* (2000), Pandey (2003), Pandey and Verma (2003), Watt (1972).

***Myrtus communis* L.**

Family: Myrtaceae

Eng., H. and Sans. Names : *Eng.*: Common Myrtle, Greek Myrtle, Swedish Myrtle, True Myrtle;

H.: Murad, Vilayatimehndi;

Sans.: Gandhamalati.

Distribution	:	Indigenous from the Mediterranean-N.W. Himalayas; often grown in gardens throughout India.
Reproductive Cycle	:	Flowering in July-August.
Part Used	:	Leaf.
Folk Use	:	Powdered leaves used as an astringent especially for sores in children in Malta.
Reference	:	Kirtikar and Basu (1984).

Myxopyrum serratulum **A. W. Hill.**

Syn.: *M. smilacifalium* C. B. Clarke.

Family: Oleaceae

H. and Sans. Names	:	*H.*: Pilacameli;
		Sans.: Caturdharalata, Hemamalati.
Distribution	:	Throughout Kerala, in evergreen forests between 600-900m.
Reproductive Cycle	:	Throughout the year.
Part Used	:	Root.
Folk Uses	:	Roots useful in scabies and prurigo in children.
Reference	:	Prajapati *et al.* (2000).

Naravelia zeylanica **(L.) DC.**

Syn.: *N. pilulifera* var. *yunnanensis* Y. Fei.

Family: Ranunculaceae

H. and Sans. Names	:	*H.*: Chagulbati, Vatanasini;
		Sans.: Dhanavalli.
Distribution	:	Found in the tropical forests of Himalayas from E. Nepal eastwards, Bengal, Assam, E. and W. Peninsula, Sri Lanka.
Reproductive Cycle	:	Flowering in rainy season and fruiting in winters.
Part Used	:	Root.
Folk Use	:	Root powder used for tonsillitis.
Reference	:	Thomas and Britto (2003).

Nauclea orientalis **(L.) L.**

Syn.: *N. lutea* Blanco.; *Sarcocephalus cordatus* (Roxb.) Miq.

Family: Rubiaceae

Eng. and Sans. Names	:	*Eng.*: Burr Tree, Canary Cheesewood, Cape York

		Leichardt, Cheesewood, Leichhardt Pine, Leichhardt Tree, Soft Leichhardt, Yellow Cheesewood; *Sans.*: Bhantriya.
Distribution	:	Native to S.E. Asia, New Guinea, and Australia.
Reproductive Cycle	:	June-January.
Part Used	:	Whole Plant.
Folk Uses	:	Used in dysentery and blood dysentery.
Reference	:	Pal *et al.* (2000).

Nelsonia canescens (Lam.) Spreng.

Syn.: *N. campestris* R. Br.; *N. lamiifolia* (Roxb.) Spreng.; *N. origanoides* (Vahl.) Roem. and Sch.; *N. tomentosa* A. Dietr.

Family: Acanthaceae

Eng. Names	:	*Eng.*: Blue Pussyleaf, Reinwardtia.
Distribution	:	Throughout India, upto 1,350m.
Reproductive Cycle	:	April-June.
Parts Used	:	Inflorescence and Leaf.
Folk Uses	:	Paste of its leaf and inflorescence given orally for diarrhoea in children by Rabhas of Jalpaiguri district in W. Bengal.
Reference	:	Molla and Roy (1985).

Nelumbo nucifera Gaertn.

Syn.: *Nelumbium nuciferum* Gaertn.; *N. rheedii* C. Presl.; *Nelumbo speciosa* Willd.; *Nymphaea nelumbo* L.

Family: Nelumbonaceae

Eng., H. and Sans. Names	:	*Eng.*: Chinese Water Lily, Egyptian Beans, Indian Lotus, Pythagorean Bean, Sacred Lotus; *H.*: Ambuj, Kamal, Kanval; *Sans.*: Ambhoja, Asyapatra, Drishopadma, Indiralaya, Padama, Pushkara, Varisona.
Distribution	:	Distributed throughout India in the ponds of warmer regions, upto 1,800m
Reproductive Cycle	:	April-October.
Parts Used	:	Rhizome, Root, Flower, Fruit and Seed.
Folk Uses	:	Rhizomes yield a kind of nutritious arrowroot given to children in diarrhoea and dysentery. Roots used in

smallpox. Flowers used in diarrhoea. One crushed fruit given with 20ml water 3 times a day to check vomiting in children. Seeds also valuable as antiemetic, diuretic, and refrigerant in children.

References : Agarwal (2003), Ambasta (1986), Anonymous (1966), Chatterjee and Pakrashi (1997), Dhiman (2003), Khanna (2002a), Khare (2004), Kirtikar and Basu (1984), Watt (1972).

Nepeta cataria L.

Syn.: *Calamintha albiflora* Vaniot.; *Cataria vulgaris* Gaterau.; *Nepeta mollis* Salisb.; *N. vulgaris* Lam.

Family: Lamiaceae

Eng., H. and Sans. Names : *Eng.*: Catmint, Catnip;

H.: Ballari, Bheki, ;

Sans.: Bha, Bhandi.

Distribution : Found in W. temperate Himalayas from Dalhousie-Kashmir, up to 1,500m.

Reproductive Cycle : July-October.

Parts Used : Leaf, Aerial Parts and Flowering Tops.

Folk Uses : Aerial parts used in flu and fever in children. Leaves and flowering tops prescribed in colic.

References : Karnick (1994a, b), Prajapati *et al.* (2006).

Nicotiana rustica L.

Syn.: *N. pavoni* Dunal.

Family: Solanaceae

Eng. and H. Names : *Eng.*: Aztec Tobacco, Mapacho, Sacred Tobacco, Wild Tobacco;

H.: Tambakhu.

Distribution : Cultivated in the W. Punjab, Baluchistan and other parts of India.

Reproductive Cycle : Flowering during July-September.

Part Used : Leaf.

Folk Use : Snuff applied to children with cold in head as a sternutatory.

Reference : Kirtikar and Basu (1984).

Nicotiana tabacum* L. (Pl. 10A)

Syn.: *N. chinensis* Fisch. *ex* Lehm.; *N. maxicana* Schltdl.; *N. pilosa* Dunal.

Family: Solanaceae

Eng., H. and Sans. Names : *Eng.*: American Tobacco, Tobacco, Virginian Tobacco;

H.: Bujjerbhang, Tamaku, Tambaku;

Sans.: Dhumrapatrika, Kalanja, Krimighni, Tamakhuh, Vajrabhringi.

Distribution : Common throughout India, cultivated in hotter parts of India.

Reproductive Cycle : December-April.

Part Used : Leaf.

Folk Uses : Powder of dried leaves mixed with lime juice used as an external application in enlargement of spleen. Petiole of the leaf introduced into anus as a laxative. Paste of endocarp ash of coconut mixed with half dose of powdered tobacco leaves, a few ml of hubble bubble liquid ('hidak mahee') applied 2-3 times daily against scratching inflammation on skin.

References : Dey and Bahadur (1973), Singh and Singh (2003a), Watt (1972).

***Nyctanthes arbor-tristis* L.**

Syn.: *Bruschia macrocarpa* Bertol.; *Parilium arbor-tristis* (L.) Gaertn.; *Scahrita triflora* L.

Family: Oleaceae

Eng., H. and Sans. Names : *Eng.*: Coral Jasmine, Indian Mourner, Night Jasmine, Sorrowful Tree;

H.: Binari, Ganga Sedi, Har Singar, Karosli, Kuri, Nibari, Parja;

Sans.: Atyuha, Kharopatraka, Nalakunkuma, Parijatah, Rajanikasa, Sephalika.

Distribution : Native of India, cultivated in many parts of India; occurring wild, outer Himalayan ranges from Chenab-Nepal, Assam, Myanmar, Bengal, C. India, southwards to Godavari.

Reproductive Cycle : August-September.

Part Used : Leaf.

Folk Uses : Juice of leaves used as a remedy for intestinal worms (1 tsp fresh juice given to infants twice a day for 4-5 days), and as a purgative for infants. Poultice of 5 leaves tied on forehead of children to treat fever.

References : Agarwal (2003), Anonymous (1966), Chatterjee and Pakrashi (1997), Dey and Bahadur (1973), Dhiman (2003), Kirtikar and Basu (1984), Nayak *et al.* (2003c), Pandey and Verma (2003), Parrotta (2001), Thakur *et al.* (1989), Tirkey (2004).

Nymphaea alba L.

Syn.: *Castalia alba* (L.) Greene.; *Nymphaea occidentalis* Moss.

Family: Nymphaeaceae

Eng., H. and Sans. Names : *Eng.*: Common Water Lily, European White Water-lily, Lotus Lily, Nilofar, Water Rose;

H.: Kumuda;

Sans.: Kumuda, Utpalam.

Distribution : Found in lakes of Kashmir, below 1,800m.

Reproductive Cycle : May-August.

Part Used : Rhizome.

Folk Use : Paste of rhizome with honey given to children of age group 6 months-2 years on empty stomach in early morning to expel roundworms (1-3 tsp for 5 consecutive days).

Reference : Borthakur (1993).

Ocimum americanum L.

Syn.: *O. album* Roxb.; *O. brachiatum* Blume.; *O. canum* Sims.

Family: Lamiaceae

Eng., H. and Sans. Names : *Eng.*: Common Basil, Hoary Basil;

H.: Babutulsi, Bahari, Mamri, Kala Tulsi, Vantulsi;

Sans.: Ajaka, Aranyatulasi, Gambhira, Kuthera, Vanabarbarika.

Distribution : Throughout tropical and hotter India, cultivated from Punjab-Ava, Travancore, Malacca and Penang. Indigenous in the Punjab on low hills.

Reproductive Cycle : April-September.

Parts Used : Whole Plant and Leaf.

Folk Uses : Bath with decoction of fresh plant enhances immunity in children. Juice of leaves given to children for cold, catarrh and bronchitis.

References : Ambily *et al.* (2010), Anonymous (1966), Parrotta (2001).

Ocimum basilicum* L.

Syn.: *O. album* L.; *O. anisatum* Benth.; *O. medium* Mill.; *O. simile* N. E. Br.

Family: Lamiaceae

Eng., H. and Sans. Names :	*Eng.*: Common Basil, Sweet Basil; *H.*: Babui Tulsi, Ban Tulsi, Gulal Tulsi, Kali Tulsi, Marua; *Sans.*: Arjakan, Asurasa, Karahi, Munjariki, Surasa, Tungi, Varvara.
Distribution :	Wasteplaces, along railway lines, throughout tropical parts of India.
Reproductive Cycle :	June-December.
Parts Used :	Root, Leaf and Seed.
Folk Uses :	Roots used in bowel complaints of children. Decoction of leaves used against stomach troubles, inflammation and constipation. Seeds useful in diarrhoea and dysentery in children (given 4-5 grains).
References :	Chatterjee and Pakrashi (1997), Khamdram *et al.* (2011), Kirtikar and Basu (1984), Sharma (2004), Watt (1972).

***Ocimum gratissimum* L.**

Family: Lamiaceae

Eng., H. and Sans. Names :	*Eng.*: Large Basil, Lemon Basil, Shrubby Basil, Tea Bush; *H.*: Ban Tulsi, Malatulsi, Ram Tulsi; *Sans.*: Ajaka, Nidralu, Shophahari, Vriddhutulsi.
Distribution :	Weed of roadsides and wasteplaces throughout India; often cultivated.
Reproductive Cycle :	September-January.
Parts Used :	Whole Plant, Leaf, Flower and Seed.
Folk Uses :	Strong decoction of leaves and flowers (5ml twice/day) given to children for aphthae. Seeds grounded with honey given to children in vomiting. Oil used in abdominal colic.
References :	Ambasta (1986), Anonymous (1966), Chatterjee and Pakrashi (1997), Kirtikar and Basu (1984), Lalramnghinglova (2003), Parrotta (2001).

Ocimum sanctum L. (Pl. 10B)

Syn.: *O. inodurum* Burm.f.; *O. monochorum* L.; *O. tenuiflorum* L.

Family: Lamiaceae

Eng., H. and Sans. Names : *Eng.*: Holy Basil, Sacred Basil;
H.: Baranda, Kala-Tulsi, Tulsi;
Sans.: Ajaka, Brinda, Manjari, Sursa, Tulsi.

Distribution : Throughout India, upto 1,800m in the Himalayas and in the Andaman and Nicobar Islands.

Reproductive Cycle : October-March.

Parts Used : Whole Plant, Leaf and Seed.

Folk Uses : Newly stitched clothes kept overnight covering the plant before being given to children for wearing as this practice believed to kill all germs and considered antiseptic. Whole plant used in chickenpox. Leaf juice given to avoid after-effects of measles. Infusion of leaves used as stomachic in gastric disorders and laxative for children. Leaf extract mixed with hot water given in cold, cough and fever and as colic if mixed with little ginger. Juice (10ml) of fresh leaves mixed with root juice of *Boerhaavia diffusa* given twice daily for 8-10 days in chronic liver complaints. Seeds used in stomach troubles of infants and small children.

References : Agarwal (2003), Anonymous (1966), Chauhan (1999), Dhiman (2003), D' Rozario *et al.* (1999), Gogoi *et al.* (2003), Jain (1991), Khare (2004), Kirtikar and Basu (1984), Parrotta (2001), Rosakutty *et al.* (2000), Sood *et al.* (2012), Sundaravelu *et al.* (2003), Thakur *et al.* (1989), Watt (1972).

Olea europaea L.

Family: Oleaceae

Eng., H. and Sans. Names : *Eng.*: Olive, Olive Tree;
H.: Jaitun;
Sans.: Jaitun.

Distribution : Cultivated in Mediterranean region, Kashmir, H.P., Nilgiris.

Reproductive Cycle : May-June.

Part Used : Seed.

Folk Use	:	Seed oil considered an emollient for infant massage to strengthen the legs and hands.
Reference	:	Agarwal (2003).

Onosma echioides **(L.) L.**

Family: Boraginaceae

Eng., H. and Sans. Names	:	*Eng.*: Alkanet, Dyer's Bugloss, Hairy Onosma; *H.*: Lajari, Ratanjot; *Sans.*: Adhanani, Anjanakeshi, Dhamani, Nali, Nalika, Nalini, Nati, Stuya.
Distribution	:	Kashmir: 1,700-3,000m and Kumaon.
Reproductive Cycle	:	June-August.
Parts Used	:	Leaf and Whole Plant.
Folk Uses	:	Plant used in poliomyelitis. Leaves powder given to children as purgative in Harboi hills.
References	:	Kirtikar and Basu (1984), Pal *et al.* (2000).

Ophiorrhiza rugosa **Wall. var.** ***argentea*** **(Wall.** ***ex.*** **G. Don) Deb and Mondal**

Syn.: *O. argentea* Wall. *ex* G. Don.

Family: Rubiaceae

Distribution	:	Sikkim Himalayas; 1,350-2,000m, Bhutan, 1,350m.
Part Used	:	Whole Plant.
Folk Uses	:	Plant used for skin eruptions and sores.
Reference	:	Pal *et al.* (2000).

Opuntia dillenii **(Ker Gawl.) Haw. (Pl. 10C)**

Syn.: *Cactus dellenii* Ker-Gawler.

Family: Cactaceae

Eng., H. and Sans. Names	:	*Eng.*: Prickly Pear, Slipper Thorn; *H.*: Haththathoira, Nagphana, Nagphani; *Sans.*: Guda, Gula, Kandarohaka, Nagadru, Snuha, Snuhi, Snuka, Sudha, Vajra, Vajri, Vidara.
Distribution	:	Naturalized all over India.
Reproductive Cycle	:	April-December.
Part Used	:	Whole Plant.

Folk Use : Plant used in bronchitis.

Reference : Kirtikar and Basu (1984).

Orchis mascula **(L.) L.**

Syn.: *O. morio* var. *mascula* L.; *O. pinetorum* Lacaita.

Family: Orchidaceae

Eng., H. and Sans. Names : *Eng.*: Purple Orchid, Salep, The Salep Orchid;
H.: Salab-Misri, Salep Misri;
Sans.: Salabmisri.

Distribution : Grown in marshy places in S. India.

Reproductive Cycle : April-June.

Part Used : Tuber.

Folk Uses : Tubers used against diarrhoea and irritated gastro-intestinal tracts.

Reference : Prajapati *et al.* (2006).

Origanum vulgare **L.**

Syn.: *O. creticum* Lour.; *Thymus origanum* Kuntze.

Family: Lamiaceae

Eng., H. and Sans. Names : *Eng.*: Argans, Orgament, Pot Marjoram, Wild Marjoram;
H.: Mirzanjosh, Sathra;
Sans.: Damnak.

Distribution : Found in temperate Himalayas, (Kashmir-Sikkim): 2,350-4,000m.

Reproductive Cycle : August-December.

Parts Used : Aerial Parts and Root.

Folk Uses : Aerial parts used in tonsillitis. An amulet of its root pieces tied to neck of infants as protective measures against conjunctivitis.

References : Chauhan (1999), Prajapati *et al.* (2006).

Oroxylum indicum **Vent.**

Syn.: *Bignonia indica* L.; *Calosanthes indica* (L.) Blume.; *Spathodea indica* (L.) Pers.

Family: Bignoniaceae

Eng., H. and Sans. Names : *Eng.*: Indian Trumpet Flower;
H.: Arlu, Assarsauna, Kharkath. Sauna, Shyona;

		Sans.: Arala, Araluka, Bhantuka, Kandarpa, Nata, Vatu.
Distribution	:	Throughout India from Himalayas-Sri Lanka and Malacca.
Reproductive Cycle	:	June-February.
Parts Used	:	Whole Plant, Root Bark, Stem Bark and Seed.
Folk Uses	:	Plant used in scabies. Root bark and / or stem bark paste applied over whole body to get relief from the effect of smallpox. Decoction of bark used in smallpox and chickenpox. Amulet of seeds tied around neck to prevent contagious diseases from affecting children. Paste of seeds mixed with fire-soot applied on lower jaw of the person suffering from mumps and tonsillitis.
References	:	Kemp (2003), Pal *et al.* (2000), Panda (2007), Panda and Das (2000), Pande *et al.* (2006), Sen and Behera (2003), Sen and Behera (2007).

***Orthosiphon rubicundus* (D. Don.) Benth.**

Syn.: *Lumnitzera rubicunda* (D. Don.) Spreng.; *Plectranthus rubicundus* D. Don.

Family: Lamiaceae

Eng. Name	:	*Eng.*: Red Java Tea.
Distribution	:	W. sub-tropical Himalaya, from C. Nepal-Jammu, 700-1,350m and southwards to the Circars and Nilgiris.
Reproductive Cycle	:	August-December.
Part Used	:	Whole Plant.
Folk Use	:	Plant used in skin diseases.
Reference	:	Pal *et al.* (2000).

***Orthosiphon thymiflorus* (Roth) Sleesen**

Syn.: *Ocimum thymiflorus* Roth.; *Orthosiphon glabratus* Benth.; *O. viscosus* Benth.; *Plectranthus tristis* (Roth) Spreng.

Family: Lamiaceae

Sans. Name	:	*Sans.*: Pratanika.
Distribution	:	Throughout India, upto 900m in dry open forests.
Reproductive Cycle	:	Throughout the year.
Part Used	:	Leaf.
Folk Uses	:	5ml of filtered leaf juice administered orally to children against cold, phlegm and intestinal worms.
Reference	:	Ravikumar and Sankar (2003).

Oryza rufipogon **Gfiff.**

Syn.: *O. glumipatula* Steud.; *O. meridionalis* N.Q.Ng.

Family: Poaceae

Eng. and H. Names : *Eng.*: Brownbread Rice, Red Rice;
H.: Chita Dhan.

Distribution : Pallikaranai (Chennai) marshland and E. india.

Reproductive Cycle : September-November.

Part Used : Whole Plant.

Folk Use : Used for eye complaints in children.

Reference : Pal *et al.* (2000).

Oryza sativa **L.**

Syn.: *O. communissima* Lour.; *O. plena* (Prain) Chowdhury.; *O. praecox* Lour.

Family: Poaceae

Eng., H. and Sans. Names : *Eng.*: Paddy, Rice;
H.: Caval, Chaaval, Dhan;
Sans.: Dhanya, Nivara, Sali, Tandula, Vrihi.

Distribution : Widely cultivated as a crop plant throughout the plains of India.

Reproductive Cycle : September-October.

Part Used : Grains.

Folk Uses : Paste of chewed polished raw rice applied on the dog bite area. Rice gruel with a pinch of salt and a squeeze of lemon makes a good drink against smallpox and measles.

References : Dhiman (2003), Singh and Singh (2003a).

Osbeckia nepalensis **Hook.f.**

Family: Melastomataceae

Eng. Name : *Eng.*: Nepal Osbeckia.

Distribution : Sub-tropical Himalayas from Nepal eastwards, Khasia hills.

Reproductive Cycle : July-December.

Part Used : Flower.

Folk Uses : Flower paste with honey locally applied in glossitis and sores of babies.

References : Barua *et al.* (2000), Kirtikar and Basu (1984), Sharma (2004).

Osbeckia stellata **Buch.-Ham. *ex* Ker Gawl.**

Syn.: *Melastoma mairei* H. Lev.; *Osbeckia opipara* C. Y, Wu and C. Chen.; *O. rostrata* D. Don.; *O. sikkimensis* Craib.

Family: Melastomataceae

Eng. and H. Names : *Eng.*: Osbeckia;

H.: Tsulesi.

Distribution : Himalayas Terai from Kumaon-Bhutan; common, ascending the hills to 1,700m.

Reproductive Cycle : July-October.

Part Used : Leaf.

Folk Uses : Juice and paste of leaves considered good for healing minor cuts and wounds in small children.

Reference : Arya *et al.* (1999).

Osmunda regalis **L.**

Syn.: *O. capensis* C. Presl.; *O. mexicana* Fee.; *O. obtusifolia* Kaulf.; *Struthiopteris regalis* (L.) Bernh.

Family: Osmundaceae

Eng. Names : *Eng.*: Buckhorn, Fern Brake, Hartshorn Bush, Herb Christopher, King's Fern, Royal Fern, Water Fern.

Distribution : India (Assam, Meghalaya, U.P., M.P., Tamil Nadu, Kerala, Maharashtra, Karnataka).

Reproductive Cycle : Flowering occurs in April.

Parts Used : Whole Plant and Root.

Folk Use : Plant along with roots used in rickets.

References : Agarwal (2003), Ambasta (1986), Das (2003), Jha *et al.* (2003), Khullar (2001), Singh and Viswanathan (1996).

Oxalis corniculata **L.**

Syn.: *O. foliosa* Blatt.; *O. pusilla* Salisb.; *O. repens* Thunb.; *O. villosa* Bieb.

Family: Oxalidaceae

Eng., H. and Sans. Names : *Eng.*: Indian Sorrel, Yellow Oxalis;

H.: Amrul, Anboti, Chuka tripati, Seh;

Sans.: Amlika, Amlotaja, Kangeri, Shuklika.

Distribution	:	Throughout the warmer parts of India and Sri Lanka, ascending Himalayas to 2,350m.
Reproductive Cycle	:	April – November.
Parts Used	:	Whole Plant and Leaf.
Folk Uses	:	Whole plant used in rickets. Leaf juice (2 tsp, thrice a day till recovery) mixed with breast milk in equal quantity given to infants for curing cold by Jenukurubas. Crushed leaves used for treating children with mouth infections, infected navels and babies having symptoms caused by the delayed closing of their fontanelles. Leaf juice applied externally on backbone of infants in rickets. Infusion of leaves used for curing convulsion in infants. 'Saambar' (sauce) prepared from leaves given to children below 4 or 5 years to reduce swelling of belly.
References	:	Kapur and Srivastava (1996), Kshirsagar and Singh (2000b, 2007), Nadanakunjidam (2003), Pande *et al.* (2006), Prajapati *et al.* (2006), Singh (2000).

Oxystelma esculentum **(L.f.) Sm.**

Syn.: *O. wallichii* Wight.; *Periploca esculenta* L.f.; *Sarcostemma esculentum* (L.f.) R. W. Holm.

Family: Asclepiadaceae

Eng., H. and Sans. Names	:	*Eng.*: Rosy Milkweed Vine; *H.*: Dudhi, Dudhialata, Dudlutta, Dugdhica, Kyirin; *Sans.*: Dugdhi, Kachhara, Kshiravi, Kshiri, Svaduparni.
Distribution	:	Throughout the plains and lower hills of India, usually near water.
Reproductive Cycle	:	July-January.
Part Used	:	Fruit.
Folk Use	:	Juice of fruits given to children as an astringent.
Reference	:	Kirtikar and Basu (1984).

Padus cornuta **(Wall. *ex* Royle) Carriere**

Syn.: *Prunus cornuta* (Wall. *ex* Royle) Steud.

Family: Rosaceae

Eng. and H. Names	:	*Eng.*: Bird Cherry; *H.*: Jamona.
Distribution	:	Temperate Himalayas from Murree-Sikkim, 2,700-4,000m and Bhutan.

Reproductive Cycle	:	April-May.
Part Used	:	Root.
Folk Use	:	Dried root powder (about ¼ tsp) with honey or sugar given for 5 days to deworm children of age group 5-10 years.
Reference	:	Arya *et al.* (1999).

Paederia foetida **Auct.**

Syn.: *P. scandens* (Lour.) Merr.; *P. tomentosa* Blume.

Family: Rubiaceae

Eng., H. and Sans. Names	:	*Eng.*: Chinese Moon-creeper, King's Tonic, Strinking Opalberry; *H.*: Gandhali, Gundali, Somaraji; *Sans.*: Balya, Chandraparni, Gandhali, Prabala, Prasarani, Sharani.
Distribution	:	Common in tropical parts of India and in the C. and E. Himalayas upto 1,700m.
Reproductive Cycle	:	September-December.
Parts Used	:	Whole Plant and Leaf.
Folk Uses	:	Plant used in diarrhoea, bowel complaints and dysentery. Leaf juice given in diarrhoea.
References	:	Chatterjee and Pakrashi (1997), Kirtikar and Basu (1984), Pal *et al.* (2000), Rawat and Chowdhury (1998), Watt (1972).

Paeonia emodi **Wall. *ex* Royle.**

Syn.: *P. officinalis* H.f. and T.

Family: Paeoniaceae

Eng., H. and Sans. Names	:	*Eng.*: Himalayan Paeony, Paeony Rose; *H.*: Udsalap; *Sans.*: Svetamula.
Distribution	:	N.W. Himalayas from Kashmir-Kumaon: 1,600-3,000m, W. temperate Himalayas, 1,700-3,350m from Kumaon-Hazara.
Reproductive Cycle	:	May-June.
Part Used	:	Root.
Folk Use	:	Roots used as a blood purifier.

References : Agarwal (2003), Anonymous (1966), Chauhan (1999), Kapur and Singh (1996), Kirtikar and Basu (1984), Sharma (2003), Watt (1972).

***Pandanus odoratissimus* L.f.**

Syn.: *P. tectorius* auct. non Soland. *ex* Parkinson.

Family: Pandanaceae

Eng., H. and Sans. Names : *Eng.*: Screw Pine, Umbrella Tree;

H.: Gagandhul, Kedgi, Kevda;

Sans.: Chamarapushpa, Halina, Jambuka, Ketaki.

Distribution : Throughout India, along the sea coast and water courses. Prefers moist and damp places. Common in and around valley throughout N.E. region.

Reproductive Cycle : February-August.

Part Used : Leaf.

Folk Use : Leaves useful in smallpox.

References : Agarwal (2003), Dhiman (2003), Kirtikar and Basu (1984), Kumar (2002), Parrotta (2001), Prajapati *et al.* (2006), Saikia and Nath (2003b).

***Panicum antidotale* Retz.**

Syn.: *P. miliare* Lam.; *P. proliferum* Lam.

Family: Poaceae

Eng. and H. Names : *Eng.*: Blue Panic Grass;

H.: Ghemar, Gunara, Jamur.

Distribution : Abundant in plains of N. India and extends to Baluchistan and Afghanistan.

Reproductive Cycle : July-October.

Part Used : Whole Plant.

Folk Use : Smoke of burning plant said to be used as disinfectant in smallpox.

References : Anonymous (1966), Kirtikar and Basu (1984), Parrotta (2001), Watt (1972).

Papaver somniferum* L.

Family: Papaveraceae

Eng., H. and Sans. Names : *Eng.*: Joan Silverpin, Opium Poppy, White Poppy;

H.: Afim, Afyun, Kashkash, Post;

		Sans.: Ahifen, Chosa, Khasa.
Distribution	:	Cultivated throughout the warmer parts of India.
Reproductive Cycle	:	February-April.
Part Used	:	Fruit.
Folk Use	:	Camphorated opium used in mumps.
Reference	:	Khare (2004).

Parmelia tinctorium **Despr.**

Family: Parmeliaceae

Eng. Name	:	*Eng.*: Lichen, Stone Flower.
Distribution	:	Indian plains and hills, 1,000-3,000m Himalayas, Nepal.
Part Used	:	Whole Plant.
Folk Use	:	Plant used in polio.
Reference	:	Pal *et al.* (2000).

Parochetus communis **D. Don.**

Syn.: *Cosmiusa repens* Alef.; *Parochetus maculata* R. Br.; *P. major* D. Don.

Family: Fabaceae

Eng. Names	:	*Eng.*: Blue Oxalis, Shamrock Pea.
Distribution	:	Alpine temperate and sub-tropical Himalayas, from Shimla and Garhwal-Assam, 1,330-4,350m, Nilgiris, Sri Lanka, Myanmar, Tenasserim.
Reproductive Cycle	:	Flowering in May-November.
Part Used	:	Leaf.
Folk Use	:	Used in stomach disorder of babies.
Reference	:	Jain (1991).

Passiflora edulis **Sims.**

Syn.: *P. gratissima* A. St.-Hil.; *P. incarnata* L.; *P. rubricaulis* Jacq.; *P. vernicosa* Barb. Rodr.

Family: Passifloraceae

Eng. and H. Names	:	*Eng.*: Maypop, Passion Flower, True Passion Flower, Wild Apricot, Wild Passion Vine;
		H.: Krishna Fal.
Distribution	:	Grown in Indian Gardens, also found in Nilgiris and in various parts of N. India.

Reproductive Cycle : July-October.

Part Used : Leaf.

Folk Use : Used in tetanus.

Reference : Prajapati *et al.* (2006).

***Pavetta crassicaulis* Bremek**

Syn.: *P. stocksii* Bremek.

Family: Rubiaceae

Eng., H. and Sans. Names : *Eng.*: Indian Pellet Shrub, White Pavetta;
H.: Kankra, Karnikara, Papri;
Sans.: Kakachedi, Papata, Triakphala.

Distribution : Throughout India, upto 1,500m.

Reproductive Cycle : July-October.

Part Used : Fruit.

Folk Use : Fruits cooked with other vegetables taken as curry to act as an anthelmintic for roundworms in children.

Reference : Kothari *et al.* (2003).

***Pavetta indica* L.var. *tomentosum* (Roxb. *ex* Sm.) Hook.f.**

Syn.: *Ixora tomentosa* Roxb. *ex* Sm.

Family: Rubiaceae

Eng., H. and Sans. Names : *Eng.*: Indian Pellet Shrub, White Pavetta;
H.: Kankra, Karnikara, Kathachampa, Papari;
Sans.: Kakachedi, Papata, Tiriakphala.

Distribution : Throughout India, from W. Himalayas in Garhwal ascending to 1,350m-Bhutan and Myanmar also southwards-Sri Lanka, Malacca and Penang.

Reproductive Cycle : January-May.

Parts Used : Flower and Root.

Folk Uses : Inhabitants of Manipur prescribe extract of its flower for eradicating roundworms in children. Roots given in powder form to children in visceral obstructions.

References : Kirtikar and Basu (1984), Sinha (1987), Watt (1972).

***Pedalium murex* L.**

Syn.: *P. microcarpum* Decne.; *P. muricatum* Salisb.; *Rogeria microcarpa* Klotzsch.

Family: Pedaliaceae

Eng., H. and Sans. Names : *Eng.*: Burra Gokhru;
H.: Bara Gokhru, Faridbuti, Kadvagokhru;
Sans.: Gokshura, Sthulagoksurah, Tiktagoksurah.

Distribution : Kathiawar, Gujarat, Konkan, Deccan Peninsula.

Reproductive Cycle : March-December.

Part Used : Whole Plant.

Folk Use : Plant used in poliomyelitis.

Reference : Pal *et al.* (2000).

***Pegolettia senegalensis* Cass.**

Syn.: *P. dubiefiana* Quezel.

Family: Asteraceae

Eng. Name : *Eng.*: Okambodi or Oumbodi.

Distribution : Occurs in Rajasthan and dry region of India.

Reproductive Cycle : September-October.

Part Used : Whole Plant.

Folk Uses : Plant with ghee given to children to make them strong and hardy.

References : Agarwal (2003), Anonymous (1966).

***Pentanema indicum* (L.) Ling.**

Syn.: *Vicoa auriculata* Cass.; *V. indica* (L.) DC.

Family: Asteraceae

Eng. and H. Names : *Eng.*: Sonkadi;
H.: Bichhloo, Sonkadi.

Distribution : Throughout the drier parts of India, from Punjab to Myanmar and southwards, ascending W. Himalayas to 1,350m and to 1,700m in Sri Lanka.

Reproductive Cycle : Throughout year.

Part Used : Root.

Folk Use : Two tsp root decoction given once a day to cure fever.

References : Girach (1992), Parrotta (2001).

Pentapetes phoenicea **L.**

Family: Sterculiaceae

Eng., H. and Sans. Names	:	*Eng.*: Noon Plant; *H.*: Dopahariya; *Sans.*: Arkavallabhah, Bandhukah.
Distribution	:	Cultivated throughout hotter parts of India, from Punjab-Myanmar and W. Peninsula.
Reproductive Cycle	:	August-December.
Part Used	:	Whole Plant.
Folk Use	:	Plant used for gastropathy in children.
Reference	:	Prajapati *et al.* (2006).

Pepromia pellucida **(L.) Kunth.**

Syn.: *P. concinna* (Haw.) A. Dietr.; *Piper concinnum* Haw.; *P. pellucidum* L.

Family: Piperaceae

Eng. Names	:	*Eng.*: Pepper Elder, Rat Ear, Shiny Bush, Silverbush.
Distribution	:	Naturalized in W. Bengal, Arunachal Pradesh, Tamil Nadu, Sikkim and Bhutan.
Reproductive Cycle	:	November-February.
Part Used	:	Whole Plant.
Folk Use	:	Plant used as refrigerant for children.
Reference	:	Anonymous (1966).

Pergularia daemia **(Forsk.) Chiov.**

Syn.: *Asclepias daemia* Forsk.; *Daemia cordifolia* K. Schum *ex* Engl.; *D extensa* R.Br.; *D. glabra* Schult.

Family: Asclepiadaceae

Eng., H. and Sans. Names	:	*Eng.*: Pergularia; *H.*: Jutuk, Sadowani, Sagovani, Utranajutuka; *Sans.*: Uttaravaruni, Yugaphala, Yugmaphalika.
Distribution	:	Throughout hotter parts of India, ascending to 1,000m.
Reproductive Cycle	:	September-January.
Parts Used	:	Leaf and Seed.
Folk Uses	:	Decoction of leaves given to children as an anthelmintic in dose not exceeding 3 tsp. Leaf juice prescribed in asthma and infantile diarrhoea, also smeared on body

to bring down fever by Jenukurubas. 2-3g seed ash mixed with a spoon of honey given in whooping cough.

References : Agarwal (2003), Ambasta (1986), Bhattacharjee (2001), Chatterjee and Pakrashi (1997), Farooq (2005), Jain (1994), Kirtikar and Basu (1984), Kshirsagar and Singh (2007), Painuli and Maheshwari (1996), Parrotta (200), Watt (1972).

Persicaria bistorta **(L.) Samp.**

Syn.: *Polygonum bistorta* L.; *P. plumosum* Small.

Family: Polygonaceae

Eng. and H. Names : *Eng.*: Bistort, Snakeroot;

H.: Sarpmool.

Distribution : Himalayas from Kashmir-Sikkim and in the hills of Assam at 2,700-4,500m.

Reproductive Cycle : April-November.

Part Used : Rhizome.

Folk Use : Used in diarrhoea.

Reference : Karnick (1994a).

Peucedanum dhana **Buch.-Ham. *ex* C.B. Clarke.**

Family: Apiaceae

Distribution : N.E. India, sub-shiwaliks upto 1,500m, Kumaon, in sub-tropical region, N. Bengal.

Reproductive Cycle : August-September.

Part Used : Whole Plant.

Folk Uses : Plant used in bowel complaints, blood dysentery and as an anthelmintic.

Reference : Pal *et al.* (2000).

Phaseolus lunatus **L.**

Syn.: *P. ilocanus* Blanco.; *P. viridis* Piper.; *P. xuarezii* Zuccagni.

Family: Fabaceae

Eng. and H. Names : *Eng.*: Burma Bean, Double Bean, Lima Bean, Rangoon Bean;

H.: Sem.

Distribution : Cultivated throughout India.

Reproductive Cycle : July-November.

Parts Used : Leaf and Seed.

Folk Uses : Crushed juice of leaves with or without macerated mass applied on forehead, chest, back of pinna and neck of children in case of high body temperature. Poultice of seeds applied to the abdomen for curing stomachache in small children.

References : Anonymous (1969), Singh and Singh (2003a, 2003b).

***Phlogacanthus thyrsiflorus* Nees.**

Syn.: *Justicia thyrsiformis* Hardw.

Family: Acanthaceae

H. Name : *H.*: Vasaka.

Distribution : In sub-tropical Himalayas and N.E. India.

Reproductive Cycle : January-May.

Parts Used : Inflorescence and Leaf.

Folk Uses : Leaves good against relief from flu. Juice of fresh leaves with or without macerated mass applied to heal burns and scalds. Two tsp of leaf extract given one to three times daily for a few days against chickenpox and cough. Inflorescence used as an antidote to pox.

References : Barua *et al.* (2000), Salam *et al.* (2011), Singh and Singh (2003a), Sinha (1987).

***Phyla nodiflora* (L.) Greene.**

Syn.: *Lippia nodiflora* Cham.; *Verbenia nodiflora* L.; *Zapania nodiflora* (L.) Lam.

Family: Verbenaceae

Eng., H. and Sans. Names : *Eng.*: Purple Lippia;

H.: Bhuiokra, Jalapapili, Ludra;

Sans.: Agnijvala, Bahushikha, Chitrapatri, Langali, Pranada, Sharadi, Trinashita, Vashira.

Distribution : Throughout India, distributed to all tropical and warm-temperate regions.

Reproductive Cycle : April-July.

Parts Used : Whole Plant, Leaf and Stalk.

Folk Uses : Plant used in indigestion after dysentery. Leaf juice used in diarrhoea. Tender bitter stalks prescribed in the form of infusion to children suffering from indigestion.

References : Chatterjee and Pakrashi (1997), Devi and Singh (2008), Kirtikar and Basu (1984), Murugesan *et al.* (2011), Parrotta (2001), Srivastava *et al.* (2003a), Watt (1972).

Phyllanthus acidus **(L.) Skeels.**

Syn.: *Averrhoa acida* L.; *Cicca disticha* L.;
***Phyllanthus distichus* (L.) Mull. Arg.**

Family: Euphorbiaceae

Eng., H. and Sans. Names : *Eng.*: The Country Gooseberry;

H.: Chalmeri, Harfarauri;

Sans.: Ghana, Lavalee.

Distribution : Widely cultivated in India, native to the coastal region of N.E. Brazil.

Reproductive Cycle : March-April.

Part Used : Leaf.

Folk Uses : Bath with decoction of leaves of *Phyllanthus*, 'neem', *Clerodendron viscosum* and *Glycosmis pentaphylla* prescribed to patients suffering from measles.

Reference : Nath and Maiti (2003).

Phyllanthus airy-shawii **Jean F. Brunal and J.P. Roux.**

Family: Euphorbiaceae

Distribution : From the Sikkim-W. and E. Himalayas, Assam, Meghalaya, in Mizoram, scattered here and there in tropical secondary forests.

Reproductive Cycle : April-July.

Part Used : Leaf.

Folk Use : Crushed leaves applied externally on measles.

Reference : Lalramnghinglova (2003).

Phyllanthus amarus **Schum. and Thonn.**

Syn.: *Diasperus nanus* (Hook.f.) Kuntze.;
***Phyllanthus nanus* Hook.f.; *P. swartzii* Kostel.**

Family: Euphorbiaceae

Eng., H. and Sans. Names : *Eng.*: India Water Lily;

H.: Jangali Amla;

Sans.: Bahupatra.

Distribution	:	Throughout Indian plains, more common in Peninsular India, ascending to 1,000m in hills.
Reproductive Cycle	:	March-April.
Part Used	:	Whole Plant.
Folk Use	:	Poultice of tender plant thoroughly grounded with milk applied to cure chickenpox.
Reference	:	Nisha and Sivadasan (2007).

***Phyllanthus emblica* L. (Pl. 10D)**

Syn.: *Emblica officinalis* Gaertn.

Family: Euphorbiaceae

Eng., H. and Sans. Names	:	*Eng.*: Emblic Myrobalan, Indian Goseberry;
		H.: Amala, Anola, Anvula, Aungra, Aura;
		Sans.: Adiphala, Akara, Amlika, Amri, Dhatri, Jatiphala, Kamu, Rochni, Shanta, Shiva, Tishya, Triphala, Umrita.
Distribution	:	Found both in wild and cultivated state; common in mixed forests in India, ascending to 1,300m on hills.
Reproductive Cycle	:	April-June.
Part Used	:	Fruit.
Folk Use	:	Two tsp mixture paste of bark of *Terminalia bellirica, Spondias pinnata* and *Psidium guajava* in equal quantities with 3 fruits each of *Terminalia chebula* and *Phyllanthus emblica* as well as few drops of alkali solution given once daily to children for 3 days to cure cholera.
Reference	:	Gogoi and Borthakur (2001).

***Phyllanthus fraternus* G.L. Webst.**

Syn.: *P. lonphali* Wall.

Family: Euphorbiaceae

Eng., H. and Sans. Names	:	*Eng.*: Carry-Me-Seed, Phyllanthus;
		H.: Jangli Amla, Jangli Amli, Jar-Amla;
		Sans.: Bhudhatri, Bhumyamalaki.
Distribution	:	Common throughout waste places and shady gardens in hotter parts of India.
Reproductive Cycle	:	August-September.
Parts Used	:	Whole Plant and Root.
Folk Uses	:	5ml juice of plant with a little honey given to children for both infective hepatitis and chronic liver problems.

Root paste given to children below 2 years, twice daily for 3 days in diarrhoea, skin eruptions due to allergy, bowel complaints and in dysentery.

References : Borthakur *et al.* (2004), Gogoi *et al.* (2003), Mohanty *et al.* (1996), Pal *et al.* (2000), Parrotta (2001), Shivaprasad and Chandrashekar (2003).

Phyllanthus reticulatus **Poir.**

Syn.: *Anisonema reticulatum* (Poir.) A. Juss.; *Kirganelia reticulata* (Poir.) Baill.

Family: Euphorbiaceae

Eng., H. and Sans. Names : *Eng.*: Potato Plant, Roast Potato Plant;

H.: Buinowla, Makhi, Panjoli, Panjuli;

Sans.: Krishna-Kamboji.

Distribution : Grows almost throughout tropical India, ascending to 2,000m in Khasi and Jaintia hills and Andaman Islands.

Reproductive Cycle : April-May.

Parts Used : Whole Plant, Root and Leaf.

Folk Uses : Plant used in smallpox. Decoction of roots given to children for cough, catarrh and asthma. Leaves used in infantile diarrhoea.

References : Agarwal (2003), Ambasta (1986), Anonymous (1959), Parrotta (2001).

Phyllanthus simplex **Retz.**

Syn.: *P. anceps* Vahl.; *P. marginatus* Herb.Heyne *ex* Wall.

Family: Euphorbiaceae

Eng. and H. Names : *Eng.*: Seeded Under Leaf, Virgate Leaf-Flower;

H.: Bhui aonla.

Distribution : Throughout India, from Kumaon-Assam and southwards-Travancore, Malacca and Sri Lanka, ascending to 2,000m.

Reproductive Cycle : Flowering and Fruiting in October.

Part Used : Leaf.

Folk Use : Extract of fresh leaves mixed with butter milk, make a good wash to cure itch in children.

References : Kirtikar and Basu (1984), Watt (1972).

Phyllanthus urinaria L.

Syn.: *Diasperus urinaria* (L.) Kuntze.; *Phyllanthus cantoniensis* Schweigg.

Family: Euphorbiaceae

Eng., H. and Sans. Names	:	*Eng.*: Shatterstone, Stone Breaker; *H.*: Hazarmani, Lal-Bhuin Anvalah; *Sans.*: Adhyand, Ajata, Ajuta, Amala, Aphala, Aruha, Bahupatra, Charati, Mala, Shiva.
Distribution	:	Found throughout India in wastelands and as weed in gardens and moist places 900m.
Reproductive Cycle	:	August-January.
Parts Used	:	Leaf, Root, Seed and Whole Plant.
Folk Uses	:	Plant used for infantile cheek eczema. Roots and seeds given to sleepless children in Chota Nagpur. Juice of leaves prescribed with coconut milk as an appetizer to children. Seeds also used in urinogenital diseases.
References	:	Ambasta (1986), Anonymous (1969), Chatterjee and Pakrashi (1997), Farooq (2005), Jain (1991), Kirtikar and Basu (1984), Parrotta (2001), Prajapati *et al.* (2006).

Physalis angulata L.

Syn.: *Boberella angulata* (L.) E.H.L. Krause.

Family: Solanaceae

Eng. Names	:	*Eng.*: Camapu, Cutleaf Groundcherry, Wild Tomato, Winter Cherry.
Distribution	:	Throughout India.
Reproductive Cycle	:	July -November.
Part Used	:	Leaf.
Folk Use	:	Juice from leaves squeezed into eyes and nose to regain consciousness in a fainting child.
Reference	:	Kirtikar and Basu (1984).

Phytolacca acinosa Roxb.

Syn.: *P. latbenia* (Moq.) Walter.; *Pircunia latbenia* Moq.

Family: Phytolaccaceae

Eng. and H. Names	:	*Eng.*: Indian Poke, Sweet Belladona; *H.*: Matazor.
Distribution	:	Temperate Himalayas, from Hazara, Kashmir-Bhutan.

Reproductive Cycle	:	June-September.
Part Used	:	Root.
Folk Uses	:	Dried roots used in tonsillitis and mumps.
Reference	:	Karnick (1994a).

Phytolacca americana **L.**

Syn.: *P. decandra* L.

Family: Phytolaccaceae

Eng. Names	:	*Eng.*: American Night Shade, Poke Root.
Distribution	:	Temperate Himalayas, wild or cultivated from Hazara and Kashmir-Bhutan, 1,700-3,000m.
Reproductive Cycle	:	Flowering in May-October.
Part Used	:	Root.
Folk Uses	:	Roots used in tonsillitis and mumps.
References	:	Karnick (1994b), Prajapati *et al.* (2006).

Pimpinella anisum **L.**

Syn.: *Anisum odoratum* Raf.; *A. officinale* DC.; *Sison anisum* (L.) Spreng.

Family: Apiaceae

Eng., H. and Sans. Names	:	*Eng.*: Anise, Aniseed;
		H.: Bandian, Saunf, Sawolf;
		Sans.: Shetapushapa.
Distribution	:	Native of E. Mediterranean, W. Asia and N. Africa, grown to a small extent as a culinary herb in India.
Reproductive Cycle	:	August-December.
Part Used	:	Seed.
Folk Use	:	Seeds given to infants and children to relieve colic.
Reference	:	Prajapati *et al.* (2006).

Pinellia pedatisecta **Schott.**

Syn.: *P. wawrae* Engl.

Family: Araceae

Eng. Names	:	*Eng.*: Palmata-leaf Pinellia, Pedatisect Pinellia, Tiger Palm.
Distribution	:	Damp shady slopes and under trees in China, Japan and Korea.

Reproductive Cycle : May-October.

Part Used : Tubers.

Folk Uses : Used in infantile convulsions and tetanus.

Reference : Prajapati *et al.* (2006).

Pinus roxburghii **Sargent (Pl. 11A)***

Syn.: *P. lonifolia* Roxb. *ex* Lamb.

Family: Pinaceae

Eng., H. and Sans. Names : *Eng.*: Chir Pine, Himalayan Long-leaved Pine, Three-leaved Pine;

H.: Chir, Cid, Cil, Sarala;

Sans.: Bhadradaru, Dhupavrikshaka, Manojna, Putikashtha, Saralan, Surabhidaruka.

Distribution : Himalayas from Kashmir-Bhutan and in the Siwalik hills between 450-2,400m.

Reproductive Cycle : September-December.

Part Used : Resin.

Folk Use : Resin used as an ingredient of a plaster for infantile paralysis.

Reference : Khare (2004).

Piper betle **L.**

Syn.: *Chavica betle* Miq.; *C. chuvya* Miq.; *C. siriboa* L.; *Piper chavya* Ham.; *P. peepuloides* Wall.; *P. siriboa* L.

Family: Piperaceae

Eng., H. and Sans. Names : *Eng.*: Betel Leaf Pepper, Betel Leaf Vine, Betel Pepper;

H.: Pan, Tambuli;

Sans.: Bhakshyapatra, Divabhishta, Nagavalli, Saptashira, Tambula, Vitika.

Distribution : Found in hotter and damper regions of India.

Reproductive Cycle : May-September.

Part Used : Leaf.

Folk Uses : Juice of leaves valuable in cough, dysponea and indigestion. Warm juice of leaves prescribed for indigestion of children. Leaves warmed and smeared with oil applied in layers over chest in catarrhal and pulmonary affections. Leaves toasted in castor oil placed over mumps. Stalk of leaf smeared with oil introduced into rectum in constipation and tympanitis of children.

References : Dhiman (2003), Satapathy (2008), Watt (1972).

***Piper hongkongense* C. DC.**

Syn.: *P. puberulum* (Benth) Maxim.

Family: Piperaceae

Distribution : Common in disturbed forests, along streams, beach, thickets, forest margins, plantation and waste areas from sea level to over 1,000m.

Reproductive Cycle : March-July.

Parts Used : Stem and Leaf.

Folk Use : Preparation made from leaf and stem used for convulsion in children.

Reference : Prajapati *et al.* (2006).

***Piper longum* L.**

Syn.: *P. latifolium* Hunter.; *P. sarmentosa* Wall.

Family: Piperaceae

Eng., H. and Sans. Names : *Eng.*: Indian Long Pepper;

H.: Gazpipal, Pipal, Piplamul, Pipli;

Sans.: Kana, Pippali, Shaundi, Upakalya.

Distribution : Distributed in hotter parts of India from C. Himalayas to Assam, Khari and Mikir hills, lower hills of W. Bengal and evergreen forests of W. Ghats from Konkan-Kerala; also recorded from Nicobar islands.

Reproductive Cycle : Flowering in rainy season and fruiting in Autumn.

Part Used : Fruit.

Folk Uses : Fruits used as vermifuge and for colic in children.

Reference : Chatterjee and Pakrashi (1997).

***Pisonia aculeata* L.**

Syn.: *P. villosa* Poir.

Family: Nyctaginaceae

Eng. Names : *Eng.*: Cockspur, Devil's Claws, Pullback.

Distribution : Scandent shrubs found along the sea coasts in Peninsular India, from Karnataka-Orissa and Andamans.

Reproductive Cycle : Flowering in January-June.

Part Used : Leaf.

Folk Use : Leaf juice mixed with pepper given to children for pulmonary diseases.

References : Agarwal (2003), Ambasta (1986), Kirtikar and Basu (1984).

***Pistacia integerrima* Stewart**

Syn.: *Rhus integerrima* Wall.

Family: Anacardiaceae

Eng., H. and Sans. Names : *Eng.*: Crab's Claw;

H.: Kakra, Kakrashingi;

Sans.: Ghosha, Karkata, Karkata sringi, Shiknari, Vakra.

Distribution : Himalayas, Garhwal, Kumaon, Bengal, Assam from 350- 2,400m.

Reproductive Cycle : March-May.

Part Used : Galls.

Folk Uses : Powdered galls and seeds of *Raphanus sativus* prescribed for whooping cough and asthma in children. Galls also used in bronchial troubles, vomiting, teething and fever problems in children.

References : Chatterjee and Pakrashi (1997), Khare (2004), Kirtikar and Basu (1984), Prajapati *et al.* (2006), Watt (1972).

***Plantago major* L.**

Syn.: *P. asiatica* L.; *P. erosa* L.; *P. longiscopa* Jacq.; *P. maxima* Ruching.; *P. officinarum* Crantz.

Family: Plantaginaceae

Eng., H. and Sans. Names : *Eng.*: Englishman's Foot, Great Plantain, Ribgrass, Waybroad;

H.: Isbgol, Luhuriya;

Sans.: Ashvagola.

Distribution : Temperate and alpine Himalayas, from Peshawar and Kashmir-Bhutan, 2,700m; W. Tibet, 3,350-4,000m, Khasia mountains, 1,350-1,700m, Myanmar and Malacca.

Reproductive Cycle : April-October.

Part Used : Leaf.

Folk Use : Paste of leaves applied on chest of 3-4 months child to cure cold.

Reference : Hosagoudar and Henry (1996c).

Plantago ovata Forsk.

Syn.: *P. decumbens* Forsk.; *P. ispaghul* Roxb.; *P. lanata* Wall.

Family: Plantaginaceae

Eng., H. and Sans. Names : *Eng.*: Blond Psyllium, Ispaghula, Spogel Seeds;

H.: Isabgal, Isabgul, Issufgul;

Sans.: Ishadgola, Snigdhabijah, Snigdhajirakah.

Distribution : Cutivated in plains of N. Gujarat, Punjab and U.P.

Reproductive Cycle : January-April.

Part Used : Seed.

Folk Use : Roasted seeds useful in irritation of bowels, dysentery, and intestinal ulcers of children. Seed husks used in diarrhoea and constipation in children.

References : Chatterjee and Pakrashi (1997), Dey and Bahadur (1973), Kirtikar and Basu (1984), Watt (1972).

Platycodon grandiflorum A. DC.

Syn.: *Campanula grandiflora* Jacq.

Family: Campanulaceae

Eng. Names : *Eng.*: Ballon Flower, Chinese Bell Flower, Japanese Bell Flower.

Distribution : Native of China and Japan; edges of ponds and streams or in swamps in India, China, Japan, and Korea.

Reproductive Cycle : July-September.

Part Used : Rhizome.

Folk Use : Used in fever due to measles.

Reference : Prajapati *et al.* (2006).

Plectranthus amboinicus (Lour.) Spreng

Syn.: *Coleus amboinicus* Lour.; *C. carnosus* Hassk.; *C. crassifolius* Benth.

Family: Lamiaceae

Eng., H. and Sans. Names : *Eng.*: Country Borage, Indian Borage;

H.: Amroda, Pathorchur, Pattaajavain;

Sans.: Parnayavanu, Pashanabhedi.

Distribution : Cultivated throughout India, specially in Rajasthan and S. India; also wild.

Reproductive Cycle : August-November.

Part Used : Leaf.

Folk Uses : Extract of leaves given as a cure for cough, cold and pneumonia. Juice of leaves given with sugar to children in colic.

References : Chatterjee and Pakrashi (1997), Jadhav (2009), Kirtikar and Basu (1984), Watt (1972).

Plectranthus barbatus **Andrews**

Syn.: *Coleus barbatus* Benth.

Family: Lamiaceae

Eng., H. and Sans. Names : *Eng.*: Indian Coleus;

H.: Makandi;

Sans.: Forskohlii.

Distribution : Sub-tropical Himalayas, Nepal and Kumaon, ascending to 2,650m, Bihar, top of Parusnath.

Reproductive Cycle : Flowering during August-October.

Part Used : Leaf.

Folk Uses : Leaf extract taken internally with pepper powder for cold, cough and continuous vomiting to children. Of the 21 tablets made from its seven leaves grounded with seven black peppers, two tablets given to children suffering from diarrhoea with vomiting tendency.

References : Satapathy (2008), Subramani and Goraya (2003).

Pleurospermum angelicoides **(Wall. *ex* DC.) Benth. *ex* C.B.Clarke**

Syn.: *Pterocyclus angelicoides* Wall *ex* DC.

Family: Apiaceae

Distribution : Kashmir, Kumaon, Rutan river, Nepal.

Reproductive Cycle : June-September.

Part Used : Root.

Folk Use : Powdered roots, cumin seeds and black pepper given to children suffering from dysentery caused due to cold.

Reference : Nautiyal *et al.* (2003).

Pogostemon benghalense **(Burm.f.) Kuntze.**

Syn.: *Mentha fruticulosa* Roxb.; *Origanum benghalense* Burm.

Family: Lamiaceae

Eng. and H. Names	:	*Eng.*: Bengal Pogostemone; *H.*: Juilata, Phangla.
Distribution	:	Found throughout most of Peninsular India, Bihar, W. Bengal and in W. Himalayas to about 1,500m.
Reproductive Cycle	:	February-April.
Part Used	:	Root.
Folk Use	:	Root paste applied on body of children to alleviate fever.
References	:	Aminuddin and Girach (1996), Parrotta (2001).

Polianthes tuberosa **L.**

Syn.: *Agave polianthes* Thiede and Eggli.; *Polianthes gracilis* Link.; *Tuberosa amica* Medik.

Family: Agavaceae

Eng., H. and Sans. Names	:	*Eng.*: Tuberose; *H.*: Gulcheri, Gulshabbo; *Sans.*: Rajnigandha.
Distribution	:	Cultivated in gardens almost throughout India.
Reproductive Cycle	:	Throughout greater parts of the year.
Part Used	:	Bulbs.
Folk Use	:	Paste of bulbs rubbed with turmeric and butter applied over red pimples in infants.
References	:	Anonymous (1969), Kirtikar and Basu (1984), Watt (1972).

Polycarpon prostratum **(Forsk.) Asch. and Schweinf.**

Syn.: *Alsine prostrata* Forssk.; *Spergula prostrata* (Forssk.) D. Dietr.

Family: Caryophyllaceae

H. Names	:	*H.*: Sureta.
Distribution	:	Fields and wasteplaces throughout hotter parts of India.
Reproductive Cycle	:	April-June.
Part Used	:	Leaf.
Folk Use	:	Infusion of roasted leaves given in measles.
References	:	Agarwal (2003), Ambasta (1986), Anonymous (1969).

***Polygonum aviculare* L.**

Syn.: *Avicularia vulgaris* Didr.

Family: Polygonaceae

Eng., H. and Sans. Names : *Eng.*: Allseed, Bian Xu, Blood Wort, Doorweed, Hogweed, Iron-grass, Knotgrass, Knot Wort, Mantie, Pig-weed, Red Legs, Sparrow;

H.: Bannalia, Ban-natia, Hunraj, Machoti;

Sans.: Nisomali.

Distribution : Kashmir-Kumaon, 2,000-4,000m.

Reproductive Cycle : June-October.

Parts Used : Whole Plant and Rhizome.

Folk Uses : Infusion of plant helps in summer complaints of children. Rhizome used in diarrhoea.

References : Karnick (1994a), Kirtikar and Basu (1984).

***Polygonum chinensis* L.**

Syn.: *Coccoloba crispata* Ham.; *Polygonum auriculatum* Meissn.; *P. brachiatum* Poir.; *P. corymbosum* Wild.; *P. cymosum* Roxb.; *P. ovatum* Heyne.; *P. patens* Don.

Family: Polygonaceae

Eng. and H. Names : *Eng.*: Chinese Knotweed;

H.: Ameta.

Distribution : All over India, ascending upto 3,100m in Himalayas.

Reproductive Cycle : January-May.

Part Used : Leaf.

Folk Use : Leaves taken internally in smallpox.

Reference : Sharma (2004).

***Polyscias fruticosa* (L.) Harms.**

Syn.: *Aralia tripinnata* Blanco.; *Panax fissus* W. Bull.; *P. fruticosus* L.

Family: Araliaceae

Eng. Names : *Eng.*: Aralia Ming, Narrwo Leaf Arelia.

Distribution : Cultivated in gardens and possibly naturalized in some areas from sea level to 500m.

Reproductive Cycle : August-December.

Part Used : Leaf.

Folk Use	:	Decoction of leaves used in tonsillitis.
Reference	:	Prajapati *et al.* (2006).

Pometia pinnata **J.R. Forst. and G. Forster**

Syn.: *Irina tomentosa* Blume.; *Pometia tomentosa* (Blume) Teijsm. and Binn.

Family: Sapindaceae

Eng. and H. Names	:	*Eng.*: Fijian longan, Island lychee; *H.*: Kasai.
Distribution	:	Distributed in Andamans and all over Malaysia, common in lowland forest, forest edges, open woodlands, lava flows and often cultivated in villages.
Reproductive Cycle	:	Flowering: July-October, and Fruiting 3 or 4 months later.
Part Used	:	Leaf.
Folk Use	:	Infusion of leaves rubbed onto head of infants or given internally to treat unclosed fontanelles.
Reference	:	Prajapati *et al.* (2006).

Pongamia pinnata **(L.) Pierre**

Syn.: *Dalbergia arborea* Willd.; *Derris indica* (Lamk.) Bennet.; *Pongamia glabra* Vent.

Family: Fabaceae

Eng., H. and Sans. Names	:	*Eng.*: Indian Beech, Karanj, Pongam Oil Tree; *H.*: Kanji, Karanji, Papar; *Sans.*: Angaravalli, Badhaphala, Chirabilva, Gaura, Karanjah, Vishari, Vrittapornia.
Distribution	:	C. and E. Himalayas-Sri Lanka and Malacca, especially near the coast.
Reproductive Cycle	:	March-June.
Parts Used	:	Whole Plant, Root, Bark, Fruit and Seed.
Folk Uses	:	Plant used in itching. Roots tied as an amulet to the waist of children to cure syphilis. Paste of basal portion of tender leaves of 'Ananas', mature leaves of *Musa balbisiana*, leaves of *Erythrina variegata* and bark of *Pongamia pinnata* in equal quantities given for roundworms in children. Fruits tied to the waist of children as a remedy for diphtheria and to neck or waist as an amulet for relief from severe cough. Paste of seeds

smeared with castor oil applied externally on throat region for treating whooping cough.

References : Chatterjee and Pakrashi (1997), Dwarakan and Ansari (1996), Gogoi and Borthakur (2001), Hosagoudar and Henry (1996a), Pal *et al.* (2000), Ravikumar and Sankar (2003), Sen and Behera (2007), Subramaniam (2000).

***Populus alba* L.**

Syn.: *P. major* Mill.

Family: Salicaceae

Eng. Names : *Eng.*: Able, White Poplar.

Distribution : N.W. Himalayas (1,200-3,000m), from Kashmir and Pakistan.

Reproductive Cycle : May-September.

Part Used : Leaf Buds.

Folk Uses : Used in laryngitis with aphonia respiratory affections.

Reference : Karnick (1994a).

***Portulaca oleracea* L.**

Syn.: *P. consanguinea* Schltdl.; *P. fosbergii* Poelln.; *P. retusa* Engelm.

Family: Portulacaceae

Eng., H. and Sans. Names : *Eng.*: Common Purslane, Garden Purslane, Pot Purslane, Purslane;

H.: Baralunia, Chhotalunia, Khurfa, Khursa, Kurfa, Lunia;

Sans.: Lona, Loni, Lonika, Lunia.

Distribution : All over India, upto 1,700m in the Himalayas.

Reproductive Cycle : July- August.

Part Used : Leaf.

Folk Use : Leaves used in stomatitis of children.

Reference : Kirtikar and Basu (1984).

***Pothos scandens* L.**

Syn.: *Batis hermaphrodita* Blanco.; *Podospadix angustifolia* Raf.; *Tapanava indica* Raf.

Family: Araceae

Eng. and Sans. Names : *Eng.*: Peacock's Tail;

Sans.: Vrikshadani.

Distribution	:	Throughout India as a creeping herb, gardens and houses for decoration.
Reproductive Cycle	:	March-April.
Part Used	:	Leaf.
Folk Use	:	Powdered leaves applied on smallpox pustules.
References	:	Agarwal (2003), Ambasta (1986), Anonymous (1969), Jain (1991), Kirtikar and Basu (1984), Mudgal *et al.* (1999), Rawat and Chowdhury (1998).

Pouzolzia zeylanica **(L.) Benn. and R.Br.**

Syn.: *P. indica* (L.) Gaud. Benn.; *Urtica glomerata* Klein *ex* Willd.

Family: Urticaceae

Eng. and H. Names	:	*Eng.*: Graceful Pouzolzsbush; *H.*: Dumpa, Piparisari.
Distribution	:	Throughout tropical and sub-tropical India, from Indus eastwards to Myanmar and southward to Malacca.
Reproductive Cycle	:	July-December.
Parts Used	:	Whole Plant and Leaf.
Folk Uses	:	Plant used for inflammation of eyes and convulscence in children. Three leaves pounded with little amount of salt applied locally in sprain of babies.
References	:	Barua *et al.* (2000), Pal *et al.* (2000).

Premna tomentosa **Willd.**

Syn.: *P. cordata* Blanco.; *P. flavescens* Juss.

Family: Lamiaceae

Eng. Name	:	*Eng.*: Bastard Teak.
Distribution	:	Deccan peninsula and Sri Lanka, from the Circas and C. regions southwards; Peninsular India and Bihar, upto 1,200m.
Reproductive Cycle	:	January-April (Flowering) and Fruiting from February onwards.
Part Used	:	Leaf.
Folk Use	:	Paste of tender leaves grounded with garlic applied on face to relieve nervous disorders in children.
Reference	:	Balasubramanian and Prasad (1996).

Primula veris **L.**

Family: Primulaceae

Eng. Names	:	*Eng.*: Cooslip, Cowslip.
Distribution	:	Grown in Indian gardens.
Reproductive Cycle	:	Flowering during April-May.
Part Used	:	Flower.
Folk Uses	:	Flowers said to be sedative and recommended for overactivity and sleeplessness in children.
Reference	:	Prajapati *et al.* (2006).

Prinsepia utilis **Royle (Pl. 11B)***

Family: Rosaceae

Eng. and H. Names	:	*Eng.*: Himalayan Prinsepia;
		H.: Bekkra, Bhekal, Cherara, Dhatela, Jhatela, Karanga, Krungora.
Distribution	:	Dry rocky hills of temperate Himalayas: 1,350-2,700m, from Hazara-Sikkim ascending to 3,000m and Bhutan; Khasia mountains, 1,700-2,000m.
Reproductive Cycle	:	January-February.
Part Used	:	Seed.
Folk Use	:	Mild heated oil used for massaging newly born child.
Reference	:	Nautiyal *et al.* (2003).

Prosopis cineraria **(L.) Druce.**

Syn.: *Mimosa cineraria* L.; *Prosopis spicata* Burm.; *P. spicigera* L.

Family: Fabaceae

Eng., H. and Sans. Names	:	*Eng.*: Algarobo, King of Desert, Wonder Tree;
		H.: Chhikura, Chhokara, Jand, Khejri, Sami;
		Sans.: Ishana, Ishta, Kananari, Lakshmi, Medhya, Surabhi.
Distribution	:	Punjab, Rajputana, Gujarat.
Reproductive Cycle	:	November-April.
Part Used	:	Whole Plant.
Folk Uses	:	Plant used in stomach pain, dysentery and colic pain.
Reference	:	Pal *et al.* (2000).

Prosopis farcta **(Banks and Sol.) J.F. Macbr.**

Syn.: *Acacia stephaniniana* Willd.; *Prosopis aculeata* Koenig.

Family: Fabaceae

Eng. Names	:	*Eng.*: Dwarf Mesquite, Syrian Mesquite.
Distribution	:	Punjab.
Reproductive Cycle	:	Flowering in April-September.
Part Used	:	Leaf.
Folk Uses	:	Juice of leaves along with boiled wheat given to children in colic, indigestion and flatulence.
Reference	:	Mitaliya and Bhatt (2003).

Prosopis glandulosa **Torr.**

Syn.: *Algarobia glandulosa* (Torr.) Cooper.; *Prosopis chilensis* Sensu auct.

Family: Fabaceae

Eng. Name	:	*Eng.*: Honey Mesquite.
Distribution	:	Introduced in India, naturalized in Punjab and Rajasthan.
Reproductive Cycle	:	January-April.
Part Used	:	Root.
Folk Use	:	A tea made from root given for umbilical hernia in children.
Reference	:	Bhattacharjee (2001).

Prunus persica **(L.) Stokes. (Pl. 11C)***

Family: Rosaceae

Eng., H. and Sans. Names	:	*Eng.*: Nectarine, Peach Tree, Wild Peach; *H.*: Aru; *Sans.*: Aaluka, Aru.
Distribution	:	Cultivated in cooler parts of India, upto 3,350m in N.W. Himalayas.
Reproductive Cycle	:	March-September.
Parts Used	:	Leaf, Flower, Seed and Bark.
Folk Uses	:	Bark, leaf, flowers and oil from seeds used in gastric irritation of children.
References	:	Kirtikar and Basu (1984), Parmar and Kaushal (1982), Rajan *et al.* (2000).

Prunus serotina Ehrh.

Syn.: *Cerasus capollin* Ser *ex* DC.

Family: Rosaceae

Eng., H. and Sans. Names : *Eng.*: Dwarf Cherry, Wild Black Cherry, Wild Cherry; *H.*: Bare (Bayr); *Sans.*: Ajamoda, Padmaka.

Distribution : Found in temperate regions.

Reproductive Cycle : May-September.

Part Used : Bark.

Folk Use : Powdered bark administered against measles.

Reference : Bhattacharjee (2001).

Psidium guajava L. (Pl. 11D)

Syn.: *P. pyriferum* L.

Family: Myrtaceae

Eng., H. and Sans. Names : *Eng.*: Common Guava, Guava; *H.*: Am, Amrud, Amrut, Bihi, Safed Safari; *Sans.*: Aprithaktvacha, Mausala, Parevata.

Distribution : Naturalized throughout India, excepting N.W. parts of Punjab.

Reproductive Cycle : April-July.

Parts Used : Shoot, Leaf, Flower, Root, Stem Bark and Root Bark.

Folk Uses : Paste of tender shoots pounded with that of *Punica granatum* in 1:1 proportion given 3-4 times a day by Girijanas to cure cough. Tincture of leaves rubbed on spine of children suffering from convulsions. Also, decoction of leaves root bark used in prolapsed anus of children. Flowers used in infantile diarrhoea. Root, root bark and stem bark all used as an astringent against diarrhoea. Two tsp mixture paste of bark of *Terminalia bellirica, Spondias pinnata* and *Psidium guajava* in equal quantities with 3 fruits each of *Terminalia chebula* and *Emblica officinalis* along with a few drops of alkali solution given daily to children for 3 days to cure cholera.

References : Anonymous (1969), Borthakur (1993), Chatterjee and Pakrashi (1997), Dey and Bahadur (1973), Farooq (2005), Gogoi and Borthakur (2001), Kirtikar and Basu (1984),

Kshirsagar and Singh (2007), Parrotta (2001), Rawat and Chowdhury (1998), Subramaniam (2000), Upadhyay and Singh (2005), Watt (1972).

***Psilotum nudum* (L.) P. Beauv.**

Syn.: *P. triquetrum* Sw.

Family: Psilotaceae

Eng. Names : *Eng.*: Skeleton Fork Fern, Whisk Fern.

Distribution : Throughout India, ascending upto 2,000m in hilly regions and Andaman and Nicobar Islands.

Parts Used : Herb and Spores.

Folk Uses : Decoction of herb administered to children suffering from thrush. Oily spores given to infants to arrest diarrhoea.

References : Agarwal (2003), Ambasta (1986), Anonymous (1969), Das (2003), Jha *et al.* (2003).

***Pteris multifida* Poir.**

Syn.: *P. serrulata* L.f.

Family: Pteridaceae

Eng. Names : *Eng.*: Spider Brake, Spider Fern.

Distribution : Found in N.W. Himalayas, Gangetic plains.

Parts Used : Fronds and Rhizome.

Folk Uses : Paste of fronds and rhizomes along with sesame oil applied on skin affections of infants.

References : Anonymous (1969), Singh (2000), Singh and Viswanathan (1996).

***Pterolobium hexapetalum* (Roth) Sant. and Wagh.**

Syn.: *Caesalpinia lingulata* DC.; *Pterolobium indicum* A. Rich.

Family: Fabaceae

Eng. Names : *Eng.*: Camp Siege, Indian Redwing.

Distribution : W. Peninsula, S. India.

Reproductive Cycle : Flowering in March-April.

Part Used : Stem Bark.

Folk Use : 1 tsp stem bark decoction given orally twice a day till cure to infants for whooping cough.

Reference : Rao and Henry (1996).

Pterospermum acerifolium **(L.) Willd.**

Syn.: *P. acerifolia* L.

Family: Malvaceae

Eng., H. and Sans. Names : *Eng.*: Maple-Leaved Bayur Tree;

H.: Kanakchampa, Kaniar, Kathachampa;

Sans.: Karnikara, Mushukunda, Padapotpala.

Distribution : Sub-Himalayan tracts up to 1,200m, extending southwards to W. Ghats. Also, cultivated along roadsides.

Reproductive Cycle : March-August.

Parts Used : Whole Plant, Flower and Bark.

Folk Uses : Plant used in diarrhoea, dysentery, bowel complaints and as anthelmintic. Ash of its bark and flowers mixed with that of *Mallotus philippensis* applied externally to smallpox eruptions.

References : Agarwal (2003), Anonymous (1969), Khare (2004), Kirtikar and Basu (1984), Lalramnghinglova (1996), Pal *et al.* (2000), Tiwari and Tiwari (1996), Watt (1972).

Pterospermum canescens **Roxb.**

Syn.: *P. suberifolium* (L.) Lamk. Non Willd.

Family: Sterculiaceae

Eng., H. and Sans. Names : *Eng.*: Cork-Leaved Bayur;

H.: Muchkand;

Sans.: Bahupatra, Chitraka, Sudala, Supushpa, Muchukunda.

Distribution : W. Peninsula, Chennai, Pondicherry, N. Circas, Deccan, Coimbatore.

Reproductive Cycle : Fruiting during August-April.

Parts Used : Bark and Flower.

Folk Use : Ash of its bark and flowers mixed with 'Kamala' powder used on smallpox eruptions in Konkan regions.

References : Anonymous (1969), Kirtikar and Basu (1984).

Pueraria tuberosa **(Roxb. *ex* Willd.) DC.**

Syn.: *Hedysarum tuberosum* Roxb.

Family: Fabaceae

Eng., H. and Sans. Names : *Eng.*: Indian Kudzu;

		H.: Patalkand, Sural, Tirra, Vidari Kaud;
		Sans.: Bhumi Kusmanda, Gajeshta, Triparna, Vidari.
Distribution	:	W. Himalayas-Sikkim, lower hills of Punjab, hilly tracts of Bengal, S. India.
Reproductive Cycle	:	February-August.
Part Used	:	Whole Plant.
Folk Use	:	Plant used in polio.
Reference	:	Pal *et al.* (2000).

Punica granatum **L. (Pl. 12A)***

Syn.: *P. nana* L.

Family: Punicaceae

Eng., H. and Sans. Names	:	*Eng.*: Pomegranate Tree;
		H.: Anar, Dhalim, Dharimb;
		Sans.: Bijapura, Dadima, Granat, Pindapushpa, Vrittaphala.
Distribution	:	Cultivated throughout India, from Balkans- Himalayas, probably wild in N.W.
Reproductive Cycle	:	January-March.
Parts Used	:	Root, Leaf, Flower Buds and Fruit.
Folk Uses	:	Decoction of roots used to expel tape-worm in children. Chronic diarrhoea can be cured by 'sharbat' of green leaves. Dried flower buds used in infantile diarrhoea and dysentery. Juice extracted from mature fruit given to children to cure diarrhoea. Fruit rind rubbed with 50ml old rice meal water consumed once a day for 3 days to control loose motion in infants. Paste of peel of its fruit and root of 'Kashmal' (*Berberis lycium*) applied on palate to ease process of emergence of teeth. 0.2g powder of one small-sized seed of opium kept in fruit of *Punica granatum* and charred on low fire made of coal given to children for treating dysentery.
References	:	Ansarali and Sivadasan (2009), Bhogaonkar and Ahmad (2007), Chatterjee and Pakrashi (1997), Chauhan (1999), Dhiman (2003), Kirtikar and Basu (1984), Tripathi (2001), Viswanathan *et al.* (2006).

Quercus incana Bartram

Syn.: *Dryopsila cinerea* (Michx.) Raf.; *Quercus oligodes* Raf.

Family: Fagaceae

Eng. and H. Names : *Eng.*: Ban Oak, Grey Oak, Himalayan White Oak, Kumaon Oak;

H.: Ban, Banj.

Distribution : Found in Kashmir and W. Himalayas upto Nepal between 1,000-2,400m.

Reproductive Cycle : April-August.

Parts Used : Bark, Fruit and Seed.

Folk Uses : Bark used in tonsillitis. Fruit and seeds used for indigestion and diarrhoea in children.

References : Ambasta (1986), Anonymous (1969), Karnick (1994a), Kirtikar and Basu (1984), Nadeem *et al.* (2001), Watt (1997).

Quercus infectoria G. Oliver.

Family: Fagaceae

Eng., H. and Sans. Names : *Eng.*: The Gall or Dyers Oak;

H.: Maja, Majuphul, Masuphal;

Sans.: Majuphul.

Distribution : Garhwal Himalayas and Nepal.

Reproductive Cycle : Flowering occurs in May.

Part Used : Galls.

Folk Use : Used as astringent in bowel complaints of children.

Reference : Watt (1972).

Quercus robur L.*

Syn.: *Q. pendunculata* Ehrh.

Family: Fagaceae

Eng. Names : *Eng.*: Common Oak, English or European Oak.

Distribution : Grows throughout the northern hemisphere in woods; introduced into India in hills, Nilgiris and H.P.

Reproductive Cycle : April-August.

Part Used : Bark.

Folk Use : Decoction of bark used in tonsillitis.

References : Karnick (1994b), Prajapati *et al.* (2006).

Radermachera xylocarpa **(Roxb.) Roxb.** ***ex*** **K. Schum.**

Syn.: *Bignonia xylocarpa* Roxb.

Family: Bignoniaceae

Eng. and H. Names	:	*Eng.*: Padri Tree; *H.*: Jaimangal, Kharsing, Paral.
Distribution	:	Throughout Deccan peninsula upto 1,500m, extending into parts of Gujarat, M.P., Bihar and Orissa.
Reproductive Cycle	:	Flowering during April-May and Fruiting after nearly one year.
Parts Used	:	Root and Leaf.
Folk Uses	:	Root juice given twice a day for 2-3 days (1/2 cup for children) to treat snake bite. Leaves smeared with castor oil tied around mumps.
References	:	Hosagoudar and Henry (1996a), Kumar and Jain (1998).

Ranunculus muricatus **L.**

Family: Ranunculaceae

Eng. Names	:	*Eng.*: Spiny Fruit Buttercup, Spiny Fruited Crowcup, Rough Fruited Buttercup.
Distribution	:	Swampy places, at low elevations in the Punjab, Himalayas and in the plains at Peshwar and Ambala.
Reproductive Cycle	:	Flowering in October.
Part Used	:	Flower.
Folk Use	:	Used in tonsils.
Reference	:	Jain (1991).

Rauvolfia serpentina **Benth.** ***ex*** **Kurz.**

Syn.: *Ophioxylon album* Gaertn.; *O. serpentium* L.; *O. trifoliatum* Gaertn.; *Rauvolfia trifoliata* (Gaertn.) Boill.

Family: Apocynaceae

Eng., H. and Sans. Names	:	*Eng.*: Rauvolfia Root, Serpentine, Serpentine Root; *H.*: Chandrabhaga, Chota-chand; *Sans.*: Chundrika, Sarpagandha.
Distribution	:	Found in tropical Himalayas, wasteplaces and shady forest in different parts of India from Punjab eastwards to Sikkim and Assam, also in some parts of C. India and W. Ghats.

Reproductive Cycle : June-August.

Part Used : Root.

Folk Uses : 2-3g of root decoction given to children to expel thread worms. Also, aqueous extract of roots given as an effective remedy against dysentery and stomachache.

References : Lalramnghinglova (2003), Panda (2007).

Remusatia vivipara (Roxb.) Schott.

Syn.: *Arum viviparum* Roxb.; *Caladium viviparum* Lodd.; *Remusatia bulbifera* Vilm.

Family: Araceae

Eng. Name : *Eng.*: Shield Leaf Ariopsis.

Distribution : Sub-tropical Himalayas, (700-1,000m) from Kumaon-Sikkim, Khasia hills, Myanmar, Bihar, Parasnath, W. Ghats and Sri Lanka.

Reproductive Cycle : Flowering in May-June.

Part Used : Tuber.

Folk Use : Used in mumps.

Reference : Jain (1991).

Rhazya stricta Dcne.

Family: Apocynaceae

Eng. and H. Names : *Eng.*: Dogbane;

H.: Sundwar.

Distribution : Found in Punjab, cultivated in gardens, Sind, trans-Indian territory, Peshawar.

Reproductive Cycle : December-May.

Part Used : Leaf.

Folk Use : Juice of leaves given to children with milk in skin eruptions.

References : Agarwal (2003), Kirtikar and Basu (1984).

Rheum australe D. Don.

Syn.: *R. emodi* Wall *ex* Meissn.

Family: Polygonaceae

Eng., H. and Sans. Names : *Eng.*: Himalayan Rhubarb, Indian Rhubarb;

H.: Dolu, Hindirevandchini, Revatchini;

Sans.: Gandhini, Pitamulika, Revatchini.

Distribution : Distributed in Himalayas from Kashmir-Sikkim between 3,300-5,200m.

Reproductive Cycle : August-September.

Part Used : Root.

Folk Uses : Powdered roots used in mumps, tonsillitis and for infant's stomach problems.

References : Chauhan (1999), Sharma (2003), Prajapati *et al.* (2006).

Rheum officinalis Baillon

Family: Polygonaceae

Eng. and Sans. Names : *Eng.*: Chinese Rhubarb, Indian Rhubarb;

Sans.: Amlavetasa.

Distribution : Native of China and Tibet, cultivated in Europe, India and Pakistan.

Reproductive Cycle : June-August.

Part Used : Rhizome.

Folk Uses : Rhizome used especially during dentition and for body smelling sour even after a bath in children.

Reference : Prajapati *et al.* (2006).

Rheum spiciforme Royle.

Syn.: *R. moorcroftianum* Royle.; *R. reticulatum* Losinsk.; *R. scaberrimum* Lingelsh.

Family: Polygonaceae

English and Sanskrit Names: *Eng.*: Spiked Rhubarb;

Sans.: Amlaparni.

Distribution : Growing commonly between 2,800-3,500m on alpine zones at Chandrakhani, Malana, Mantalai slopes.

Reproductive Cycle : June-August.

Part Used : Root.

Folk Uses : Root paste used against mumps and tonsils.

Reference : Sharma *et al.* (2003b).

Rhododendron arboreum Sm. (Pl. 12B)*

Family: Ericaceae

Eng. and H. Names : *Eng.*: Rose Tree, Tree Rhododendron;

H.: Burans.

Distribution	:	Temperate Himalayas, 1,500-3,600m, Manipur, Meghalaya and Nilgiri hills.
Reproductive Cycle	:	Flowering during March-April/June-September.
Part Used	:	Flower.
Folk Use	:	Flowers used for infantile diarrhoea in Sikkim.
Reference	:	Srivastava *et al.* (1987).

Rhus coriaria **L.**

Family: Anacardiaceae

Eng., H. and Sans. Names	:	*Eng.*: The Sumach Tree of Europe; *H.*: Mahi, Mutchli, Shumuk, Tatrak; *Sans.*: Tandidik.
Distribution	:	Native of Persia, Syria, Palestine and S. Europe.
Reproductive Cycle	:	April-July.
Part Used	:	Leaf.
Folk Uses	:	Poultice of leaves applied to abdomen in children suffering from diarrhoea or dysentery.
Reference	:	Watt (1972).

Rhus semialata **Murr.**

Syn.: *Schinus indicus* Burm.f.

Family: Anacardiaceae

Eng. and H. Names	:	*Eng.*: Chinese Sumac; *H.*: Tatri.
Distribution	:	Temperate Himalayas, from Banahal-Sikkim, 1,000-2,000m, Khasia mountains, 1,000-1,700m.
Reproductive Cycle	:	April-August.
Part Used	:	Root.
Folk Use	:	1/2 tsp of powdered roots mixed with double the amount of water given twice daily after food for children in colic.
Reference	:	Ahmed and Borthakur (2005).

Rhus succedanea **L.**

Syn.: *Rhus succedanea L.*

Family: Anacardiaceae

Eng., H. and Sans. Names	:	*Eng.*: Wax Tree, Wild Varnish Tree;

		H.: Kakra-singi;
		Sans.: Karkata Sringee.
Distribution	:	Found in mixed forests of Himalayas upto 2,500m from Punjab-Assam and Khasi hills.
Reproductive Cycle	:	May-September.
Parts Used	:	Galls and Root.
Folk Uses	:	Galls prescribed to children suffering from diarrhoea, dysentery and vomiting. Roots used in urinary-incontinence.
References	:	Chatterjee and Pakrashi (1997), Farooq (2005), Karnick (1994a), Kirtikar and Basu (1984).

Ricinus communis L. (Pl. 12C)*

Syn.: *Cataputia major* Ludw.; *Ricinus africanus* Mill.; *R. armatus* Andr.; *R. badius* Rchb.

Family: Euphorbiaceae

Eng., H. and Sans. Names	:	*Eng.*: Castor, Castor Oil Plant, Castor Seed, Palma Christi;
		H.: Arand, Erandi;
		Sans.: Erand, Eranda, Vatari.
Distribution	:	Cultivated almost throughout India in gardens and fields, also runs wild in wastelands, ascending to 1,500m.
Reproductive Cycle	:	Whole year.
Parts Used	:	Whole Plant, Leaf and Seed.
Folk Uses	:	Plant used in nightblindness of children. Leaves with warm ghee tied on belly in stomach swelling of children, also given as a mild laxative. Paste of seed kernels, 'neem' leaves and turmeric powder applied externally to cure smallpox and chickenpox after bath to avoid pustule formation. Oil obtained from crushed seeds used in diarrhoea and as mild laxative for children.
References	:	Dey and Bahadur (1973), Nadeem *et al.* (2001), Pal *et al.* (200), Parrotta (2001), Subramaniam (2000), Verma and Chauhan (2006).

Rorippa sarmentosa (Soland. *ex* Forst.) J.F.Macbr.

Syn.: *Nasturtium sarmentosum* O.E. Schulz.; *Rorippa elstonii* Hochr.

Family: Brassicaceae

Eng. Name	:	*Eng.*: Polynesian Cress.

Distribution : Common in damp wasteareas from sea-level-1,000m elevation.

Reproductive Cycle : Throughout the year.

Part Used : Whole Plant.

Folk Use : Plant used for curing convulsion in children.

Reference : Prajapati *et al.* (2006).

Rosa centifolia **L.**

Family: Rosaceae

Eng., H. and Sans. Names : *Eng.*: Cabbage Rose, Hundred-leaved Rose, Pale Rose;
H.: Gulab, Gulabzurdi;
Sans.: Atikesara, Kharva, Satapatri, Saumyagandha.

Distribution : Cultivated throughout India.

Reproductive Cycle : June-July.

Part Used : Petals.

Folk Use : Syrup of petals given as a laxative to infants.

References : Kirtikar and Basu (1984), Watt (1972).

Rosa damascena **Mill. (Pl. 12D)**

Family: Rosaceae

Eng., H. and Sans. Names : *Eng.*: Damask Rose, Otto of Roses;
H.: Fasli Gulab, Gulab, Sudharg;
Sans.: Atimanjula, Lakshapushpa, Shatapatri.

Distribution : Common throughout India.

Reproductive Cycle : April-June.

Part Used : Flower.

Folk Uses : 'Gulkand' (preserve) made from petals considered laxative and also used for enlarged tonsils.

Reference : Ambasta (1986).

Roylea cinerea **(D.Don) Bail**

Syn.: *Ballota cinerea* D. Don.; *Phlomis calycina* Roxb.; *Roylea elegans* Wall.

Family: Lamiaceae

Eng. and H. Names : *Eng.*: Ashy Roylea;
H.: Patkarru, Titpatti.

Distribution : Subtropical W. Himalayas from Kashmir – Kumaon: 700-1,700m.

Reproductive Cycle	:	Flowering in September.
Part Used	:	Leaf.
Folk Use	:	Used as snuff in tonsils.
Reference	:	Jain (1991).

Rubia cordifolia L. Sensu Hook.f.

Syn.: *R. alata* Wall.; *R. chinensis* Regel and Maack.; *R. javana* DC.; *R. mitis* Miq.; *R. mungisth* Desv.; *R. munjista* Roxb.

Family: Rubiaceae

Eng., H. and Sans. Names	:	*Eng.*: Indian Madder; *H.*: Manjit, Manjith; *Sans.*: Aruna, Bhandi, Chhatra, Gandiri, Harini, Jingi, Kala-meshika, Manjistha, Rohini, Samanga, Tamrika, Vapra, Vikasa.
Distribution	:	Throughout hilly district of India from N.W. Himalayas eastwards, ascending to 2,650m and southwards to Sri Lanka and Malacca.
Reproductive Cycle	:	July-November.
Part Used	:	Root.
Folk Use	:	Root paste used against fever in children.
Reference	:	Prusti (2007).

Rubus idaeus L.

Family: Rosaceae

Eng., H. and Sans. Names	:	*Eng.*: European Raspberry, Indian Summer Raspberry; *H.*: Rasabharii; *Sans.*: Gauriphal.
Distribution	:	Cultivated in hill stations of S. India.
Reproductive Cycle	:	Flowering in March.
Parts Used	:	Leaf and Fruit.
Folk Use	:	Both used in tonsillitis.
Reference	:	Karnick (1994b).

Rubus moluccanus L.

Family: Rosaceae

Eng. and H. Names	:	*Eng.*: Black Berry, Broad-leaf bramble, Native Raspberry; *H.*: Katsol.

Distribution	:	W. Ghats, C. and E. and tropical Himalayas, Nepal, Sikkim, Myanmar, Assam, Sri Lanka and Malaya.
Reproductive Cycle	:	May-November.
Part Used	:	Fruit.
Folk Use	:	Considered a valuable remedy for nocturnal micturition in children by Malayans.
References	:	Kirtikar and Basu (1984), Rawat and Chowdhury (1998), Watt (1972).

Rubus niveus **Thunb. (Pl. 13A)**

Syn.: *R. bonatii* H.Lev.; *R. distans* D. Don.; *R. lasiocarpus* Sm.; *R. pinnatus* D. Don.

Family: Rosaceae

Eng., H. and Sans. Names	:	*Eng.*: Mahabaleshwar Raspberry, Mysore Raspberry, Raspberry;
		H.: Kala Hinsalu, Kale Hinure;
		Sans.: Gauriphal.
Distribution	:	Temperate Himalayas from Murree and Kashmir, 1,700-2,650m (ascending to 1,500m in Kumaon)-Sikkim, 1,300-3,350m.
Reproductive Cycle	:	April-June.
Part Used	:	Fruit.
Folk Use	:	Used in nocturnal micturition.
Reference	:	Kapur (1996b).

Rungia pectinata **(L.) Nees.**

Family: Acanthaceae

Eng. and Sans. Names	:	*Eng.*: Comb Rungia;
		Sans.: Pindi.
Distribution	:	Found throughout warmer parts of India upto 1,200m in Himalayas.
Reproductive Cycle	:	Flowering in February-November.
Parts Used	:	Whole Plant and Leaf.
Folk Use	:	Plant along with juice of leaves given 2 tsp daily in smallpox.
References	:	Agarwal (2003), Farooq (2005), Jain (1991), Parrotta (2001), Upadhyay and Singh (2005).

Rungia repens **(L.) Nees.**

Syn.: *Justicia repens* L.

Family: Acanthaceae

Eng., H. and Sans. Names : *Eng.*: Creeping Rungia;
H.: Kharmar, Kharmor;
Sans.: Parpata, Parpatakah, Parpatha.

Distribution : Plains and low hills throughout India, usually as a weed in moist habitats such as riverbanks, ditches, channels and bunds of paddy fields.

Reproductive Cycle : Throughout the year.

Part Used : Leaf.

Folk Use : Fresh bruised leaves mixed with castor oil applied to scalp to cure a scaly fungoid infection (*Tinea capitis*) that occurs in children.

Reference : Parrotta (2001).

Ruta chalepensis **Wall.**

Syn.: *R. angustifolia* Pers.

Family: Rutaceae

Eng., H. and Sans. Names : *Eng.*: Common Rue, Fringed Rue, Garden Rue;
H.: Pismaram, Sadab, Satari;
Sans.: Gucchapatra.

Distribution : Cultivated in Indian gardens.

Reproductive Cycle : Flowering in March-April.

Part Used : Leaf.

Folk Uses : Juice of leaves given to children for helminthic infections. Decoction of leaves used as fumigant in infant catarrh.

References : Agarwal (2003), Prajapati *et al.* (2006).

Ruta graveolens **L.**

Syn.: *R. hortensis* Mill.

Family: Rutaceae

Eng., H. and Sans. Names : *Eng.*: Ave Grace, Herb of Grace, Rue Plant;
H.: Pismarum, Sadab, Satari;
Sans.: Guchhapatra, Somalata, Vishopaha.

Distribution : Cultivated throughout India westwards to Canaries.

Reproductive Cycle : May-September.

Parts Used : Whole Plant and Leaf.

Folk Uses : Juice of plant given to children as a remedy for worms, bronchial catarrh and acute bronchitis. Bruised leaves given in infantile diarrhoea. Dried leaves used as a fumigatory for children suffering from catarrh. Leaf juice given to infants and children with convulsion and fits. Leaf juice mixed with breast milk in equal proportion given to infants 1 tsp thrice daily till recovery for curing coughs by Jenukurubas.

References : Kirtikar and Basu (1984), Kshirsagar and Singh (2007), Parrotta (2001), Watt (1972).

***Saccharum bengalense* Retz.**

Syn.: *S. ciliare* Anders.; *S. munja* Roxb.

Family: Poaceae

Eng., H. and Sans. Names : *Eng.*: Munj Sweetcane;
H.: Munja, Ramsar, Sarkanda;
Sans.: Bahupraja, Bana, Munja.

Distribution : Throughout the plains and low hills of India.

Reproductive Cycle : October-May.

Part Used : Bark.

Folk Use : Decoction of bark used in tonsillitis.

Reference : Chauhan *et al.* (2007).

***Salomonia cantoniensis* Lour.**

Syn.: *S. edentula* DC.

Family: Polygonaceae

Distribution : E. Bengal, Assam, Khasia mountains 0-1,350m, E. Peninsula.

Reproductive Cycle : August-November.

Part Used : Whole Plant.

Folk Use : Plant pulp rubbed around mouth against sprue in children.

Reference : Agarwal (2003).

***Salvadora persica* L.**

Syn.: *Cissus arborea* Forsk.; *Salvadora indica* Wight.; *S. wightiana* Planck.

Family: Salvadoraceae

Eng., H. and Sans. Names	:	*Eng.*: Mustard Tree, Salt Bush, Tooth Brush Tree; *H.*: Jhak, Kharjal, Pilu; *Sans.*: Brihatpilu, Maha vriksha, Pilu, Piluh.
Distribution	:	Drier parts of India, up to 500m.
Reproductive Cycle	:	March-July.
Part Used	:	Seed.
Folk Use	:	Seed oil used for massaging polio-affected limbs.
Reference	:	Kalpesh *et al.* (2004).

***Salvia officinalis* L.**

Family: Lamiaceae

Eng. and H. Names	:	*Eng.*: Garden Sage, Red Sage, Sage; *H.*: Salbiasetakuss.
Distribution	:	Native to Mediterranean region; introduced into some Indian Gardens.
Reproductive Cycle	:	Flowering in April-June.
Part Used	:	Leaf.
Folk Use	:	Leaves used for tonsillitis.
Reference	:	Kirtikar and Basu (1984).

***Sambucus nigra* L.**

Syn.: *S. graveolens* Willd.; *S. peruviana* Kunth.

Family: Caprifoliaceae

Eng. Names	:	*Eng.*: Common Elder, European Elder.
Distribution	:	Parks and gardens throughout India; cultivated to a small extent in botanical gardens in India, occur wild in Kangra district.
Reproductive Cycle	:	February-June.
Parts Used	:	Leaf and Flower.
Folk Uses	:	Leaves used in dry coryza of infants. Flowers used as a laxative to infants.
Reference	:	Rajan *et al.* (2000).

Sansevieria roxburghiana **Schult and Schult.f.**

Syn.: *S. zeylanica* Roxb.

Family: Dracaenaceae

Eng., H. and Sans. Names	:	*Eng.*: Indian Bowstring Hemp; *H.*: Marul, Murva; *Sans.*: Marura, Maurvi, Murva.
Distribution	:	Grows in coromandel coast of India.
Reproductive Cycle	:	December-September.
Part Used	:	Shoot.
Folk Use	:	Juice of tender shoots given to children to clear viscid phlegm in throat.
References	:	Agarwal (2003), Ambasta (1986), Kirtikar and Basu (1984), Subramaniam (2000), Watt (1972).

Santalum album **L.**

Syn.: *Sandalum album* Rumph.; *S. myrtifolium* Roxb.; *S. verum* L.

Family: Santalaceae

Eng., H. and Sans. Names	:	*Eng.*: Sandal Tree, White Sandal Tree; *H.*: Chandal, Safed Chandan, Santal; *Sans.*: Bhadrasri, Candanah, Malayaja, Pitachandana, Srikhandam.
Distribution	:	S. regions of India, especially in Karnataka, Tamil Nadu and Kerala, ascending up to 1,200m; scattered in deciduous forests, Deccan Peninsula.
Reproductive Cycle	:	July-March.
Parts Used	:	Heartwood and Stem.
Folk Uses	:	Heartwood paste mixed with rice powder, mixture applied on mumps by Jenukurubas, also considered a coolant and applied as such over spots of pox, measles. Poultice of stem on forehead applied to check severe headache due to inflammation in Manipur caused by smallpox vaccination.
References	:	Hosagoudar and Henry (1996a), Kirtikar and Basu (1984), Kshirsagar and Singh (2000b, 2007), Pal *et al.* (2000), Sinha (1986, 1987), Vihari (1995).

Sapindus laurifolius **Balb. *ex* DC.**

Syn.: *S. trifoliatus* auct. non L.

Family: Sapindaceae

Eng., H. and Sans. Names :	*Eng.*: South India Soapnut, Three-Leaf Soapberry, Trijugate-Leaved Sop-nut; *H.*: Phenil, Reetha, Risht, Rishtak; *Sans.*: Arishta, Phenila, Rithakaranja.
Distribution :	Very common on hill slopes and in W. Peninsula; also cultivated in Bengal and N. India.
Reproductive Cycle :	Flowering during November-January.
Part Used :	Fruit.
Folk Uses :	Fruit paste applied locally to cure mumps, also used for epileptic fits in children.
References :	Kirtikar and Basu (1984), Kshirsagar and Singh (2003, 2007).

Sapindus mukorossi **L.**

Syn.: *S. abruptus* Lour.; *S. forsythii* DC.; *S. rigidus* Mill.; *S. saponaria* L.

Family: Sapindaceae

Eng., H. and Sans. Names :	*Eng.*: Soap Nut Tree of North India; *H.*: Areetha, Reetha, Ritha; *Sans.*: Arista, Arishtak, Phenila.
Distribution :	Cultivated throughout India, N.W. India and Bengal, Kumaon, Silhet, Assam.
Reproductive Cycle :	May-September.
Part Used :	Fruit.
Folk Uses :	Fruit with that of clove, Indian long pepper and ginger used for tonsillitis in Assam. Paste of fleshy part of fruits applied on forehead for worm affections in children. Foam produced due to stirring of epicarp and mesocarp of fruit along with little water smeared over the forehead, chest, back of pinna and neck of children to check high body temperature. 2g pill powder of nut shell mixed with honey and warm milk given twice daily in pneumonia.
References :	Borthakur (1993), Karnick *et al.* (1981), Salam *et al.* (2011), Singh and Singh (2003a, 2003b).

Sauropus androgynus **(L.) Merr.**

Syn.: *S. albicans* Blume.; *S. indicus* Wight.

Family: Euphorbiaceae

Eng., H. and Sans. Names	:	*Eng.*: Star Gooseberry; *H.*: Kutchumku, Surasarabi; *Sans.*: Aruni.
Distribution	:	Cultivated everywhere as a vegetable.
Part Used	:	Leaf.
Folk Use	:	A mouth wash made of juice of fresh leaves and honey applied inside mouth to cure tongue thrush in infants.
Reference	:	Prajapati *et al.* (2006).

Sauropus quadrangularis **(Willd.) Muell.-Arg.**

Syn.: *Phyllanthus rhamnoides* Roxb.

Family: Euphorbiaceae

Distribution	:	Bihar, Chota Nagpur, W. Peninsula, Myanmar.
Reproductive Cycle	:	June-September.
Part Used	:	Leaf.
Folk Use	:	Dried leaves smoked as a cure for tonsillitis.
References	:	Ambasta (1986), Kirtikar and Basu (1984).

Schefflera octophylla **(Lour.) Harms.**

Family: Araliaceae

Eng. Name	:	*Eng.*: Octophyllous Aralia.
Distribution	:	Grows wild in secondary forests in mountainous regions.
Reproductive Cycle	:	September-January.
Part Used	:	Bark.
Folk Use	:	Used in infantile rickets.
Reference	:	Prajapati *et al.* (2006).

Schleichera oleosa **(Lour.) Oken.**

Syn.: *S. trijuga* Willd.

Family: Sapindaceae

Eng., H. and Sans. Names	:	*Eng.*: Ceyolon Oak, Lac Tree; *H.*: Kausum, Kusum;

		Sans.: Mukulakah, Laksavrksah, Raktamrah.
Distribution	:	Throughout the plains of India.
Reproductive Cycle	:	June-October.
Part Used	:	Whole Plant.
Folk Use	:	Plant used in skin diseases.
Reference	:	Pal *et al.* (2000).

Schrebera swieteniodes Roxb.

Syn.: *Nathusia swietenioides* (Roxb.) Kuntze.; *Schrebera pubescens* Kurz.

Family: Oleaceae

Eng., H. and Sans. Names	:	*Eng.*: Weaver's Beam Tree;
		H.: Banpalas, Moka;
		Sans.: Ksarasresthah, Muskakah.
Distribution	:	Deciduous forests throughout India, upto 1,200m.
Reproductive Cycle	:	February-May.
Part Used	:	Whole Plant.
Folk Uses	:	Plant used in diarrhoea, stomach heat and colic pain.
Reference	:	Pal *et al.* (2000).

Scleria lithosperma (L.) Sw.

Syn.: *Carex lithosperma* J.F. Gmel.; *Hypoporum lithospermum* (L.) Nees.

Family: Cyperaceae

Eng. Names	:	*Eng.*: Florida Keys Nutrush, Mountain Grass.
Distribution	:	Found in Kerala and Malabar in India.
Reproductive Cycle	:	June-October.
Part Used	:	Aerial parts.
Folk Use	:	Given to children with enlarged stomach.
Reference	:	Ambasta (1986).

Scoparia dulcis L.

Syn.: *Ambulia micrantha* Raf.; *Capraria dulcis* (L.) Kuntze.; *Gratiola micrantha* Nutt.

Family: Scrophulariaceae

Eng., H. and Sans. Names	:	*Eng.*: Sweet Broom, Toothbrush;
		H.: Ghoda Tulsi, Mithi Patti;
		Sans.: Asmaghni, Pashanabheda.

Distribution	:	Common as a wild plant in abandoned fields, roadsides and river banks, plains of Bengal.
Reproductive Cycle	:	March-June.
Parts Used	:	Whole Plant, Root and Leaf.
Folk Uses	:	Plant decoction used for washing rashes among children. 1/2 teaspoonful juice of root and leaves given thrice daily in malaria. Fresh leaf juice drops poured in nostrils of children during convulsion.
References	:	Molla and Roy (1996), Mudgal *et al.* (1999), Prajapati *et al.* (2006), Prusti (2007), Sikarwar (2002).

***Scrophularia ningpoensis* Hemsl.**

Syn.: *S. microdonta* Franch.; *S. silvestrii* Bonati and Pamp.

Family: Scrophulariaceae

Eng. Names	:	*Eng.*: Chinese Figwort, Ning-Po-Figwort.
Distribution	:	Naturalized and widespread in the mountains, midlands and plains.
Reproductive Cycle	:	June-November.
Part Used	:	Root.
Folk Use	:	Roots used in tonsillitis.
Reference	:	Prajapati *et al.* (2006).

***Selaginella bryopteris* (SW.) Spring.**

Family: Selaginellaceae

Eng., H. and Sans. Names	:	*Eng.*: Life Giving Herb, Resurrecting Plant, Selaginella; *H.*: Sanjivini; *Sans.*: Sanjeevani, Sanjeevani Bhoothi.
Distribution	:	Throughout India.
Part Used	:	Whole Plant.
Folk Use	:	Plant used in poliomyelitis.
Reference	:	Pal *et al.* (2000).

***Semecarpus anacardium* L.f.**

Syn.: *S. latifolius* Pers.

Family: Anacardiaceae

Eng., H. and Sans. Names	:	*Eng.*: Marking Nut Tree, Oriental Cashew, Varnish Tree; *H.*: Belatak, Bhela, Bhilara;

		Sans.: Arusharah, Bhallatakah, Shailabija, Tapona, Vatori.
Distribution	:	Tropical outer Himalayas, Sirmour-Sikkim, ascending to 1,200m and throughout hotter parts of India, as far east as Assam.
Reproductive Cycle	:	October-April.
Parts Used	:	Whole Plant and Seed.
Folk Uses	:	Plant used in skin eruptions. Seed oil applied to check inflammation of tonsils.
References	:	Dash and Misra (2000), Pal *et al.* (2000).

***Senna alexandrina* Mill.**

Syn.: *Cassia angustifolia* M. Vahl.; *C. senna* L.; *Senna angustifolia* (Vahl.) Batka.

Family: Fabaceae

Eng., H. and Sans. Names	:	*Eng.*: Bombay Senna, Indian Senna, Lotus Senna, Medicinal Senna, Senna;
		H.: Bhuikhakhasa, Hindisana, Sanai, Sanay, Sonamukhi, Swarnapatri;
		Sans.: Bhumiari, Bhupadma, Hemapatri, Markandi, Markundi, Pitapushpi, Swargistha, Swarnpatri.
Distribution	:	Occurs throughout S. India and cultivated, also found in Punjab, Sind, W. Peninsula.
Reproductive Cycle	:	August-October.
Parts Used	:	Fruit and Leaf.
Folk Use	:	Leaves and fruits used as 'ghooty', a general tonic preparations for babies.
Reference	:	Agarwal (2003).

***Senna auriculata* (L.) Roxb.**

Syn.: *Cassia auriculata* L.; *C. densistipulata* Taub.

Family: Fabaceae

Eng., H. and Sans. Names	:	*Eng.*: Tanner's Cassia;
		H.: Anwal, Aval, Tarwar;
		Sans.: Ahula, Mandari, Pitakalika.
Distribution	:	Wild in the C. regions, W. Peninsula and Sri Lanka; often planted elsewhere.
Reproductive Cycle	:	Flowering in early and late Monsoon.

Part Used : Leaf.

Folk Use : Leaves used in chickenpox.

Reference : Murugesan *et al.* (2011).

Senna occidentalis **(L.) Link**

Syn.: *Cassia caroliniana* Walter.; *C. ciliata* Raf.; *C. falcata* L.; *C. foetida* Pers.

Family: Fabaceae

Eng., H. and Sans. Names : *Eng.*: Coffee Senna, Negro Coffee, Stinking Weed;

H.: Barikasaumdi, Chakunda, Kasondi;

Sans.: Arimarda, Jarana, Kasmarda.

Distribution : Common weed throughout India along roadsides, scattered from Himalayas to S. India.

Reproductive Cycle : February-December.

Parts Used : Whole Plant, Leaf, Fruit and Seed.

Folk Uses : Paste of plant applied externally on mumps by Jenukurubas, and also valuable for itching, scabies, ringworms and fever. Leaves used for curing stomach troubles, fever, and eye complaints. Decoction of leaves used as a mild febrifuge and anthelmintic in children. In Konkan, seeds used as a cure for convulsions of children. Fruit pulp along with jaggery given to children suffering from abnormally excessive saliva secretion.

References : Katewa and Arora (1997), Kirtikar and Basu (1984), Kshirsagar and Singh (2000b, 2007), Pal *et al.* (2000), Parrotta (2001).

Senna tora **(L.) Roxb.**

Syn.: *Cassia tora* L.

Family: Fabaceae

Eng., H. and Sans. Names : *Eng.*: Chinese Senna, Foetid Cassia, Sicklepod, Sickle Senna;

H.: Chakramarda, Chakunda, Chakavat, Pamad, Panevar;

Sans.: Ayudham, Chakramarda, Dadamari, Praputrata, Taga.

Distribution : Throughout India, Sri Lanka and the tropics generally, and common weed occurring at Chimtipur of Chhindwara district of M.P.

Reproductive Cycle : Flowering and fruiting throughout the year, most commonly September-December.

Parts Used : Whole Plant and Leaf.

Folk Uses : Plant used for skin diseases and ringworm. Decoction of its mucilaginous and foetid-smelling leaves prescribed in dose of about two ounces for children suffering from feverish attacks while teething and intestinal troubles. Leaf paste applied on whole body for a week against measles.

References : Bhattacharyya (1996), Kirtikar and Basu (1984), Kumar and Pullaiah (1998), Pal *et al.* (2000), Watt (1972).

Seriphidium brevifolium **(Wall. *ex* DC.) Ling and Y.R. Ling.**

Syn.: *Artemisia brevifolia* Wall. *ex* DC.

Family: Asteraceae

Eng. Name : *Eng.*: Short-Leaved Worm Seed.

Distribution : W. Himalayas, from Kashmir-Kumaon, 2,300-3,000m.

Reproductive Cycle : August-December.

Part Used : Leaf.

Folk Uses : Leaf extract given twice or thrice a day against stomach problems and intestinal worms.

Reference : Ballabh and Chaurasia (2006).

Seriphidium thomsonianum **(C.B. Clarke) Ling and Y.R. Ling.**

Syn.: *Artemisia maritima* L. var. *thomsoniana* C.B. Clarke.; *A. thomsoniana* (C.B. Clarke) Filatova

Family: Asteraceae

Eng., H. and Sans. Names : *Eng.*: Drooping Sea Wormweed, Roman Wormweed, Wormseed;

H.: Ajavayana, Chhuari, Kirmala;

Sans.: Gadadhara, Gandha.

Distribution : Distributed chiefly from Kashmir-Kumaon: 2,400-3,000m.

Reproductive Cycle : July-November.

Parts Used : Whole Plant, Leaf and Flower.

Folk Uses : Plant used in fever. Decoction of leaves and flower given orally to remove abdominal parasites. Flowers used in infantile spasms and whooping cough.

References : Chatterjee and Pakrashi (1997), Pal *et al.* (2000), Sood *et al.* (2001).

Sesamum indicum **L.**

Syn.: *S. luteum* Retz.; *S. occidentale* Heer and Regel.; *S. orientale* L.

Family: Pedaliaceae

Eng., H. and Sans. Names : *Eng.*: Gingelly, Sesame;

H.: Bariktel, Til, Tilkatel;

Sans.: Homadhanya, Pavitra, Snehaphala, Tila, Tilah, Vanodbhava.

Distribution : Mainly grown in M.P., U.P., A.P., Rajasthan, Tamil Nadu and Maharashtra; often wild on roadsides and wastelands.

Reproductive Cycle : July-September.

Parts Used : Leaf and Seed.

Folk Uses : Infusion of leaves used as demulcent in infantile diarrhoea. Paste of 'sesame' oil mixed with puffy mass of 'lupa hidak' (a salt available in market) and equal amount of 'nungshin' (a salt available in market) applied 2-3 times daily to scratching inflammation on skin. Oil cake of seeds given to weak children for improving their health.

References : Sharma (2003), Singh and Singh (2003a), Pande and Joshi (2001).

Sesbania grandiflora **Pers.**

Syn.: *Agati grandiflora* Desv.; *Coronilla grandiflora* Willd.

Family: Fabaceae

Eng., H. and Sans. Names : *Eng.*: Sesban, Swamp Pea;

H.: Agasti, Bak, Basna, Hatiya;

Sans.: Agasti, Agastyah, Agati, Anari.

Distribution : Cultivated in many parts of India, in S.E. India, the Ganges, and Myanmar.

Reproductive Cycle : Flowering throughout the year and fruiting during winters.

Parts Used : Leaf and Bark.

Folk Uses : Leaves used in mumps. Infusion of bark given during the first stage of smallpox and eruptive fevers.

References : Chatterjee and Pakrashi (1997), Dey and Bahadur (1973), Dhiman (2003), Jain (1991), Khare (2004), Kirtikar and Basu (1984), Parrotta (2001), Watt (1972).

Sesbania sesban **(L.) Merr.**

Syn.: *Aeschynomene sesban* L.; *Sesbania aegyptiaca* Pers.

Family: Fabaceae

Eng., H. and Sans. Names : *Eng.*: Common Sesban, Egyptian Rattle Pod;

H.: Jainti, Jait, Rawasan;

Sans.: Jayanti, Jayantika, Nadayi, Raja.

Distribution : Plains from the Himalayas-Sri Lanka and Slam, ascending to 1,330m in N.W. Himalayas.

Reproductive Cycle : August-December.

Part Used : Seed.

Folk Uses : Paste of seeds along with stems of *Coculus villosus* applied on ribs to cure pneumonia of children. Seeds also effective for smallpox eruptions.

References : Kirtikar and Basu (1984), Mishra (2008).

Setaria italica **(L.) P. Beauv.**

Syn.: *Panicum chinense* Trin.; *P. germanicum* Mill.; *P. italicum* L.

Family: Poaceae

Eng., H. and Sans. Names : *Eng.*: Italian Millet;

H.: Chena, Kakni, Kakun, Kanni;

Sans.: Chinaka, Kangu, Kanguni.

Distribution : Cultivated throughout India, upto 2,000m in Himalayas.

Part Used : Grains.

Folk Use : Cooked grains given to children recovering from measles.

Reference : Pande and Joshi (2001).

Shorea robusta **Gaertn.f.**

Syn.: *Vatica robusta* Steud.

Family: Dipterocarpaceae

Eng., H. and Sans. Names : *Eng.*: Indian Dammer, Sal Tree;

H.: Damar, Dhuna, Ral, Sal;

Sans.: Asvakarna, Guggilam, Koushi-kaha, Salah.

Distribution : Tropical Himalayas and along its base from Assam-Satluj, eastern districts of C. India, W. Bengal hills.

Reproductive Cycle : March-July.

Parts Used : Whole Plant and Resin.

Folk Uses : Plant used in itching and eye diseases ('kajal'). Resin with honey or sugar given in dysentery and bleeding piles to children in dose of 0.6-1.3g. Smoke from resin used against chickenpox.

References : Dhiman (2003), Pal *et al.* (2000), Rao and Henry (1996).

Sida acuta **Burm.f.**

Syn.: *Sida carpinifolia* Mast.; *S. lanceolata* Roxb.; *S. scoparia* Lour.; *S. stauntoniana* DC.; *S. stipulata* Cav.

Family: Malvaceae

Eng., H. and Sans. Names : *Eng.*: Hornbean Leaved Sida;

H.: Bariara, Kareta, Kharenta;

Sans.: Bala, Pitberela, Rajbala.

Distribution : Hotter parts of India, tropics generally.

Reproductive Cycle : September-March.

Part Used : Leaf.

Folk Use : Infusion of leaves in water used as an enema for paralysed children to help them to walk.

Reference : Kirtikar and Basu (1984).

Sida rhombifolia **L.***

Syn.: *S. canariensis* Willd.; *S. compressa* Wall.

Family: Malvaceae

Eng., H. and Sans. Names : *Eng.*: Broom Jute Sida, Country Mallow, Mahot, Sida, Teaweed;

H.: Bhiunli, Sahadabi, Swetbarela;

Sans.: Ahikhanda, Brihadagala, Lalbarila, Sahadeva, Vatya.

Distribution : Throughout India, upto 200m in Himalayas.

Reproductive Cycle : July-September.

Part Used : Whole Plant.

Folk Uses : Plant extract used as antipyretic for babies. Macerated plant used against measles.

References : Barua *et al.* (2000), Keller *et al.* (2011).

***Skimmia laureola* (DC.) Sieb. and Zucc. *ex* Walp**

Syn.: *Limonia laureola* DC.

Family: Rutaceae

Eng. and H. Names	:	*Eng.*: Ner; *H.*: Kasturchara, Kedarpatta, Nora.
Distribution	:	Found in E. Himalayas.
Reproductive Cycle	:	April-October.
Part Used	:	Leaf.
Folk Use	:	Leaves used in smallpox by Monpas in Arunachal Pradesh.
References	:	Dam and Hajra (1981), Farooq (2005), Jain (1991), Kapur and Srivastava (1996), Pande *et al.* (2006), Samant *et al.* (2001b), Watt (1972).

***Smilax perfoliata* Lour.**

Syn.: *S. annamensis* Rendle.; *S. prolifera* Roxb.

Family: Smilacaceae

Distribution	:	Almost all over the tropical parts of India.
Reproductive Cycle	:	April-August.
Parts Used	:	Whole Plant and Leaf.
Folk Uses	:	Plant used in diarrhoea, stomach pain, bowel complaints, dysentery, dyspepsia/indigestion and anthelmintic. Children suffering from bed wetting cured by serving their food on leaves of this plant.
References	:	Alagesboopathi *et al.* (2000), Jain (1991), Pal *et al.* (2000).

***Smilax zeylanica* L.**

Syn.: *S. collina* Kunth.; *S. elliptica* Desv.; *S. indica* Burm.f.

Family: Smilacaceae

Eng., H. and Sans. Names	:	*Eng.*: Indian Smilax; *H.*: Bhitura, Jangli Aushbah, Kumarika; *Sans.*: Nanamadhusnahi.
Distribution	:	Throughout hilly tropical districts of India, from E. Himalayas and Myanmar to Malay Peninsula and from Concan-Sri Lanka.
Reproductive Cycle	:	Flowering during January-April.

Parts Used	:	Whole Plant and Seed.
Folk Uses	:	Plant used in stomach pain, dysentery, dyspepsia/ indigestion and as anthelmintic. Seed paste applied locally to children for curing itching due to scabies.
References	:	Pal *et al.* (2000), Srivastava *et al.* (2003b).

Solanum americanum Mill.

Syn.: *S. adventitium* Polgar.; *S. caribaeum* Dunal.; *S. depilatum* Bitter.; *S. gollmeri* Bitter.

Family: Solanaceae

Eng. Names	:	*Eng.*: American Black Nightshade, Small Flower Nightshade, White Nightshade.
Distribution	:	Found in tropical and temperate regions in wastegrounds, disturbed sites, open pastures, roadsides and railroads.
Reproductive Cycle	:	May-November.
Part Used	:	Bud.
Folk Use	:	Buds used for measles.
Reference	:	Keller *et al.* (2011).

Solanum anguivi Lam.

Syn.: *S. indicum* L.; *S. lividum* Link.; *S. sodomeum* L.; *S. violaceum* Jacq.

Family: Solanaceae

Eng., H. and Sans. Names	:	*Eng.*: Indian Nightshade, Poison Berry; *H.*: Barhanta, Barikateri, Kateli; *Sans.*: Brhati, Brihatee, Simhi.
Distribution	:	Throughout tropical India, ascending to 1,700m.
Reproductive Cycle	:	Flowering during January-April.
Parts Used	:	Root and Fruit.
Folk Uses	:	Extract of roots given to infants to check vomiting following intake of milk among Kathodias and Raibaris of Rajasthan. 2 tsp fruit extract given thrice daily to children for expelling worms from stomach by Jenukurubas and Kondhs of S.W. Orissa.
References	:	Kshirsagar and Singh (2007), Parrotta (2001).

Solanum dulcamara L.

Syn.: *S. laxum* Royle.; *S. lyratum* Thunb.; *S. persicum* Willd.

Family: Solanaceae

Eng., H. and Sans. Names : *Eng.*: Bitter Sweet, Dulcamara, Woody Nightshade;
H.: Bhalumas;
Sans.: Kakmachi.

Distribution : Temperate W. Himalayas, wild in Kashmir, H.P. and U.P., 1,219-2,438m.

Reproductive Cycle : July-August.

Part Used : Whole Plant before flowering.

Folk Use : Used in measles.

Reference : Rajan *et al.* (2000).

Solanum erianthum D. Don.

Syn.: *S. adulterinum* Buch.-Ham. *ex* Wall.

Family: Solanaceae

Eng., H. and Sans. Names : *Eng.*: Big Egg Plant, Potato Tree, Tobacco Tree;
H.: Akra, Ban Tamakhu;
Sans.: Gandhira, Pathi, Pryamkari, Vidari.

Distribution : Throughout tropical and subtropical zone in India.

Reproductive Cycle : July-October.

Part Used : Fruit.

Folk Use : Fruit juice used as nasal drops (twice daily for 2 days) to expel worms.

References : Farooq (2005), Hosagoudar and Henry (1996c).

Solanum giganteum Jacq.

Syn.: *S. niveum* Thunb.

Family: Solanaceae

Eng. Names : *Eng.*: Healing-Leaf Tree, Red Bitter-Apple, Thorny Bug-Tree.

Distribution : Mountains of Deccan and Sri Lanka, 1,000-2,350m.

Reproductive Cycle : January-April.

Part Used : Root.

Folk Uses	:	Paste of roots grounded with garlic applied all over the body of children before bath to cure dysentery and diarrhoea.
Reference	:	Hosagoudar and Henry (1996c).

Solanum nigrum **L. (Pl. 13B)***

Syn.: *S. pupureilineatum* Sabins and Bhatt.; *S. rubrum* Mill.

Family: Solanaceae

Eng., H. and Sans. Names	:	*Eng.*: Black Nightshade, Hound's Berry, Morelle, Petty Morel;
		H.: Gurkkamai, Kabaiya, Makoi;
		Sans.: Dhvankshamachi, Ghanaghana, Kakamachi, Rasayani, Vayasi.
Distribution	:	Wasteplaces throughout India.
Reproductive Cycle	:	February-June.
Parts Used	:	Whole Plant and Leaf.
Folk Uses	:	Plant infusion useful in anaemia, smallpox, tetanus and as an enema to infants with abdominal upsets. Extract of leaves fried and consumed 1/4 tsp, once daily for 4 days given for stomach problems in children.
References	:	Agarwal (2003), Gogoi and Das (2003), Kirtikar and Basu (1984), Rajan *et al.* (2000), Subramaniam (2000), Viswanathan *et al.*(2006).

Solanum stramoniifolium **Jacq.**

Syn.: *S. coconilla* Huber.; *S. demerarense* Poepp.; *S. orinocense* Kunth.

Family: Solanaceae

Eng. and Sans. Names	:	*Eng.*: Hairy-Fruited Pea-Eggplant, Red-Fruited Nightshade;
		Sans.: Garbhada.
Distribution	:	Found occasionally in tropical parts of India.
Reproductive Cycle	:	April-July.
Part Used	:	Root.
Folk Use	:	Roots pounded with lime juice given orally to children for constipation.
Reference	:	Bhandary *et al.* (1996).

Solanum surattense **Burm.f. (Pl. 13C)**

Syn.: *S. armatum* Br.; *S. virginianum* L.; *S. xanthocarpum* Schrad.

Family: Solanaceae

Eng., H. and Sans. Names	:	*Eng.*: Dwarf Lid Brinjal, Yellow-berried Nightshade; *H.*: Kateli, Remgani, Satyanasi; *Sans.*: Dhavanika, Kautakari, Nidigdhika, Putrapada, Sinhi.
Distribution	:	As a weed on roadsides and wastelands in dry situation throughout India; common at Isvalee of Uttar Kashi between 1,000-2,000m and at Harvalachar of Chindwara district of M.P.
Reproductive Cycle	:	May-October.
Parts Used	:	Root, Flower and Fruit.
Folk Uses	:	Root decoction along with paste of long pepper used as an antidote to smallpox. Crushed flowers given orally to relieve diarrhoea in children among the Bhils of Rajasthan. 5-15ml preparation comprising 30-40 fresh leaves of 'Ramtulsi', 4-5 fresh plants of 'Punghphai' and 40-50 crushed fruits of this plant, 5-10ml clear supernatant of fresh cowdung and honey prescribed orally to children twice daily for 3-4 days against sore mouth, tongue etc.
References	:	Chatterjee and Pakrashi (1997), D' Rozario *et al.* (1999), Parrotta (2001), Singh and Singh (2003a).

Solanum torvum **Swartz.**

Syn.: *S. ferrugineum* Jacq.; *S. pseudosaponaceum* Blume.; *S. stramonifolium* Roxb.; *S. wightii* Miq.

Family: Solanaceae

Eng., H. and Sans. Names	:	*Eng.*: Devil's Fig, Prickly Nightshade, Turkey Berry; *H.*: Bari, Kateli; *Sans.*: Gosthabartaku, Shwet Brihati.
Distribution	:	Common in wasteplaces, hedges, along railway line, throughout tropical parts of India.
Reproductive Cycle	:	August-December.
Part Used	:	Fruit.
Folk Uses	:	Wreath made of its fruits worn around neck of children to treat cold, cough and tonsils. Fresh fruit paste applied locally for chickenpox and measles.

References : Nath and Maiti (2003), Singh and Narain (2008), Sinha (1987).

***Solidago virgaurea* L.**

Syn.: *Aster virgaurea* (L.) Kuntze.; *Dectis decurrens* Raf.; *Solidago minor* Mill.; *S. nudiflora* Viv.

Family: Asteraceae

Eng. and H. Names : *Eng.*: European Goldenrod, Woundwort;

H.: Pinja Phool, Sonali.

Distribution : Native of Europe and Asia, naturalized in N. America; prefers open areas, wasteground and hillsides.

Reproductive Cycle : July-October.

Part Used : Aerial parts.

Folk Use : Used for treating gastroenteritis in children.

Reference : Prajapati *et al.* (2006).

***Sphaeranthus indicus* auct. non L.**

Syn.: *S. hirtus* Willd.; *S. senegalensis* DC.

Family: Asteraceae

Eng., H. and Sans. Names : *Eng.*: East Indian Globe-thistle;

H.: Gorakh Mundi, Mundi;

Sans.: Hapusa, Mahamundi, Mundi, Mundirika.

Distribution : Occurs at Chindwara district of M.P., moist damp places of tropical zones of Garhwal Himalayas.

Reproductive Cycle : January-February.

Parts Used : Stem, Leaf and Flower.

Folk Uses : Flower, leaf and stem paste used 6-7 times a day in smallpox. Leaf extract mixed with sugar and milk administered (1 tsp twice a day for 3 days) to infants for diarrhoea. A garland of flowers also worn to cure smallpox.

References : Bhattacharjee (2001), Jain (1991), Rao and Henry (1996), Sharma (2003).

***Spondias pinnata* (L.f.) Kurz.**

Syn.: *Evia amara* Comn.; *Mangifera pinnata* Koen.; *Spondias mangifera* Willd.

Family: Anacardiaceae

Eng., H. and Sans. Names : *Eng.*: Bile Tree, Hogplum, Wild Mango;

		H.: Ambodha, Amra, Jungli Aam;
		Sans.: Amratakan, Bhringiphalla, Kapitana, Rasadhya, Tanukshiri.
Distribution	:	Common in dry forests of many parts of India; often cultivated in village surroundings.
Reproductive Cycle	:	February-June.
Parts Used	:	Fruit and Bark.
Folk Uses	:	An amulet of its fruit around neck of children cure mouth ulcers. Powdered bark mixed with part of *Mangifera indica* in equal proportion and honey given to children to cure diarrhoea. Paste of its bark and that of *Terminalia bellirica, Psidium guajava* in equal quantities with 3 fruits each of *Terminalia chebula* and *Emblica officinalis* along with a few drops of alkali solution given two tsp once daily to children for 3 days to cure cholera.
References	:	Gogoi and Borthakur (2001), Satapathy and Brahmam (2000), Sen and Behera (2007).

***Stachytarpheta jamaicensis* (L.) Vahl.**

Syn.: *Abena jamaicensis* (L.) Vahl.; *Stachytarpheta mexicana* Steud.

Family: Verbenaceae

Eng., H. and Sans. Names	:	*Eng.*: Aaron's Rod, Bastard Nervain, Brazilian Tea, Jamaica False Vervain;
		H.: Karyartharani, Ramdatun;
		Sans.: Kariyartharani.
Distribution	:	Tropical India, from Punjab and Silhet-Travancore.
Reproductive Cycle	:	Throughout the year.
Part Used	:	Root.
Folk Use	:	Extract of roots grounded with that of *Ziziphus oenoplia*, pepper, garlic, cumin and salt administered orally to 1-2 year old children in fever.
Reference	:	Hosagoudar and Henry (1996a).

***Stephania japonica* (Thunb.) Miers.**

Family: Menispermaceae.

Eng., H. and Sans. Names	:	*Eng.*: Tape Vine;
		H.: Par;
		Sans.: Patha, Rajapatha, Vanatiktika.

Distribution : West and E. coast, Sikkim, E. Bengal, Assam.

Reproductive Cycle : June-August.

Part Used : Leaf.

Folk Use : Poultice of leaves applied on head or hand of children as a coolant.

References : Viswanathan *et al.* (2001, 2006).

***Stereospermum cholonoides* (L.f.) DC.**

Syn.: *Bignonia chelonoides* L.f.; *B. suaveolens* Roxb.; *Hieranthes fragrans* Raf.

Family: Bignoniaceae

Eng., H. and Sans. Names : *Eng.*: Trumpet-Flower;

H.: Padal, Padaria, Padiala, Paral, Parur, Pur, Purula;

Sans.: Abhipriya, Amova, Kokila, Patali, Patoli.

Distribution : Mixed deciduous and sal (*Shorea robusta*) dominant forests and sub-Himalayan tract upto 1,500m.

Reproductive Cycle : Flowering at the beginning of the dry season after the leaves have been shed.

Part Used : Seed.

Folk Use : Paste of seeds gently applied in throat once in the morning by fingers to cure throat swelling among children.

Reference : Maliya (2004).

***Streblus asper* Lour.**

Syn.: *Epicarpurus asper* Steud.; *E. orientalis* Bl.; *Trophis aspera* Retz.

Family: Moraceae

Eng., H. and Sans. Names : *Eng.*: Siamese Rough-bush;

H.: Cherootpathi, Daheya, Karchanua, Rusa, Siora;

Sans.: Akshadhara, Gavakshi, Patrollekhataruh, Shakhotaka.

Distribution : Drier parts of India, from Rohilkund, eastwards and southward to Travancore, Penang and Andaman islands.

Reproductive Cycle : January-July.

Part Used : Latex.

Folk Uses : 3 drops of latex mixed with breast milk given twice daily for 3 days to infants for dysentery. Latex also applied on measles.

Reference : Rao and Henry (1996).

Strychnos nux-vomica **L.**

Syn.: *S. colubrine* Wight.; *S. lucida* Wall.

Family: Loganiaceae

Eng., H. and Sans. Names : *Eng.*: Nux-vomica, Snake-wood, Strychnine Tree;

H.: Bailewa, Chibbige, Kajra, Kuchla;

Sans.: Chipita, Dirghapatra, Gerdruma, Marakatincles, Kakasphurja, Karaskara, Vishamushti.

Distribution : Throughout tropical India, 0-1,350m.

Reproductive Cycle : February-June.

Part Used : Seed.

Folk Use : Seeds used in chickenpox.

Reference : Jain (1991).

Strychnos potatorum **Willd.**

Syn.: *S. tettankotta* Retz.

Family: Loganiaceae

Eng., H. and Sans. Names : *Eng.*: Clearing Nut, Clearing-nut Tree;

H.: Neimal, Nelmal, Nirmali;

Sans.: Ambu-prasada, Chakshushya, Guchnaphala, Katakatphalam, Lekhanatmaka, Ruchishya, Shadanatmaka, Tikaphala.

Distribution : Konkan, C. India, Deccan, S. Travancore, Sri Lanka, Myanmar.

Reproductive Cycle : April-May.

Part Used : Whole Plant.

Folk Uses : Plant good against dysentery, bowel and eye complaints.

Reference : Pal *et al.* (2000).

Stylosanthes fruticosa **(Retz.) Alston.**

Syn.: *Arachis fruticosa* Retz.; *Hedysarum hamatum* Burm.f.

Family: Fabaceae

Eng. Names : *Eng.*: Shrubby Pencil Flower, Wild Lucerne.

Distribution : Grows throughout S.India and Orissa.

Reproductive Cycle : Flowering in December.

Part Used : Leaf.

Folk Use : Leaves used in infantile diarrhoea.

Reference : Agarwal (2003).

Swertia chirayita **(Roxb.** ***ex*** **Flem) Karst.**

Syn.: ***Agathotes chirata*** **D. Don.;** ***Gentiana chirayita*** **Roxb.;** ***Ophelia chirata*** **Griseb.;** ***Swertia chirata*** **Buch.-Ham.** ***ex*** **C.B. Clarke.**

Family: Gentianaceae

Eng., H. and Sans. Names : *Eng.*: Chiretta Plant;

H.: Chirayita;

Sans.: Bhunimba, Haima, Jvarantaka, Kirata-tikta, Naipala, Ramasenaka, Viktaka.

Distribution : Found in temperate Himalayas: 1,200-3,000m, (Kashmir-Bhutan), Khasia hills; in Meghalaya at 1,200-1,500m.

Reproductive Cycle : August-October.

Part Used : Whole Plant.

Folk Use : Plant used in bed wetting of children.

Reference : Khare (2004).

Syagrus romanzoffiana **(Cham.) Glassman.**

Syn.: ***Arecastrum romanzoffianum*** **(Cham.) Becc.;** ***Calappa australis*** **(Mart.) Kuntze.;** ***Cocos acrocomioides*** **Drude.**

Family: Arecaceae

Eng. Names : *Eng.*: Feathery Coconut Palm, Giriba Palm, Queen Palm.

Distribution : Introduced into Indian gardens.

Reproductive Cycle : Flowering in Spring and Summer.

Parts Used : Leaf and Seed.

Folk Uses : Wax covering tertiary leaf veins used as cicatrizant for navel of recently born children. Beetle larvae in seeds given to children to prevent cavities in teeth and intestinal parasites.

Reference : Keller *et al.* (2010).

Syzygium aromaticum **(L.) Merr. and Perry.**

Syn.: *Caryophyllus aromaticus* L.; *Eugenia aromatica* Kuntze.; *E. caryophyllata* Thunb.

Family: Myrtaceae

Eng., H. and Sans. Names	:	*Eng.*: Clove Tree; *H.*: Laung; *Sans.*: Devakusumam, Lavangam.
Distribution	:	Cultivated in South India.
Reproductive Cycle	:	January-June.
Part Used	:	Fruit.
Folk Use	:	Fruit of this plant along with that of Indian long pepper, soapnut tree and ginger used for tonsillitis in Assam.
Reference	:	Karnick *et al.* (1981).

Syzygium corynocarpum **(A. Gray) Muell. Stuttg.**

Syn.: *Eugenia corynocarpa* A. Gray.

Family: Myrtaceae

Distribution	:	Common in dense wet forest from sea level- 1,000m; elevation.
Parts Used	:	Leaf and Bark.
Folk Use	:	Infusion of bark or leaf given to babies with teething problems.
Reference	:	Prajapati *et al.* (2006).

Syzygium cumini **(L.) Skeels.**

Syn.: *Eugenia jambolana* Lam.; *E. obovata* Poir.; *E. obtusifolia* Roxb.; *E. tsoi* Merr. and Chun.; *Syzygium jambolanum* Lam. DC.

Family: Myrtaceae

Eng., H. and Sans. Names	:	*Eng.*: Black Plum, Jaman, Jambolan; *H.*: Jaman, Jambhal, Jamun, Paiman, Phalinda; *Sans.*: Jambavam, Jambuh, Rajphala.
Distribution	:	Throughout India, ascending to 1,000m on Punjab Himalayas and to 1,700m in Kumaon.
Reproductive Cycle	:	May-August.
Parts Used	:	Leaf and Bark.

Folk Uses	:	Paste of leaves given in diarrhoea and dysentery of children. Fresh juice of bark given with goat's milk in diarrhoea of children.
References	:	Banerjee (2000b), Chatterjee and Pakrashi (1997), Dhiman (2003), Kirtikar and Basu (1984), Parrotta (2001), Upadhyay and Singh (2005), Watt (1972).

Syzygium malaccense (L.) Merr. and L.M. Perry.

Syn.: *Eugenia domestica* Baill.; *E. malaccensis* L.; *E. purpurea* Roxb.; *Jambosa domestica* Blume.

Family: Myrtaceae

Eng. and H. Names	:	*Eng.*: Malay Apple, Mountain Apple, Rose Apple; *H.*: Malay jamun.
Distribution	:	Common in villages, lowland secondary forests and cultivated valleys from near sea level upto 1,000m.
Reproductive Cycle	:	April-May.
Part Used	:	Bark.
Folk Use	:	Bark used for curing mouth sores in children.
Reference	:	Prajapati *et al.* (2006).

Syzygium samarangense (Bl.) Merr. and L.M. Perry.

Syn.: *Eugenia javanica* Lamk.; *Myrtus javanica* (Lam.) Blume.; *M. samarangensis* Blume.

Family: Myrtaceae

Eng. and H. Names	:	*Eng.*: Cloud Apple, Java Apple, Mountain Apple, Water Apple, Wax Apple, Wax Jambu; *H.*: Amrool, Jamrul.
Distribution	:	Native of Andaman and Nicobar islands, found in different parts of India.
Reproductive Cycle	:	March-June.
Part Used	:	Leaf.
Folk Use	:	Paste of mixture comprising its pounded leaves mixed with pig blood smeared on body of weak children as a renovating agent.
Reference	:	Dagar and Dagar (1996).

Tabernaemontana alternifolia **L.**

Syn.: *Ervatamia heyneana* (Wall.) T. Cooke.; *Tabernaemontana heyneana* Wall.; *T. oblonga* Wall.

Family: Apocynaceae

Eng., H. and Sans. Names	:	*Eng.*: Wax Flower; *H.*: Chandni, Chandui; *Sans.*: Kshiri, Nandi.
Distribution	:	Common in W. forests of Coorg and occasional in E. forests. Upper Gangetic plains, Garhwal, E. Bengal, Khasia hills, Assam, hills of *Viz.*agapatam.
Reproductive Cycle	:	November-June.
Part Used	:	Fruit.
Folk Use	:	An amulet of its fruit tied around neck for curing cough in children by Girijanas.
References	:	Kshirsagar and Singh (2000a, 2007).

Tacca plantaginea **(Hance.) Drenth.**

Syn.: *Schizocapsa guangxiensis* P.P. Ling and C.T. Ting.; *S. plantaginea* Hance.

Family: Taccaceae

Eng. Name	:	*Eng.*: Tacca.
Distribution	:	Grows wild in wet places in mountainous regions.
Reproductive Cycle	:	October-February.
Part Used	:	Tuberous Root.
Folk Use	:	Roots used in infantile poliomyelitis.
Reference	:	Prajapati *et al.* (2006).

Tadehagi triquetrum **(L.) H. Ohashi**

Syn.: *Desmodium acrocarpum* Hance.; *D. triquetrum* (L.) DC.; *Hedysarum triquetrum* L.

Family: Fabaceae

Eng. Name	:	*Eng.*: Trefle Gros.
Distribution	:	C. and E. Himalayas through India proper-Sri Lanka, ascending to 1,300m in Kumaon, Sikkim and Khasia, Chittagong, Tavoy, Pegu.
Reproductive Cycle	:	Flowering in summer.
Part Used	:	Root.

Folk Use : 3g powdered drug of its roots along with *Terminalia bellirica's* stem bark and 5 black pepper given twice daily to children to cure rickets.

Reference : Aminuddin and Girach (1996).

Tagetes minuta **L. (Pl. 13D)***

Syn.: *T. glanduliflora* Schrank.

Family: Asteraceae

Eng., H. and Sans. Names : *Eng.*: Marigold, Stinking-roger;

H.: Ganindri, Jangli-gainda;

Sans.: Sthulpushpa, Zandu, Zendu.

Distribution : N.W. Himalayas region, cultivated as a crop in J&K, Punjab, H.P., Uttranchal, U.P., Karnataka, A.P. and Tamil Nadu.

Reproductive Cycle : September-October.

Part Used : Leaf.

Folk Use : Raw paste of its bruised leaves applied on naval of children to expel roundworms.

Reference : Chauhan (1999).

Tamarindus indica **L.**

Syn.: *T. lumbrosa* Salisb.; *T. occidentalis* Gaertn.; *T. officinalis* Hook.

Family: Fabaceae

Eng., H. and Sans. Names : *Eng.*: Tamarind Tree;

H.: Amli, Ampli, Imli, Nuli;

Sans.: Amlika, Chukra, Cinca, Sukta, Tintrini.

Distribution : Throughout India, particularly in S. India, often cultivated.

Reproductive Cycle : April-August.

Parts Used : Whole Plant, Leaf and Fruit.

Folk Uses : Lozenges made from pulp mixed with senna and chocolates used as a cathartic in infantile diseases. Tamarind pulp boiled with water and sweetening with sugar considered a good laxative for children. Leaves used in smallpox.

References : Bhattacharjee (2001), Kirtikar and Basu (1984), Prajapati *et al.* (2006), Sharma (2004), Subramaniam (2000), Watt (1972).

Tamilnadia uliginosa **(Retz.) Tirveng and Sastre**

Syn.: *Gardenia uliginosa* Retz.; *Posoqueria uliginosa* (Retz.) Roxb.; *Randia uliginosa* DC.; *Xeromphis uliginosa* (Retz.) Maheshwari.

Family: Rubiaceae

Eng., H. and Sans. Names	:	*Eng.*: Grey Emetic Nut; *H.*: Bharani, Katul, Panar, Paniah, Penaura, Pendua, Pindora; *Sans.*: Devatamalla, Gangati, Pindalu.
Distribution	:	Grows throughout India.
Reproductive Cycle	:	May-December.
Part Used	:	Root.
Folk Use	:	Roots given for boils in children.
References	:	Agarwal (2003), Anonymous (1969), Kirtikar and Basu (1984).

Tanacetum gracile **Hook.f. and Thoms.**

Syn.: *Chrysanthemum hookeri* Kitam.

Family: Asteraceae

Distribution	:	Alpine W. Himalayas (Kashmir-Kumaon) upto 4,500m.
Reproductive Cycle	:	July-August.
Parts Used	:	Leaf and Flower.
Folk Use	:	Decoction of leaves and flowers given in small dose upto one week against intestinal worms in children.
Reference	:	Ballabh and Chaurasia (2006).

Taraxacum mongolicum **Hand.-Mazz.**

Syn.: *T. huhhoticum* Z.xu and H.C.Fu.

Family: Asteraceae

Eng. Name	:	*Eng.*: Mongolian Dandelion.
Distribution	:	Hillsides, grasslands and by roadsides in China, Korea, Mongolia.
Reproductive Cycle	:	Flowering during Spring-late Summer.
Part Used	:	Whole Plant.
Folk Use	:	Plant used in tonsillitis.
Reference	:	Prajapati *et al.* (2006).

Tarenna sambucina **(G.Forst.) T. Durand *ex* Drake**

Syn.: *Chiococca sambucina* (G. Forst.) Spreng.; *Ixora sambucina* (G. Forst.) Kuntze.

Family: Rubiaceae

Distribution	:	Common in coastal and grassland thickets, disturbed, dry or dense forest and even mangrove margins from sea level to about 500m elevation.
Part Used	:	Bark.
Folk Uses	:	Infusion of grated bark used to treat children's fever, inflammation, diarrhoea and internal injuries.
Reference	:	Prajapati *et al.* (2006).

Tectaria coadunata **(Wall. *ex* Hook. and Grev.) C. Christ**

Syn.: *T. macrodonta* (Fee) C. Chr.; *T. viridifrons* Ching.

Family: Dryopteridaceae

Distribution	:	E. Himalayas.
Reproductive Cycle	:	September-February.
Part Used	:	Rhizome.
Folk Use	:	Inhabitants of Darjeeling district in W. Bengal prescribe 2-3 drops of juice of its rhizome mixed with mother's milk to children having diarrhoea.
Reference	:	Dixit *et al.* (1978).

Tephrosia purpurea **(L.) Pers.**

Syn.: *T. hamiltonii* Drumm *ex* Gamble.; *T. stricta* Grah.

Family: Fabaceae

Eng., H. and Sans. Names	:	*Eng.*: Purple Tephrosia, Wild Indigo; *H.*: Sarphonka; *Sans.*: Banah, Banapunkhah, Kriti, Nilavriksha, Sharpunkha.
Distribution	:	All over India, ascending upto 1,850m in Himalayas.
Reproductive Cycle	:	Whole year.
Parts Used	:	Whole Plant, Root and Seed.
Folk Uses	:	Plant used in diarrhoea, bowel complaints, dysentery, anthelmintic, blood dysentery and eye sore. Roots used in tonsillitis, also as an anthelmintic and blood purifier. Seeds used as laxative and for worm infestation in children.

References : Chatterjee and Pakrashi (1997), Dwivedi (2000), Karnick (1994a), Mishra (2008), Pal *et al.* (2000), Parrotta(2001), Prasad *et al.* (2003), Singh *et al.* (2007), Subramaniam (2000), Watt (1972).

Terminalia alata Roth.

Syn.: *T. tomentosa* (Roxb.) W. and A.

Family: Combretaceae

Eng., H. and Sans. Names : *Eng.*: Black Murdah;

H.: Asluna, Kauha, Sein;

Sans.: Krishnatvaka, Nissaraphalaka.

Distribution : Common throughout India, sub-Himalayan tracts, Deccan Peninsula.

Reproductive Cycle : May-February.

Part Used : Whole Plant.

Folk Uses : Plant used in diarrhoea, bowel complaints, dysentery especially blood dysentery.

Reference : Pal *et al.* (2000).

Terminalia arjuna (Roxb. *ex* DC.) Wight and Arn.

Syn.: *Pentaptera angustifolia* Roxb.; *P. glabra* Roxb.; *Terminalia berryi* W. and A.; *T. glabra* W. and A.; *T. ovalifolia* Rottl.

Family: Combretaceae

Eng., H. and Sans. Names : *Eng.*: Arjun Tree, White Murdah;

H.: Arhun, Kahu, Kahua, Kangach;

Sans.: Arjunah, Dhananjaya, Kakubhah, Karaviraka, Sachi, Savya.

Distribution : Throughout the greater part of India, banks of rivers, streams and dry water courses in sub-himalayan tract, C. and S. India and W. Bengal.

Reproductive Cycle : May-September.

Part Used : Whole Plant.

Folk Use : Powdered plant mixed with oil used against aphthae of infants.

Reference : Kirtikar and Basu (1984).

Terminalia bellirica **(Gaertn.) Roxb.**

Syn.: *Myrobalanus bellirica* Gaertn

Family: Combretaceae

Eng., H. and Sans. Names : *Eng.*: Beddanuts, Belleric Myrobalan;
H.: Bahera, Bhaira, Sagona;
Sans.: Aksah, Anilaghnaka, Bahira, Behera, Bibhitaka, Kaliyugalaya, Karshaphala.

Distribution : Throughout India, common in plains and lower hills, extending to Sri Lanka; not in dense region in W. India.

Reproductive Cycle : May-September.

Parts Used : Leaf, Fruit, Seed and Bark.

Folk Uses : Fruits, leaf and seeds used for treating pox. Fruits chewed by children to check cough and cold. Powdered dried fruit mixed with mother's milk given to children to arrest diarrhoea. 2 tsp mixture of paste of bark of *Terminalia bellirica, Spondias pinnata* and *Psidium guajava* in equal quantities with 3 fruits each of *Terminalia chebula* and *Emblica officinalis* as well as a few drops of alkali solution given once daily to children for 3 days to cure cholera.

References : Dhiman (2007), Gogoi and Borthakur (2001), Parrotta (2001), Prasad *et al.* (1996).

Terminalia catappa **L.***

Syn.: *Badamia comonersoni* Gaertn.; *Juglans catappa* Lour.; *Terminalia intermedia* Spreng.; *T. moluccana* Lamk.; *T. procera* Roxb.; *T. subcordata* Willd.

Family: Combretaceae

Eng., H. and Sans. Names : *Eng.*: Indian Almond, Malabar Almond, Malay Almond;
H.: Badami, Deshi-badam, Jangli-badam;
Sans.: Desabadama, Grahdruma, Ingudi.

Distribution : Common along beaches, rocky coast, upper margins of mangrove swamps, low land clearings and secondary forests, Andaman and Peninsular islands.

Reproductive Cycle : June-October.

Parts Used : Whole Plant, Leaf and Bark.

Folk Uses : Plant used for leprosy, scabies and other skin diseases. Leaves used as an emetic for infants. Infusion of bark used for treating internal ailments in children.

References : Pal *et al.* (2000), Prajapati *et al.* (2006).

Terminalia chebula Retz.*

Syn.: *Embryogonia arborea* Teys and Binn.; *Terminalia aruta* Ham.; *T. reticulata* Roth.

Family: Combretaceae

Eng., H. and Sans. Names : *Eng.*: Chebulic Myrobalan;

H.: Harad, Harara, Harra;

Sans.: Abhaya, Haritaki, Harra, Pathya.

Distribution : Abundantly found in N. India from Kangra-Bengal and southwards to the Deccan between 304-914m and up to 1,828m in Travancore; higher forests of Bombay Ghats, Satpuras, Bdgaum and Kanara.

Reproductive Cycle : April-October.

Parts Used : Whole Plant, Fruit, Seed and Galls.

Folk Uses : Plant used for aphthae in children. Fruit paste mixed with breast milk given orally to infants for cough and stomachache by Jenukurubas (1 tsp for an acute attack). Also, the paste applied locally on mumps and tonsillitis, by Soligas and also to check sound of snoring during breathing in children. Seeds tied to the arm to prevent smallpox. Galls used in infantile diarrhoea, the dose for a child of one year old being one grain every 3 hour.

References : Kshirsagar and Singh (2003, 2007), Rai and Sharma (2001), Rao and Henry (1996), Sen and Behera (2007), Watt (1972).

Thuja occidentalis L.

Syn.: *Juniperus ericoides* Mast.

Family: Cupressaceae

Eng. and H. Names : *Eng.*: American Arborvitae, Arbor-Vitae, White Cedar;

H.: Morpankhi, Thuja.

Distribution : Native of northeastern U.S., became a popular ornamental tree in Europe; found in wet, marshy ground and along riverbanks.

Reproductive Cycle : April-September.

Part Used : Leaf.

Folk Uses : Leaves considered diuretic and used for treating acute cystitis and bed wetting.

Reference : Prajapati *et al.* (2006).

Thunbergia fragrans **Roxb.**

Syn.: *Roxburghia rostrata* Russell *ex* Nees.

Family: Acanthaceae

Eng. and H. Names : *Eng.*: Sweet Clock Vine, White Thunbergia;
H.: Chimine.

Distribution : Occasional on other small trees, throughout India, 300-1,350m from Himalayas and Terai-Sri Lanka and Tenasserim; ascending to 2,350m.

Reproductive Cycle : April-October.

Part Used : Leaf.

Folk Use : Leaf juice, 1 tsp 3-4 times a day, given for dizziness in children by Soligas.

Reference : Kshirsagar and Singh (2007).

Thunbergia fragrans **var.** ***laevis*** **(Nees) C.B. Clarke** ***ex*** **Hook.f.**

Syn.: *T. laevis* Nees.

Family: Acanthaceae

Distribution : Deccan Peninsula, extending to Bihar and W. Bengal.

Part Used : Whole Plant.

Folk Use : An amulet of small seven pieces of the plant tied on the arm or neck of children for nine days for treatment of vesicles.

Reference : Sharma and Singh (2000).

Tinospora cordifolia **(Willd.) Miers** ***ex*** **Hook. and Thoms.**

Syn.: *Cocculus convolvulacrus* DC.; *C. cordifolia* DC.; *C. verrucosus* Wall.; *Menispermum cordifolium* Willd.

Family: Menispermaceae

Eng., H. and Sans. Names : *Eng.*: Gulancha Tinospora, Tinospora;
H.: Amrita, Giloy, Gulanca;
Sans.: Chakranai, Gurunchi, Jivanthika, Madhuparni, Nirjara, Tantri.

Distribution : Throughout tropical India, Myanmar, Andamans, Sri Lanka, from Kumaon-Assam and from Bihar and Concan-Sri Lanka and the Carnatic.

Reproductive Cycle : January-March.

Part Used : Whole Plant.

Folk Uses	:	Plant used in malarial fever, and poliomyelitis.
References	:	Kirtikar and Basu (1984), Pal *et al.* (2000).

Toddalia asiatica (L.) Lam.

Syn.: *T. aculeata* Pers.; *T. angustifolia* Miq.; *T. floribunda* Wall.; *T. rubricaulis* Willd.

Family: Rutaceae

Eng., H. and Sans. Names	:	*Eng.*: Forest-Pepper, Wild Orange-Tree, Lopez Tree; *H.*: Dehan, Jangli Kalimirch, Kanj; *Sans.*: Dahana, Kanchano, Tiksnaksah.
Distribution	:	Common in scrub Jungles, sub-tropical Himalayas from Kumaon eastwards-Bhutan, ascending to 1,700m, Khasia mountains ascending to 2,000m and throughout the W. Peninsula and Sri Lanka.
Reproductive Cycle	:	All seasons.
Parts Used	:	Root, Leaf and Bud.
Folk Uses	:	Root paste mixed with lemon juice applied locally on mumps by Jenukurubas. Cake made out of buds and leaves grounded with rice flour used against cough and cold in children.
References	:	Kshirsagar and Singh (2007), Kumar *et al.* (2000).

Toona ciliata M. Roem. (Pl. 14A)

Syn.: *Cedrela serrulata* Miq.; *C. toona* Roxb. *ex* Rottl.; *Toona kingii* (C. DC.) Harms.

Family: Meliaceae

Eng., H. and Sans. Names	:	*Eng.*: Happy Tree, Indian Mahogany, Red Cedar, Sandal Neem, Toon; *H.*: Lim, Mahanim, Toon, Tuni; *Sans.*: Apina, Kanta, Nandiviriksha, Tunah.
Distribution	:	Sub-Himalayan tract from the Indus eastwards, Chittagong, Assam, Myanmar, Chota Nagpur, Ganjam, W. Ghats of Bombay to the Nilgiri's and Anamalais and hills of W. Peninsula.
Reproductive Cycle	:	May-June.
Parts Used	:	Stem Bark and Wood.
Folk Uses	:	Decoction of stem bark with common alum used as gargle in infantile diarrhoea, dysentery and tonsillitis. Wood used in tonsillitis.

References	:	Ambasta (1986), Farooq (2005), Kapur and Singh (1996), Kirtikar and Basu (1984), Lalramnghinglova (2003), Pandey and Verma (2002), Parrotta (2001), Prajapati *et al.* (2006), Sood *et al.* (2012), Watt (1972).

Tournefortia montana **Lour.**

Syn.: *T. boniana* Gagnep.; *T. brachyantha* Merr. and Chun.; *T. sampsonii* Hance.

Family: Boraginaceae

Eng. Name	:	*Eng.*: Mountain Tournefortia.
Distribution	:	A subscandent shrub found in E. Himalayas.
Part Used	:	Root.
Folk Use	:	Bath of root infusion good for weak babies.
Reference	:	Agarwal (2003).

Trachyspermum ammi **(L.) Sprague**

Syn.: *Carum copticum* Hiern.; *Trachyspermum copticum* Link.

Family: Apiaceae

Eng., H. and Sans. Names	:	*Eng.*: Bishop's Weed, Carum, Lovage; *H.*: Ajowanj, Ajwain; *Sans.*: Yavani, Yawani.
Distribution	:	Cultivated throughout India.
Reproductive Cycle	:	Flower blooms in November and December.
Part Used	:	Whole Plant.
Folk Use	:	Plant used in polio.
Reference	:	Pal *et al.* (2000).

Trachyspermum roxburghianum **(DC.) Craib.**

Syn.: *Carum roxburghianum* Benth. *ex* Kurz.; *Trachyspermum involucratum* Wolff.

Family: Apiaceae

Eng., H. and Sans. Names	:	*Eng.*: Wild Celery; *H.*: Ajmod, Ajmot, Ajmud, Randhuni; *Sans.*: Ajamoda, Ajaoda.
Distribution	:	Cultivated throughout India.
Reproductive Cycle	:	January-April.

Part Used	:	Fruit.
Folk Use	:	Fruits prescribed for tonsillitis.
Reference	:	Chatterjee and Pakrashi (1997).

Tragia involucrata **L.**

Syn.: *T. trifida* Wall.

Family: Euphorbiaceae

Eng., H. and Sans. Names	:	*Eng.*: Indian Stinging-Nettle; *H.*: Barhanta; *Sans.*: Durlabha, Grahini, Virupa.
Distribution	:	Throughout India, ascending upto 750m.
Reproductive Cycle	:	October-May.
Part Used	:	Root.
Folk Use	:	Fresh roots grounded with rice water taken orally frequently in smallpox.
Reference	:	Ambily *et al.* (2010).

Tribulus terrestris **L.**

Syn.: *T. lanuginosus* L.

Family: Zygophyllaceae

Eng., H. and Sans. Names	:	*Eng.*: Land-caltrops, Puncture-vine; *H.*: Chhota Gokhru, Gokhru, Gokshri, Hussuk; *Sans.*: Bahukantaka, Chanadruma, Gokshura, Ikshugandha.
Distribution	:	Throughout warmer regions of India, upto 3,000m.
Reproductive Cycle	:	May-June.
Part Used	:	Whole Plant.
Folk Use	:	Plant used in polio.
Reference	:	Pal *et al.* (2000).

Trichaurus ericoides **(Rottler and Willd) Arn.**

Syn.: *Tamarix ericoides* Rottl.

Family: Tamaricaceae

Eng. and H. Names	:	*Eng.*: Erica Tamarisk; *H.*: Jhau.
Distribution	:	C. India, Bengal, W. Peninsula and Sri Lanka.

Reproductive Cycle : Flowering during September-December.

Part Used : Leaf.

Folk Uses : Decoction of leaves reported to cure enlarged spleen in children. Few leaves cooked with rice given to children against cough.

References : Agarwal (2003), Ambasta (1986).

Trichodesma indicum **(L.) Lehm.**

Syn.: *Borago indica* L.

Family: Boraginaceae

Eng., H. and Sans. Names : *Eng.*: Indian Borage;

H.: Andhahuli, Chhotakulpha, Hetamundiya, Ondhaphuli, Ratmundiya;

Sans.: Adhapushpi, Adhomukha, Darvika, Gandhapushpika.

Distribution : Weed in disturbed open habitats throughout most of India to 1,500m in Himalayas, typically on roadsides, cultivated fields, wastelands and black cotton soil in Peninsular India.

Reproductive Cycle : Most part of the year.

Part Used : Root.

Folk Use : Roots pounded with water given as a drink to children with dysentery.

References : Kirtikar and Basu (1984), Parrotta (2001).

Trichosanthes tricuspidata **Lour.**

Syn.: *T. palmata* Roxb.; *T. pubera* Blume.

Family: Cucurbitaceae

Eng., H. and Sans. Names : *Eng.*: Redball Snakegourd;

H.: Indrayan, Mahakal;

Sans.: Shvetapushpi.

Distribution : Common in all moist thickets from Himalayas-Sri Lanka and Singapore, ascending to 1,700m.

Reproductive Cycle : May-October.

Part Used : Fruit.

Folk Uses : Paste of fruit epicarp paste mixed with breast milk and honey given orally to infants for cough and fever.

Reference : Rao and Henry (1996).

***Trifolium pratense* L.**

Syn.: *T. borysthenicum* Gruner.; *T. bracteatum* Schousb.

Family: Fabaceae

Eng., H. and Sans. Names : *Eng.*: Beebread, Cow Clover, Cow Grass, Red Clover, Wild Clover;

H.: Tripatra;

Sans.: Vana-Methika.

Distribution : Kashmir-Garhwal, 1,350-2,650m.

Reproductive Cycle : May-October.

Part Used : Flower Heads.

Folk Use : Used in mumps.

Reference : Rajan *et al.* (2000).

***Trifolium repens* L.**

Syn.: *Lotodes repens* Kuntze.; *Trifolium limonium* Phil.; *T. stipitatum* Clos.

Family: Fabaceae

Eng. Names : *Eng.*: Duch White Clover, White Clover.

Distribution : Temperate and alpine Himalayas, ascending to 6,650m, Nilgiris and Sri Lanka, perhaps introduced.

Reproductive Cycle : July-October.

Part Used : Whole Plant.

Folk Use : Plant used in mumps.

Reference : Baburaj *et al.* (2000).

Trigonella foenum-graecum* L. (Pl. 14B)

Syn.: *T. tibetana* (Alef.) Vassilcz.

Family: Fabaceae

Eng., H. and Sans. Names : *Eng.*: Fenugreek, Greek Hayes;

H.: Methi, Muthi;

Sans.: Asumodhagam, Bahuparni, Chandrika, Dipani, Methika.

Distribution : Punjab, Kashmir, cultivated in many parts of India, upper Gangetic plains.

Reproductive Cycle : February-April.

Part Used : Seed.

Folk Uses : Infusion of seeds considered good cool drink for smallpox patients. Powdered seeds given with milk to check rickets and bedwetting in children.

References : Chatterjee and Pakrashi (1997), Chauhan (1999), Kirtikar and Basu (1984), Prajapati *et al.* (2006), Singh *et al.* (2007).

Tylophora asthmatica (L.f.) Wight and Arn.

Family: Asclepiadaceae

Eng., H. and Sans. Names : *Eng.*: Country Ipecacuanha, Emetic Swallow-wort, Indian Ipecacunha;

H.: Antamul, Jangli-pikvam;

Sans.: Lataksiri.

Distribution : Met within N. and E. Bengal, Assam. Kachar, Chittagong, Deccan peninsula, Myanmar-Malacca, Common in Sri Lanka.

Reproductive Cycle : July-October.

Part Used : Whole Plant.

Folk Use : Plant infusion given as emetic to those children suffering from phlegm.

Reference : Watt (1972).

Tylophora flexuosa R. Br.

Syn.: *T. tenuis* Bl.; *T. tenuissima* (Roxb.) Wt. and Arn.; *T. tetrapetala* (Dennst.) Suresh.

Family: Asclepiadaceae

Distribution : W. coast and Ghats of Chennai, Bengal.

Reproductive Cycle : April-December.

Part Used : Whole Plant.

Folk Use : Juice of plant used in smallpox.

References : Agarwal (2003), Ambasta (1986), Kirtikar and Basu (1984), Nayak *et al.* (2003b, c).

Typha australis K. Schum. and Thonner

Family: Typhaceae

Eng., H. and Sans. Names : *Eng.*: Elephant Grass;

H.: Patera;

Sans.: Eraka.

Distribution	:	Grows in marshy regions of India, upto 1,730m from U.P., Assam, Kashmir.
Reproductive Cycle	:	July-October.
Part Used	:	Rhizome.
Folk Use	:	Used in measles.
Reference	:	Agarwal (2003).

Typha elephantina **Roxb.**

Syn.: *T. maresii* Batt.; *T. schimperi* Rohrb.

Family: Typhaceae

Eng., H. and Sans. Names	:	*Eng.*: Elephant Grass, Indian Reed-Mace; *H.*: Bora, Mothitrina, Pater; *Sans.*: Eraka, Shari, Shimbi.
Distribution	:	Marshes from N.W. India-Assam and southwards.
Part Used	:	Rhizome.
Folk Use	:	Used in measles.
References	:	Ambasta (1986), Kirtikar and Basu(1984).

Typhonium trilobatum **(L.) Schott.**

Syn.: *Arum orixense* Roxb.; *A. trilobatum* L.

Family: Araceae

Eng. Names	:	*Eng.*: Bengal Arum, Greater Cat's Tail, Loked Leaf Typhonium.
Distribution	:	W. Peninsula, Bengal, Assam.
Reproductive Cycle	:	May-October.
Part Used	:	Root.
Folk Use	:	Rootstock used against tonsillitis.
Reference	:	Gogoi and Das (2003).

Uncaria rhynchophylla **(Miq.) Miq. *ex* Havil.**

Syn.: *Nauclea rhynchophylla* Miq.; *Uncaria rhynchophylloides* F.C. How.

Family: Rubiaceae

Eng. Name	:	*Eng.*: Sharp Leaved Gambir.
Distribution	:	In woods or ravines, on stream banks in China and Japan.
Reproductive Cycle	:	May-December.

Part Used : Unciferous branches.

Folk Use : Used in infantile convulsions.

Reference : Prajapati *et al.* (2006).

***Uncaria sessilifructus* Roxb.**

Syn.: *Nauclea sessilifructus* (Roxb.) D. Dietr.; *N. sessilis* Spreng.

Family: Rubiaceae

Eng. Name : *Eng.*: Gambir Vine.

Distribution : E. Bengal, from Sikkim and the Khasia mountains to Chittagong, Pegu and Tenasserim.

Reproductive Cycle : March-December.

Parts Used : Leaf and Shoot.

Folk Use : Half a cup decoction of leaves and shoot taken twice daily for tonsillitis.

Reference : Lalramnghinglova (1996).

***Uraria picta* (Jacq.) DC.**

Syn.: *Doodia picta* Roxb.; *Hedysarum pictum* Jacq.; *Uraria linearis* Hassk.

Family: Fabaceae

Eng., H. and Sans. Names : *Eng.*: Painted- Leaved Uraria;

H.: Dabra, Prisniparni;

Sans.: Guha, Prasniparni, Prishniparni, Shaliparni.

Distribution : Found in dry grasslands, wasteplaces and open forests in the sub-Himalayan tract from Kashmir-W. Bengal and Assam, upto 1,800m and all the plains of India.

Reproductive Cycle : July-September.

Parts Used : Root and Fruit.

Folk Uses : Both roots and pods employed in Bihar against prolapsed anus and sore-mouth in children.

References : Agarwal (2003), Ambasta (1986), Chatterjee and Pakrashi (1997), Chauhan (1999), Khare (2004), Kirtikar and Basu (1984), Parrotta (2001), Prajapati *et al.* (2006), Watt (1972).

Urtica dioica* L.

Family: Urticaceae

Eng., H. and Sans. Names : *Eng.*: Man's Plaything, Nacighty, Nettle, The Stinging Nettle;

		H.: Bichhu Booti;
		Sans.: Vrscikali.
Distribution	:	Commonly found between 500-2,000m at Arakot of Uttarkashi, N.W. Himalayas; 2,700-3,550m.
Reproductive Cycle	:	June-September.
Parts Used	:	Aerial Parts, Stem and Leaf.
Folk Uses	:	Aerial parts used in infantile eczema. Small young and soft stem applied locally into anus of small children in chronic constipation. Leaf juice used as nasal drops in frequent nose bleeding.
References	:	Arya and Prakash (2000), Karnick (1994b), Salam *et al.* (2011).

Vaccinium myrtillus **L.**

Syn.: *V. oreophilum* Rydb.

Family: Ericaceae

Eng. Names	:	*Eng.*: Bilberry, Farkle Berry, Sparkle Berry.
Distribution	:	Found in S. India.
Reproductive Cycle	:	Flowering in April-June.
Part Used	:	Fruit.
Folk Use	:	Decoction of fruits useful for treating diarrhoea in children.
Reference	:	Prajapati *et al.* (2006).

Valeriana jatamansi **Jones.**

Syn.: *V. spica* Vahl.; *V. villosa* Wall.; *V. wallichi* DC.

Family: Valerianaceae

Eng., H. and Sans. Names	:	*Eng.*: Indian Valerian;
		H.: Mushkbala, Sugandhbala, Tagar;
		Sans.: Chebra, Danda, Dipana, Tagara.
Distribution	:	Alpine Himalaya, 3,650-5,000m from Kumaon-Sikkim, ascending to 5,650m in Sikkim.
Reproductive Cycle	:	June-October.
Part Used	:	Leaf.
Folk Use	:	Leaf juice with honey administered orally to children having diarrhoea.
Reference	:	Mohanty *et al.* (1996).

Ventilago madraspatana **Gaertn.**

Syn.: *V. bracteata* Wall.

Family: Rhamnaceae

Eng., H. and Sans. Names	:	*Eng.*: Red Creeper; *H.*: Pitti; *Sans.*: Dinesavalli, Raktavalli.
Distribution	:	Forests throughout India.
Reproductive Cycle	:	March-June.
Part Used	:	Whole Plant.
Folk Use	:	Plant used in itching.
Reference	:	Pal *et al.* (2000).

Veratrum viride **Ait.**

Syn.: *Helonias viridis* (Aiton) Ker Gawl.

Family: Melanthiaceae

Eng. Names	:	*Eng.*: American Hellebore, Green Hellebore.
Distribution	:	Grown in Kashmir.
Reproductive Cycle	:	June-September.
Part Used	:	Rhizome.
Folk Use	:	Dried rhizome used in acute tonsillitis.
References	:	Agarwal (2003), Ambasta (1986).

Verbascum thapsus **L. (Pl. 14C)***

Syn.: *V. indicum* Wall.

Family: Scrophulariaceae

Eng., H. and Sans. Names	:	*Eng.*: Aaron's Rod, Bullock's Lungwort, Common Mullein, Cow's Lungwort, Golden Grain, High Taper, Old Man's Flannel, Velvet Dock, Virgin Mary's Candle; *H.*: Gidar Tamaku; *Sans.*: Ban Tamaku, Gidar Tamaku, Phulla.
Distribution	:	Found in H.P. between 1,200-3,000m and temperate Himalayas, 2,000-3,650m from Kashmir-Bhutan.
Reproductive Cycle	:	June-September.
Part Used	:	Flower.
Folk Use	:	Flower oil used as a remedy for enuresia afflicting children.
Reference	:	Aman and Dar (2003).

Vernonia anthelmintica (L.) Willd.

Syn.: *Centratherum anthelminticum* (L.) Kuntze. *ex* Gamble.; *Conyza anthelmintica* L.; *Vernonia stenolepis* Oliv.

Family: Asteraceae

Eng., H. and Sans. Names : *Eng.*: Purple Fleebane;
H.: Baksi, Kalijira, Samraj;
Sans.: Aranyajiraka.

Distribution : Throughout India, from Kashmir-Sri Lanka and Malacca, ascending to 1,850m in the Himalayas and Khasia mountains.

Reproductive Cycle : August-September.

Part Used : Seed.

Folk Uses : Paste of grounded seeds in coconut oil applied to cure wounds and skin eruptions.

Reference : Reddy *et al.* (2003).

Vernonia cinerea (L.) Juss.

Syn.: *Conyza cinerea* L.; *C. mollis* Willd.; *C. ovata* Wall.; *Serratula cinerea* Roxb.; *Vernonia albicans* DC.; *V. laxiflora* Less.

Family: Asteraceae

Eng., H. and Sans. Names : *Eng.*: Ash Coloured Fleabane;
H.: Dandotpala, Sahadevi, Sodori;
Sans.: Devasasha, Sahadevi.

Distribution : Found throughout India, ascending to 2,700m in Himalayas, Khasia and Peninsular mountains.

Reproductive Cycle : August-May.

Part Used : Whole Plant.

Folk Use : Plant juice given to infants in irregular painful urine.

References : Agarwal (2003), Parrotta (2001).

Vernonia divergens (DC.) Edgew.

Syn.: *Cacalia divergens* Kuntze.; *Eupatorium acuminatum* D. Don.; *E. nepalense* Spreng.

Family: Asteraceae

Eng. Name : *Eng.*: Diverging Vernonia.

Distribution : Bihar, on Parusnath, 1,350m C. India and from the Concan-Travancore mountains, Myanmar and Tenasserim.

Reproductive Cycle	:	Flowering in December.
Part Used	:	Leaf.
Folk Use	:	1 tsp shade dried powdered leaves given twice daily for 3 days to asthmatic children.
Reference	:	Viswanathan *et al.* (2006).

Vicia sativa subsp. *nigra* (L.) Ehrh.

Syn.: *V. angustifolia* L.; *V. bobartii* Koch.; *V. cuneata* Guss.; *V. sativa* L. var. *angustifolia* L.

Family: Fabaceae

Eng. and H. Names	:	*Eng.*: Narrow Leaved Vetch; *H.*: Akra, Akta, Ankra.
Distribution	:	N.W. regions, ascending from the plains of Bengal to 2,450m in Kumaon.
Reproductive Cycle	:	June-November.
Part Used	:	Seed.
Folk Use	:	Decoction of seeds used in smallpox and measles.
Reference	:	Ambasta (1986).

Vigna marina (Burm) Merr.

Syn.: *Dolichos luteus* Sw.; *Vigna lutea* (Sw.) A. Grey.; *V. retusa* (E.Mey.) Walp.

Family: Fabaceae

Eng. Names	:	*Eng.*: Beach Bean, Noched Cow-Pea.
Distribution	:	Common on sandy seashores and among coastal vegetation and plantation.
Reproductive Cycle	:	Fruiting in October.
Part Used	:	Leaf.
Folk Use	:	Leaf infusion given for treating fever in children.
Reference	:	Prajapati *et al.* (2006).

Vigna umbellata (Thunb.) Ohwi and H. Ohashi

Syn.: *Dolichos umbellatus* Thunb.; *Vigna calcarata* (Roxb.) Kurz.; *V. papuana* Baker.f.

Family: Fabaceae

Eng. and H. Names	:	*Eng.*: Climbing Mountain Bean; *H.*: Ghurush, Sutri.

Distribution	:	Throughout Indian plains, ascending to 1,500m in hills; cultivated or wild.
Reproductive Cycle	:	May-September.
Part Used	:	Seed.
Folk Use	:	Seeds cooked as pulse and given to children suffering from measles.
Reference	:	Pande and Joshi (2001).

Viola biflora **L.**

Syn.: *Chrysion biflorum* (L.) Spach.; *Dischidium biflorum* (L.) Opiz.

Family: Violaceae

Eng., H. and Sans. Names	:	*Eng.*: Twin Flowered Violet; *H.*: Banfsa; *Sans.*: Banfsa.
Distribution	:	Temperate Himalayas from Kashmir-Sikkim, 2,000-3,650m.
Reproductive Cycle	:	May-October.
Part Used	:	Whole Plant.
Folk Use	:	Decoction of plant used in chickenpox.
Reference	:	Kumar (2002).

Viola odorata **L.**

Family: Violaceae

Eng., H. and Sans. Names	:	*Eng.*: Appel-leaf, Bairwort, March Violet, Sweet Violet; *H.*: Banaphasha; *Sans.*: Jvarapaha, Nilapushpa, Sukshmapatra.
Distribution	:	N.W. Himalayas, Kashmir, 1,700-2,000m.
Reproductive Cycle	:	Flowering in March-July.
Part Used	:	Petals.
Folk Use	:	Extract of petals mixed with sugar given as a remedy for infantile disorders.
Reference	:	Kaul (1997).

Viola pilosa **Blume. (Pl. 14D)**

Syn.: *V. pogonantha* W.W. Sn.; *V. serpens*Wall *ex* Ging.

Family: Violaceae

Eng., H. and Sans. Names	:	*Eng.*: Smooth-Leaf White Voilet;

		H.: Banafshah;
		Sans.: Nilapushpa.
Distribution	:	Throughout temperate Himalayas, Khasia hills, Pulney and Nilgiri mountains and Sri Lanka, 1,700-2,650m.
Reproductive Cycle	:	March-July.
Part Used	:	Flower.
Folk Use	:	Used in infantile affections.
Reference	:	Singh (1999).

Vitex leucoxylon **L.f.**

Syn.: *V. rheedei* Kostel.; *V. saligna* Roxb.

Family: Lamiaceae

Eng. and Sans. Names	:	*Eng.*: Five Leaved Chaste Tree, White-Wood Chaste Tree, White-Wooded Wallrothia;
		Sans.: Paravatapadi.
Distribution	:	From S. Assam and Cachar-Malacca, W. Peninsula, throughout Deccan Peninsula.
Reproductive Cycle	:	September-December.
Part Used	:	Leaf.
Folk Use	:	Vapours emanating from boiling water containing its leaves inhaled to cure infant's cold.
Reference	:	Kumar (1996).

Vitex negundo **L. (Pl. 15A)**

Family: Lamiaceae

Eng., H. and Sans. Names	:	*Eng.*: Chinese Chaste Tree, Indian Privet;
		H.: Mewri, Nisinda, Panikisambhala, Sambhalu, Shambalu;
		Sans.: Indrani, Nilapushpa, Nirgundi, Suvaha.
Distribution	:	Throughout the greater parts of India, ascending to 1,500m in the outer Himalayas.
Reproductive Cycle	:	March-June.
Parts Used	:	Whole Plant, Root and Leaf.
Folk Uses	:	Extract of plant used for expelling worms in children. Plant also used in painful teething and various skin diseases in children. Roots used in tonsillitis and to overcome teething problems. Juice of leaves given to children in tonsils and eradication of worms.

References : Chauhan (1999), Kirtikar and Basu (1984), Pal *et al.* (2000), Sharma (2003), Swami and Gupta (1996), Thakor (2009).

Vitex trifolia **L.**

Syn.: *V. indica* Mill.

Family: Lamiaceae

Eng., H. and Sans. Names : *Eng.*: Fox Grape, Indian Wild Pepper;

H.: Amalbel, Pani ki Sambhalu, Ramchana;

Sans.: Aranyavasini, Atyamlaporni, Neelanirgundi.

Distribution : Throughout India in tropical and sub-tropical regions.

Reproductive Cycle : July-November.

Part Used : Leaf.

Folk Use : Bath (2-3 times in a week) with decoction of 100-200 leaves boiled with about 5L of water, till reduced to 1/3rd useful for scabies in children.

Reference : Singh and Singh (2003a).

Vitis vinifera **L. (Pl. 15B)**

Family: Vitaceae

Eng., H. and Sans. Names : *Eng.*: Common Grape Vine, European Grape, Wine-grape;

H.: Angur, Dakh;

Sans.: Amrutaphala, Draksha, Mrdvika, Yakshmaghni.

Distribution : N.W. Himalayas, cultivated extensively in N.W. India and rarely in the Peninsula and Sri Lanka.

Reproductive Cycle : October-April.

Part Used : Berries.

Folk Uses : Used against thrush and constipation during teething in children.

References : Chauhan *et al.* (2007), Chatterjee and Pakrashi (1997), Khare (2004).

Volutarella divaricata **Benth. and Hook.f.**

Syn.: *Amberboia divaricata* Kuntze.

Family: Asteraceae

Distribution : C. and S. India, from Bihar and upper Gangetic plains-Lahore and from Sind-Mysore and Deccan, ascending to 1,000m in the N.W. Himalayas.

Reproductive Cycle : January-March.

Part Used : Whole Plant.

Folk Use : Plant used in fever.

Reference : Pal *et al.* (2000).

***Withania somnifera* (L.) Dunal.**

Syn.: *Physalis flexuosa* L.; *P. somnifera* Link.; *Withania microphysalis* Suess.

Family: Solanaceae

Eng., H. and Sans. Names : *Eng.*: Winter Cherry;

H.: Asgandh, Punir;

Sans.: Ashwagandha, Gandhapatri, Hoya, Priyakari, Turangi-gandha, Varahakareni.

Distribution : Drier parts of India, ascending upto 1,700m in the Himalayas.

Reproductive Cycle : December-June.

Parts Used : Whole Plant, Tuber and Root.

Folk Uses : Plant used in poliomyelitis. Tubers good for marasmus. Roots used in emaciation, smallpox and in marasmus.

References : Chauhan (1999), Kirtikar and Basu (1984), Pal *et al.* (2000), Sharma (2003), Srivastava *et al.* (2003a), Thakur *et al.* (1989).

Woodfordia fruticosa* (L.) Kurz.

Syn.: *W. floribunda* Salisb.; *W. tomentosa* Bedd.

Family: Lythraceae

Eng., H. and Sans. Names : *Eng.*: Fire-flame Bush, Shiranjitia;

H.: Dawi, Dhai, Dhaula, Dhaura, Santha, Thawi;

Sans.: Agnijwala, Dhataki, Madaniyahetu.

Distribution : Throughout N. India, ascending to 1,500m in Himalayas.

Reproductive Cycle : January-June.

Part Used : Flower.

Folk Uses : Flowers used in smallpox. Paste of flowers, long pepper and fruits of *Phyllanthus emblica* rubbed on gums of infants during teething troubles. A sweet bolus of its flowers parched paddy and sugar given to children suffering from diarrhoea.

References : Agarwal (2003), Borthakur (1993), Khare (2004), Kothari and Rao (2000), Pande *et al.* (2006), Thakur *et al.* (1989).

Wrightia arborea **(Dennst) Mabb.**

Syn.: *W. hamiltoniana* Wall.; *W. wallichi* A. DC.

Family: Apocynaceae

Eng., H. and Sans. Names : *Eng.*: Wooly Dyeing Rosebay;

H.: Daira, Dharauli, Dudhi;

Sans.: Kutaja, Kutajah, Kutajastri.

Distribution : Throughout warmer regions of India, upto 600m in the Himalayas and upto 1,200m in Nilgiris in S. India.

Reproductive Cycle : March-July.

Part Used : Roots.

Folk Use : Roots used in fever of babies.

Reference : Jain (1991).

Wrightia tinctoria **(Roxb.) R. Br.**

Syn.: *Nerium tinctorium* Roxb.

Family: Apocynaceae

Eng., H. and Sans. Names : *Eng.*: Pala Indigo Plant;

H.: Dudhi, Khirni, Mitha;

Sans.: Hayamaraka, Madhuindrayava, Svetakutaja.

Distribution : C. India and throughout the W. Peninsula, Myanmar, widely distributed from Rajasthan-M.P. southwards to Tamil Nadu in deciduous forests on dry, sandy soils, often along streams.

Reproductive Cycle : May-September.

Parts Used : Root, Fruit and Bark.

Folk Uses : Paste of bark, root and fruits in equal proportion given twice a day to nursing mothers to expel thread worms from breast fed infants.

References : Brahmam and Saxena (1990), Parrotta (2001).

Xanthium strumarium **L.**

Syn.: *X. indicum* DC.; *X. roxburghii* Wall.

Family: Asteraceae

Eng., H. and Sans. Names : *Eng.*: Burweed, Cocklebur, Ditch Bur, Louse-bur;

		H.: Adhasisi, Banokra, Chota-dhatura, Chota-gokhru, Gokhru;
		Sans.: Arishta, Chanda, Itora, Shwetakusuma, Vanamalini.
Distribution	:	Naturalized throughout hotter parts of India, ascending upto 1,828m in Himalayas.
Reproductive Cycle	:	August-October.
Parts Used	:	Whole Plant, Root, Fruit and Seed.
Folk Uses	:	Plant good against smallpox, bowel complaints, dysentery, blood dysentery and as anthelmintic. Roots tied around neck to check continous crying in newborn child. Fruits used in smallpox. Wreath of seeds worn around neck of children on Thursday or Sunday as a treatment of stye, chronic conjunctivitis and inflammation.
References	:	Bhalla *et al.* (1996), Chauhan (1999), Farooq (2005), Kirtikar and Basu (1984), Nayak *et al.* (2003a), Pal *et al.* (2000), Pande *et al.* (2006), Pandey and Verma (2003), Parrotta (2001), Prajapati *et al.* (2006), Sandhya *et al.* (2011), Sharma (2003, 2004), Singh and Narain (2008), Singh and Srivastava (2000), Watt (1972).

***Ximenia americana* L.**

Syn.: *X. aculeata* Crantz.; *X. spinosa* Salisb.

Family: Olacaceae

Eng. Names	:	*Eng.*: False Sandle Wood, Mountain Plum, Tallow Wood, Wild Plum.
Distribution	:	Deccan Peninsula and coastal regions on Andamans.
Reproductive Cycle	:	March-June.
Part Used	:	Bark.
Folk Use	:	Paste of bark used as a post-measle lotion.
Reference	:	Agarwal (2003).

***Xylocarpus granatum* J. Koen.**

Syn.: *X. obovatus* Juss.

Family: Meliaceae

Eng., H. and Sans. Names	:	*Eng.*: Apple Mangrove, Cannonball Mangrove, Cedar Mangrove, Puzzle Nut.
Distribution	:	Muddy seacoast throughout India.

Reproductive Cycle : April-November.

Part Used : Seed.

Folk Use : Seed paste applied on mumps.

References : Agarwal (2003), Jain (1991).

Xylosma longifolia **Clos.**

Family: Flacourtiaceae

H. Names : *H.*: Dandal, Kandhara, Katari, Katpatra, Sialu.

Distribution : W. Himalayas on the outer ranges, ascending to 1,700m from Kumaon-Marri, Assam, at Nowgong and Gowahatty.

Reproductive Cycle : February-May.

Part Used : Leaf.

Folk Use : Leaves used for scabies in children.

Reference : Singh and Singh (2003a).

Zanthoxylum armatum **DC. (Pl. 15C)**

Syn.: *Fagara armata* Thunb.; *Zanthoxylum planispinum* Sieb and Zucc.

Family: Rutaceae

Eng., H. and Sans. Names : *Eng.*: Toothache Tree;

H.: Darmar, Nepali Dhaniya, Tej-phal, Tumru;

Sans.: Dhiva, Gandhalu, Tumburu.

Distribution : In Himalayas and N.E. India.

Reproductive Cycle : March-November.

Parts Used : Whole Plant and Bark.

Folk Uses : Plant used in diarrhoea, dysentery, blood dysentery and as an anthelmintic. Bark with root of *Aristolochia cathartii* and *Cyclea peltata* used against smallpox by Garos of Meghalaya.

References : Jain (1991), Kumar *et al.* (1980), Pal *et al.* (2000).

Zingiber officinale **Ross. (Pl. 15D)***

Syn.: *Amomum zingiber* L.; *Curcuma longifolia* Wall.

Family: Zingiberaceae

Eng., H. and Sans. Names : *Eng.*: Ginger;

H.: Ada, Adrak, Sonth (when dry);

Sans.: Ardrakam, Gulmamula, Mahija, Shunthi.

Distribution : Cultivated in tropical regions widely cultivated in India upto 1,500m.

Reproductive Cycle : August-September.

Part Used : Rhizome.

Folk Use : Rhizome with fruit of clove, *Piper longum* and soap nut tree or along with rice used for tonsils in Assam.

References : Boissya and Majumdar (1980), Karnick *et al.* (1981).

Ziziphus abyssinica **Hochst.** ***ex*** **A. Rich.**

Syn.: *Z. xylopyrus* Hochst. *ex* A. Rich.

Family: Rhamnaceae

Eng., H. and Sans. Names : *Eng.*: Catch Thorn, Jujube;

H.: Bhander, Ghont, Kakor ;

Sans.: Ghonta.

Distribution : N.W. India, Nepal, Banda, Rajputana, ascending to 700m.

Reproductive Cycle : January-October.

Part Used : Stem Bark.

Folk Uses : Medicated oil prepared by boiling its stem bark in mustard oil prescribed for massaging body of an infant every 8 hours to treat fever/cold.

References : Girach and Aminuddin (1995), Parrotta (2001).

Ziziphus jujuba **Mill.***

Syn.: *Z. sativa* Gaertn.; *Z. vulgaris* Lam.

Family: Rhamnaceae

Eng., H. and Sans. Names : *Eng.*: Chinese Date, Common Jujube, Indian Jujube, Jujube;

H.: Ban Ber, Kandiari, Kandika, Pitni-ber, Singli;

Sans.: Soubira.

Distribution : Throughout India and in the outer Himalayas up to 1,350m.

Reproductive Cycle : September-February.

Parts Used : Leaf, Stem Bark and Fruit.

Folk Uses : Leaf juice used in typhoid. Infusion of stem bark mixed with asafoetida leaves given orally once a day for 3 days for constipation in children. Fruits used against smallpox, measles and chickenpox.

References : Agarwal (2003), Parrotta (2001), Ram *et al.* (2002), Thakur *et al.* (1989).

Zornia diphylla **(L.) Pers.**

Syn.: *Hedysarum conjugatum* Willd.; *Zornia conjugata* (Willd.) Sm.

Family: Fabaceae

Distribution : Throughout plains of India from Himalayas-Sri Lanka and Myanmar, ascending to 1,350m in Kumaon.

Reproductive Cycle : Flowering during August-November.

Part Used : Root.

Folk Uses : Roots given to children to induce sleep, and as a soporific.

References : Agarwal (2003), Ambasta (1986), Kirtikar and Basu (1984).

Zornia gibbosa **Span.**

Syn.: *Z. angustifolia* Sensu. auct.; *Z. graminea* Span.

Family: Fabaceae

Eng. and H. Names : *Eng.*: Grassllike Zornia;

H.: Gewani, Somrapani.

Distribution : Widespread in India in open, grassy habitats or as a weed in cultivated fields to an elevation of 1,500m.

Reproductive Cycle : August-September.

Part Used : Root.

Folk Use : Root used as a tranquillizer for inducing sleep in children.

References : Banerjee (2000), Deshmukh *et al.* (2000), Nayak *et al.* (2003a), Parrotta (2001).

Epilogue

From the foregoing description and analysis of data on the plants used for child health care in India, the following important conclusions can be drawn:

1. Usage of 800 species belonging to 568 genera under 163 families have been delineated for the purpose, of which a large majority are herbaceous taxa **(Figure 1; Appendix I)**.

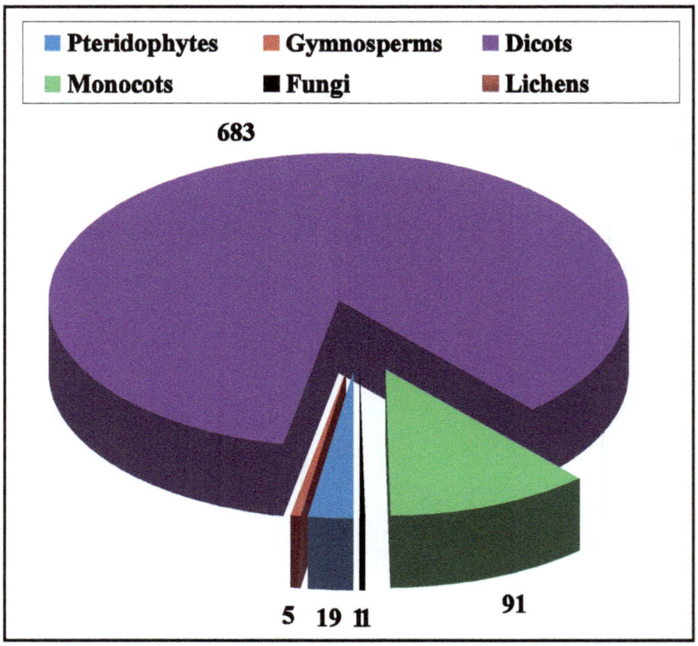

Figure 1: Various Divisions of Ethnic Indian Plants Used for Child Health Care.

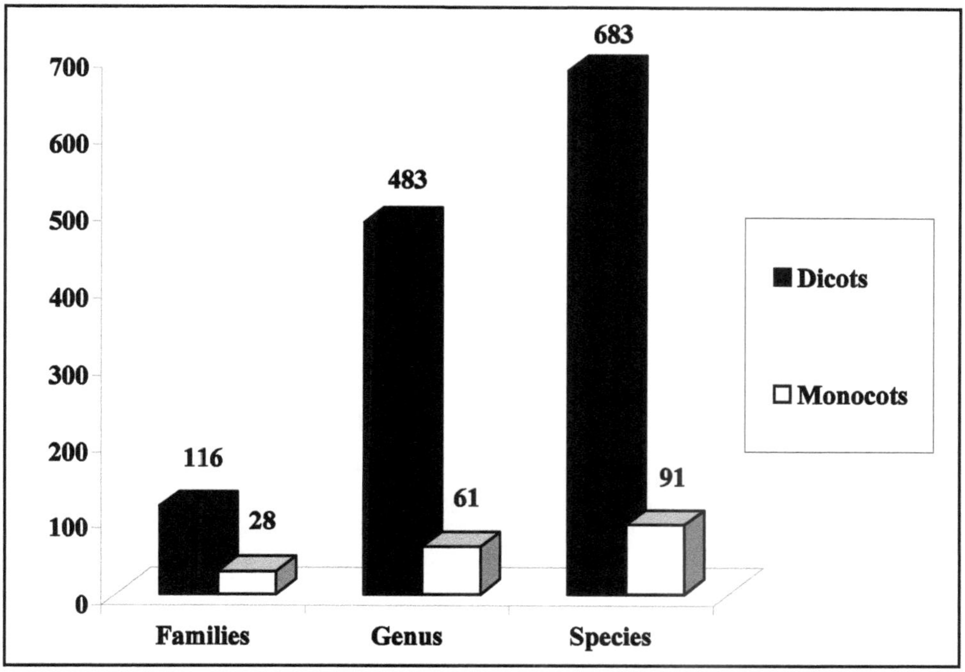

Figure 2: Families, Genera and Species of Plants for Child Health Care.

2. Plant category-wise analysis of the available data shows the usage of 1 species (1 genus) belonging to fungi; 1 species (1 genus) to lichens; 5 species (5 genera) to gymnosperms; 19 species (17 genera) to pteridophytes and the remaining 774 species (544 genera) to angiosperms of which 683 species are dicots and the remaining 91 as monocots **(Appendix I)**.
3. Statistically, the ratio of species, genera and families between monocots and dicots is 1:7.50; 1:7.91 and 1:4.14 **(Figure 2).** Imperatively, the dicotyledonous taxa predominate in their usage.
4. Habitat-wise analysis of the presently collated data reveals that plant species having scattered or throughout distribution (382 species) predominate in usage, followed by Peninsular India (Ghats, Coastal Plains, Deccan Plateau; 97 species), N.E. and E. Himalayas (90 species), Entire Himalayas (71 species), N.W. and W. Himalayas (63 species), Central Highlands and Plains of India (39 species), Tropical forests of India (39 species) and Miscellaneous (19 species) **(Figure 3; Table 1)**. Evidently, herbs employed for child health care by Indian tribals are located in various phytogeographical regions.
5. Most frequently used ethnobotanical families for child health care in India are represented by Fabaceae (82 species, 50 genera), Asteraceae (60 species, 44 genera), Euphorbiaceae (44 species, 22 genera), Lamiaceae (39 species, 22 genera), Rubiaceae (30 species, 23 genera), Acanthaceae (22 species, 16 genera), Solanaceae (17 species, 7 genera), Apiaceae (16 species, 13 genera),

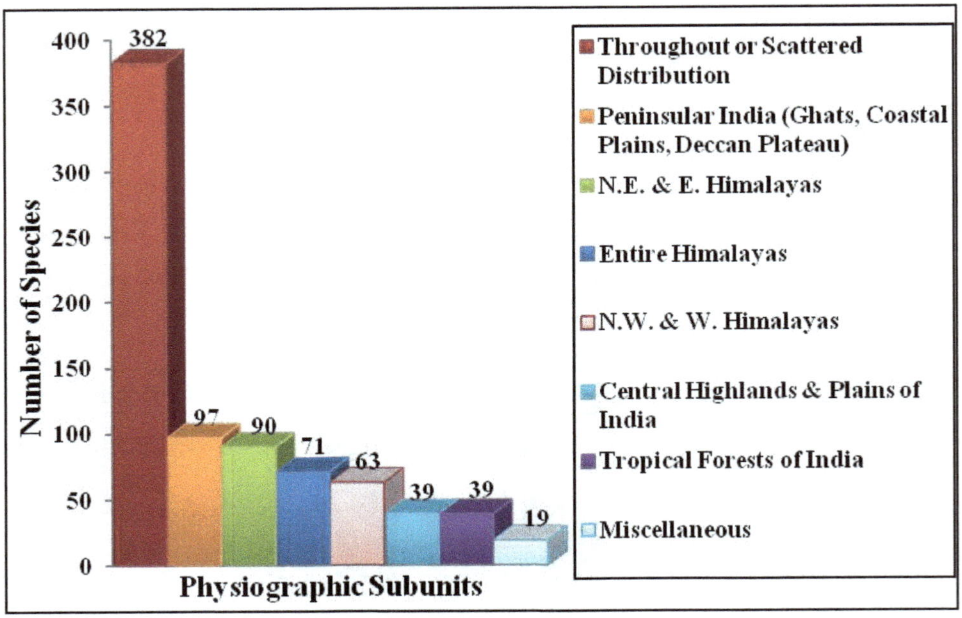

Figure 3: Histogram Showing Distribution Patterns of Indian Plants for Child Health Care.

Rutaceae (15 species, 12 genera), Asclepiadaceae (14 species, 11 genera) etc. amongst dicots **(Figure 4; Table 2; Appendix II)** whereas these are Poaceae (19 species, 13 genera), Zingiberaceae (14 species, 6 genera), Araceae (7 species, 7 genera), Arecaceae (6 species, 5 genera), Liliaceae (6 species, 2 genera), Asparagaceae (5 species, 1 genus), Cyperaceae (4 species, 3 genera), Iridaceae (3 species, 2 genera), Orchidaceae (3 species, 3 genera) and others for monocots **(Figure 5; Table 3; Appendix II)**. Factually, the size of families has a strong bearing to the above statistics.

Table 1: Distribution Patterns of Plants for Child Health Care

Distribution	*No. of Species*
Throughout or Scattered Distribution	382
N.W. and W. Himalayas	63
Peninsular India (Ghats, Coastal Plains, Deccan Plateau)	97
Entire Himalayas	71
Central Highlands and Plains of India	39
N.E. and E. Himalayas	90
Miscellaneous	19
Tropical Forests of India	39

Table 2: Dicot Families Employed for Child Health Care

Name of Family	*Total No. of Species*	*Percentage of Taxa*
Fabaceae	82	10.25
Asteraceae	60	7.5
Euphorbiaceae	44	5.5
Lamiaceae	39	4.87
Rubiaceae	30	3.75
Acanthaceae	22	2.75
Solanaceae	17	2.12
Apiaceae	16	2
Rutaceae	15	1.87
Asclepiadaceae	14	1.75
Rosaceae	14	1.75
Apocynaceae	13	1.62
Cucurbitaceae	13	1.62
Malvaceae	12	1.5
Ranunculaceae	12	1.5
Moraceae	11	1.37
Myrtaceae	11	1.37
Anacardiaceae	10	1.25
Oleaceae	10	1.25
Combretaceae	8	1
Polygonaceae	8	1
Sapindaceae	7	0.87
Amaranthaceae	6	0.75
Boraginaceae	6	0.75
Convolvulaceae	6	0.75
Lauraceae	6	0.75
Menispermaceae	6	0.75
Verbenaceae	6	0.75
Capparaceae	5	0.62
Scrophulariaceae	5	0.62
Araliaceae	4	0.5
Aristolochiaceae	4	0.5
Bignoniaceae	4	0.5
Flacourtiaceae	4	0.5
Myrsinaceae	4	0.5
Papaveraceae	4	0.5
Piperaceae	4	0.5
Rhamnaceae	4	0.5

Contd...

Table 2–*Contd...*

Name of Family	*Total No. of Species*	*Percentage of Taxa*
Sterculiaceae	4	0.5
Urticaceae	4	0.5
Violaceae	4	0.5
Vitaceae	4	0.5
Zygophyllaceae	4	0.5
Berberidaceae	3	0.37
Bombacaceae	3	0.37
Brassicaceae	3	0.37
Celastraceae	3	0.37
Clusiaceae	3	0.37
Fagaceae	3	0.37
Loganiaceae	3	0.37
Melastomataceae	3	0.37
Meliaceae	3	0.37
Oxalidaceae	3	0.37
Sapotaceae	3	0.37
Simaroubaceae	3	0.37
Tiliaceae	3	0.37
Annonaceae	2	0.25
Burseraceae	2	0.25
Campanulaceae	2	0.25
Caprifoliaceae	2	0.25
Caryophyllaceae	2	0.25
Chenopodiaceae	2	0.25
Crassulaceae	2	0.25
Ebenaceae	2	0.25
Ericaceae	2	0.25
Gentianaceae	2	0.25
Lythraceae	2	0.25
Nyctaginaceae	2	0.25
Olacaceae	2	0.25
Pedaliaceae	2	0.25
Phytolaccaceae	2	0.25
Plantaginaceae	2	0.25
Salvadoraceae	2	0.25
Saxifragaceae	2	0.25
Alangiaceae	1	0.12
Aquifoliaceae	1	0.12

Contd...

Table 2–*Contd...*

Name of Family	*Total No. of Species*	*Percentage of Taxa*
Balanitaceae	1	0.12
Basellaceae	1	0.12
Bixaceae	1	0.12
Cactaceae	1	0.12
Cannabinaceae	1	0.12
Caricaceae	1	0.12
Dilleniaceae	1	0.12
Dipterocarpaceae	1	0.12
Droseraceae	1	0.12
Elaeocarpaceae	1	0.12
Geraniaceae	1	0.12
Gesneriaceae	1	0.12
Icacinaceae	1	0.12
Juglandaceae	1	0.12
Lecythidaceae	1	0.12
Loranthaceae	1	0.12
Martyniaceae	1	0.12
Molluginaceae	1	0.12
Moringaceae	1	0.12
Myricaceae	1	0.12
Myristicaceae	1	0.12
Nelumbonaceae	1	0.12
Nymphaeaceae	1	0.12
Onagraceae	1	0.12
Opiliaceae	1	0.12
Orobanchaceae	1	0.12
Paeoniaceae	1	0.12
Pandanaceae	1	0.12
Passifloraceae	1	0.12
Portulacaceae	1	0.12
Primulaceae	1	0.12
Punicaceae	1	0.12
Salicaceae	1	0.12
Samydaceae	1	0.12
Santalaceae	1	0.12
Saururaceae	1	0.12
Tamaricaceae	1	0.12
Thymelaeaceae	1	0.12
Ulmaceae	1	0.12
Valerianaceae	1	0.12

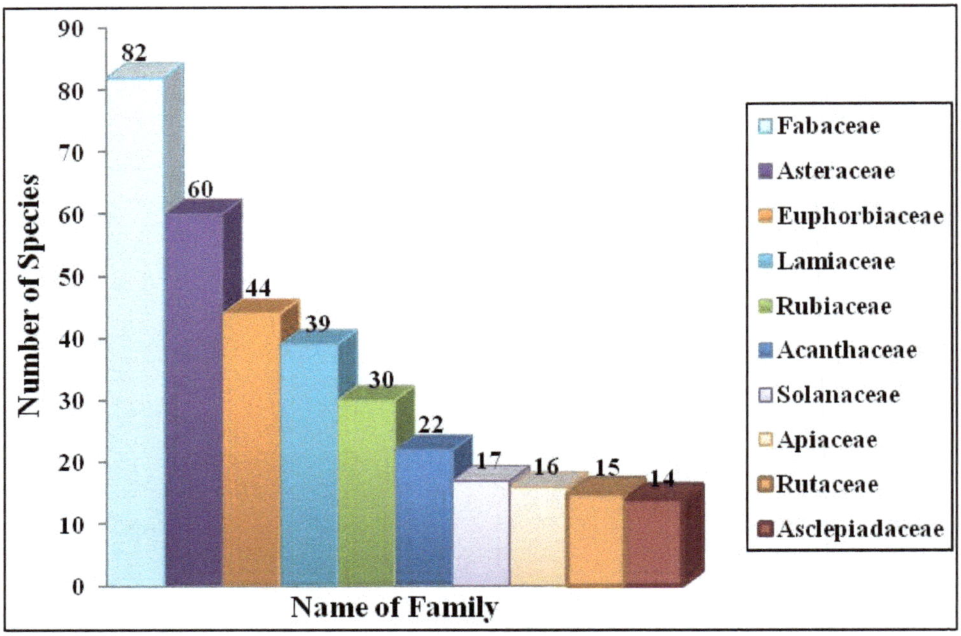

Figure 4: Predominant Dicotyledonous Plant Families for Child Health Care.

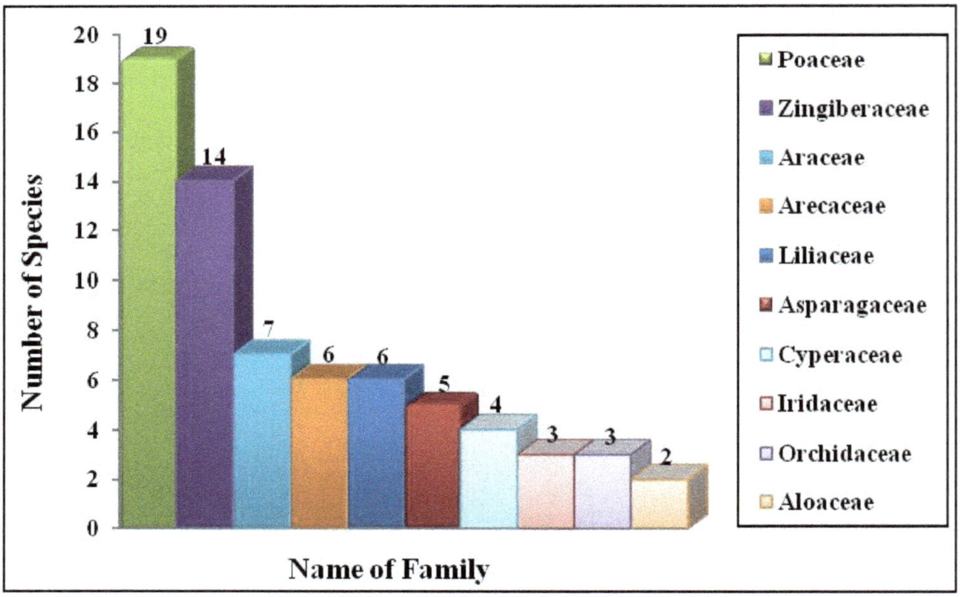

Figure 5: Predominant Monocotyledonous Plant Families for Child Health Care.

Table 3: Monocot Plant Families Employed for Child Health Care

Name of Family	*Total No. of Species*	*Percentage of Taxa*
Poaceae	19	2.37
Zingiberaceae	14	1.75
Araceae	7	0.87
Arecaceae	6	0.75
Liliaceae	6	0.75
Asparagaceae	5	0.62
Cyperaceae	4	0.5
Iridaceae	3	0.37
Orchidaceae	3	0.37
Aloaceae	2	0.25
Bambusaceae	2	0.25
Pontederiaceae	2	0.25
Smilacaceae	2	0.25
Typhaceae	2	0.25
Acoraceae	1	0.12
Agavaceae	1	0.12
Bromeliaceae	1	0.12
Cannaceae	1	0.12
Colchicaceae	1	0.12
Commelinaceae	1	0.12
Costaceae	1	0.12
Dioscoreaceae	1	0.12
Dracaenaceae	1	0.12
Flagellariaceae	1	0.12
Hypoxidaceae	1	0.12
Melanthiaceae	1	0.12
Musaceae	1	0.12
Taccaceae	1	0.12

6. Comparatively, the relative per cent in taxa for predominant Fabaceae amongst dicots is 10.25 **(Table 2)** and likewise it is 2.37 for Poaceae (monocots) **(Table 3)**. Relatively, when usages are contrasted, Fabaceae have been found to be employed over 4 times more than the Poaceae.
7. Organ-wise analysis revealed that the most frequently employed usages of plant part/s for child health care in order of preference in India are: whole plant (311 species), leaf (285 species), root (151 species), fruit (97 species), seed (90 species), stem (89 species), bark (76 species), inflorescence (67 species), aerial parts (35 species) and wood (7 species) **(Figure 6; Appendix III)**.

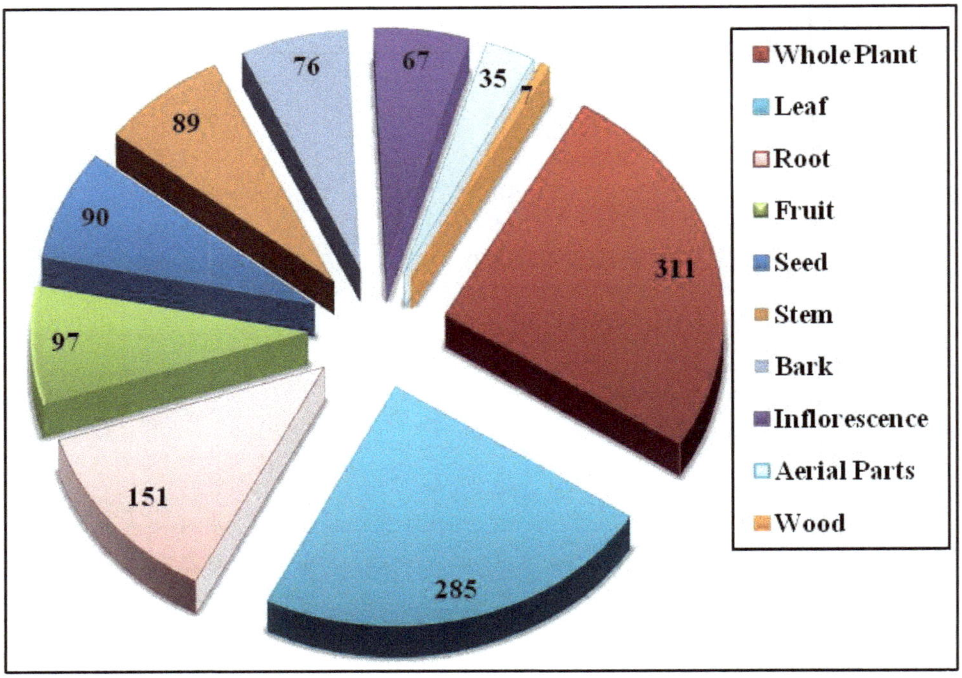

Figure 6: Predominant Plant Parts Used for Child Health Care.

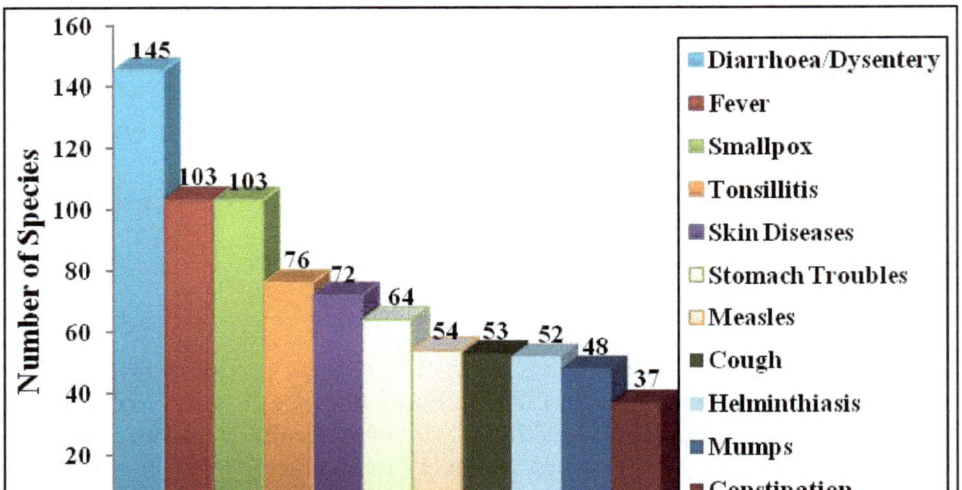

Figure 7: Plants Used in Various Ailments of Children.

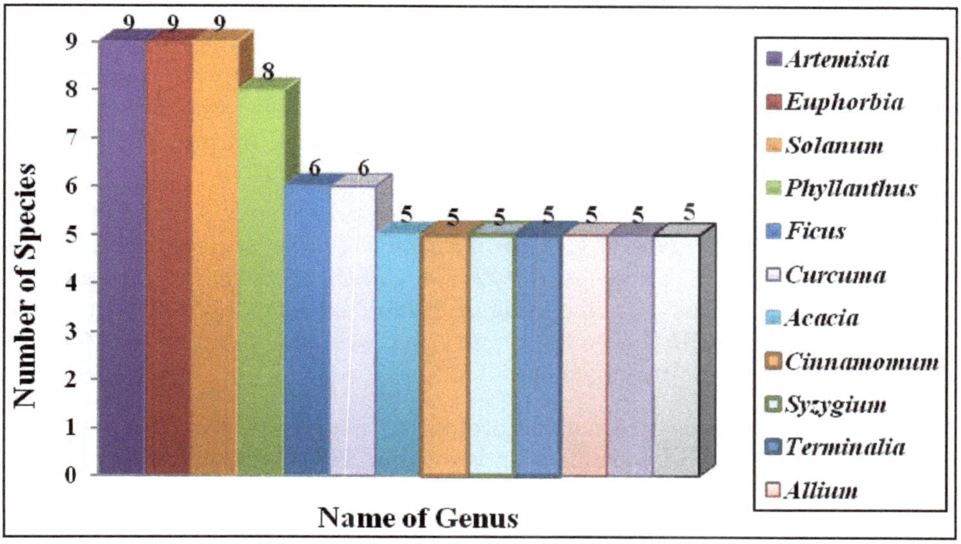

Figure 8: Promising Genera of Childcare Plants.

Table 4: Predominant Genera of Plants For Child Health Care

Name of Genus	*Number of Species*	*Name of Genus*	*Number of Species*
Artemisia	9	*Ailanthus*	3
Euphorbia	9	*Andrographis*	3
Solanum	9	*Bauhinia*	3
Phyllanthus	8	*Berberis*	3
Ficus	6	*Blumea*	3
Curcuma	6	*Clerodendrum*	3
Acacia	5	*Combretum*	3
Cinnamomum	5	*Elsholtzia*	3
Syzygium	5	*Embelia*	3
Terminalia	5	*Erythrina*	3
Allium	5	*Gardenia*	3
Asparagus	5	*Ipomoea*	3
Cymbopogon	5	*Launaea*	3
Aconitum	4	*Piper*	3
Aristolochia	4	*Prosopis*	3
Crotalaria	4	*Quercus*	3
Desmodium	4	*Rheum*	3
Jasminum	4	*Rhus*	3
Jatropha	4	*Rubus*	3
Mentha	4	*Vernonia*	3
Ocimum	4	*Viola*	3
Senna	4		

8. Of the children ailments in India, most prevalent ones as evidenced from the usage of maximum number of plants are diarrhoea/dysentery (145 species), fever, smallpox (103 species each), tonsillitis (76 species), skin diseases (72 species), stomach troubles (64 species), measles (54 species), cough (53 species), helminthiasis (52 species), mumps (48 species), constipation (37 species) etc. **(Figure 7; Appendix IV)**.
9. With regard to utilitarian genera, *Artemisia, Euphorbia, Solanum* (9 species each), *Phyllanthus* (8 species), *Ficus, Curcuma* (6 species each), *Acacia, Cinnamomum, Syzygium, Terminalia, Allium, Asparagus, Cymbopogon* (5 species each) etc. and others **(Figure 8; Table 4)** are predominantly employed for child health care in the country, and overall accounting for 14.43 per cent of the reported diversity.

Hopefully, the generated data on promising species be screened on scientific lines for clinical validation, and utilized in formulating effective strategies for checking child mortality on account of prevalent diseases in India.

Literature Cited

Agarwal, V.S. 1986. *Economic Plants of India*. Bishen Singh Mahendra Pal Singh, Dehradun (India).

Agarwal, V.S. 2003. *Directory of Indian Economic Plants.* Bishen Singh Mahendra Pal Singh, Dehradun (India).

Ahmed, A.A. and Borthakur, S.K. 2005. *Ethnobotanical Wisdom of Khasis of Meghalaya.* Bishen Singh Mahendra Pal Singh, Dehradun (India).

Ahuja, B.S. 1965. *Medicinal Plants of Saharanpur*. Survey of Medicinal Plants, Central Council of Ayurvedic Research Gurukul Kangri Vishwavidyalya, Hardwar.

Ainslie, W. 1813. *Materia Medica of Hindoostan*. Govt. Press Madras (India).

Ainslie, W. 1926. *Materia Medica.* Neeraj Publ. House, Delhi.

Alagesaboopathi, C., Dwarkan, P. and Balu, S. 2000. Plants used as medicine by tribals of Shevaroy Hill, Tamil Nadu: 391-393. In Maheshwari, J.K. (*eds.*): *Ethnobotany and Medicinal Plants of Indian Subcontinent*. Sci. Publ., Jodhpur (India).

Aman, N. and Dar, G.H. 2003. Medicinal plants in Scrophulariaceae from Kashmir Himalaya: 914-921. In Singh, V. and Jain, A.P. (*eds.*): *Ethnobotany and Medicinal Plants of India and Nepal*, Vol. II. Sci. Publ., Jodhpur (India).

Ambasta, S.P. 1986. *The Useful Plants of India.* C.S.I.R., New Delhi (India).

Ambasta, S.P., Ramachandran, K., Kashyapa, K. and Chand, R. 1992. *The Useful Plants of India*. P.I.D., C.S.I.R., New Delhi.

Ambily, D.V., Rajith, N.P., George, V. and Pushpangadan, P. 2010. Utilization of medicinal plants by the Paniyan tribe of Kozhikode district, Kerala. *Ethnobotany* ***22***: 14-24.

Aminuddin and Girach, R.D. 1991. Ethnobotanical studies on Bondo tribe of district Koraput (Orissa), India. *Ethnobotany* ***3***: 15-19.

Aminuddin and Girach, R.D. 1993. Observations on ethnobotany of the Bhunjia- a tribe of Sonabera plateau. *Ethnobotany* ***5***: 83-86.

Aminuddin and Girach, R.D. 1996. Native phytotherapy among the Paudi Bhuinya of Bonai hills. *Ethnobotany* ***8***: 66-70.

Anonymous, 1948-1976a. *The Wealth of India (A Dictionary of Indian Raw Materials and Industrial Products)*, Vol. I-VIII, C.S.I.R., New Delhi (India).

Anonymous, 1976b. *Medicinal Plants of India*, Vol.1. Cambridge Printing Press, New Delhi (India).

Anonymous, 1981. *Medicinal Plants of Kerala Forest.* Kerala Forest Research Institute, Peechi, Kerala.

Anonymous, 1994. *Ethnobiology in India- A Status Report.* Ministry of Environment and Forests, Govt. of India.

Anonymous, 1999. *Medicinal Plants of Andhra Pradesh.* CCRUM, New Delhi.

Ansarali, K.C. and Sivdasan, M. 2009. Ethnobotanical investigations in Lakshadweep Islands, India. *Ethnobotany* ***21***: 18-24.

Arya, K.R. and Prakash, V. 2000. Ethnomedicinal study of a remote tribal areas of Almora district: a survey report-part I: 247-252. In Maheshwari, J.K. (*ed.*): *Ethnobotany and Medicinal Plants of Indian Subcontinent.* Sci. Publ., Jodhpur (India).

Arya, K.R., Pande, P.C. and Prakash, V. 1999. Ethnobotanical study on tribal areas of Almora district-II. *Ethnobotany* ***11***: 100-104.

Ashraf, S., Rahman, A.J., Satwani, H., Naz, F., Abbas, K. and Hussan, A. 2010. Trend of complimentary therapies in paediatric age group. *J Pak Med Assoc 60*: 1015-1018.

Asolkar, L.V., Kakkar, K.K. and Chakre, O.J. 1992. *Second Supplement to Glossary of Indian Medicinal Plants with Active Principle*, Part I (A-K) (1965-81). C.S.I.R., New Delhi (India).

Aswal, B.S. 2001. Zingiberaceae of western Himalaya: 101-118. In Pande, P.C. and Samant, S.S. (*eds.*): *Plant Diversity of the Himalaya.* Gyanodaya Prakashan, Nainital (India).

Atkinson, E.T. 1982. *Economic Botany of Himalayan Region.* Cosmos Publ., New Delhi.

Augustine, J. and Sivadasan, M. 2004. Ethnobotanical plants of Periyar tiger reserve, Kerala, India. *Ethnobotany* ***16***: 44-49.

Baburaj, D.S., Britto, S.J., Mathew, G.K. and Rajan, S. 2000. Cultivated medicinal plants useful in homoeopathy found in Nilgiri district, Tamil Nadu: 381-389. In Maheshwari, J.K. (*ed.*): *Ethnobotany and Medicinal Plants of Indian Subcontinent.* Sci. Publ., Jodhpur (India).

Balasubramanian, P. and Prasad, S.N. 1996. Medicinal plants among the Irulars of Attappady and Boluvampatti forests in the Nilgiri biosphere reserve: 253-259. In Maheshwari, J.K. (*ed.*): *Ethnobotany in South Asia.* Sci. Publ., Jodhpur (India).

Ballabh, B. and Chaurasia, O.P. 2006. Ethnobotanical studies on Boto tribe in Ladakh. *Ethnobotany* ***18***: 87-95.

Banerjee, A. 2000a. Ethnobotany of a few plant species in the eroded soil of Birbhum, West Bengal: 527-530. In Maheshwari, J.K. (*ed.*): *Ethnobotany and Medicinal Plants of Indian Subcontinent.* Sci. Publ., Jodhpur (India).

Banerjee, A. 2000b. Ethnobotany of some trees in the Santhal villages of Birbhum, West Bengal: 531-534. In Maheshwari, J.K. (*ed.*): *Ethnobotany and Medicinal Plants of Indian Subcontinent.* Sci. Publ., Jodhpur (India).

Banerjee, R.N. and Ghora, C. 1996. On the domestic use of some unreported plants of West Dinajpur district West Bengal: 325-328. In Maheshwari, J.K. (*ed.*): *Ethnobotany in South Asia.* Sci. Publ., Jodhpur (India).

Baqui, A.H., Williams, E.K., Darmstadt, G.L., Kumar, V., Kiran, T.U., Panwar, D., Sharma, R.K., Ahmed, S., Sreevasta, V., Ahuja, R., Santosham, and Black, R.E. 2007. Newborn care in rural Uttar Pradesh. *Indian J Pediatr* ***74*** *(3)*: 241-247.

Barua, K.N., Barua, I.C. and Das, M. 2000. Ethnobotany of Rajbanshis of Assam: 609-614. In Maheshwari, J.K. (*ed.*): *Ethnobotany and Medicinal Plants of Indian Subcontinent.* Sci. Publ., Jodhpur (India).

Beegam, A.R. and Nayar, T.S. 2011. Plants used for natal healthcare in folk medicine of Kerala, India. *Indian J. Trad. Knowledge* ***10*** *(3)*: 523-527.

Beigh, S.Y., Nawchoo, I.A. and Iqbal, M. 2003. Plants in traditional medicine of Kashmir himalaya: 99-104. In Singh, V and Jain, A.P. (*eds.*): *Ethnobotany and Medicinal Plants of India and Nepal*, Vol. I. Sci. Publ., Jodhpur (India).

Bennet, S.S.R. 1985. Ethnobotanical studies in West Sikkim. *J. Econ. Tax. Bot.* ***7***: 317-321.

Bennet, S.S.R. 1986. *Name Changes in Flowering Plants and Adjacent Regions.* Triseas Publ., Dehradun, India.

Bhalla, S., Patel, J.R. and Bhalla, N.P. 1996. Ethnomedicinal observations on some Asteraceae of Bundelkhand region, Madhya Pradesh: 175-178. In Maheshwari, J.K. (*ed.*): *Ethnobotany in South Asia.* Sci. Publ., Jodhpur (India).

Bhandary, M.J., Chandrashekar, K.R. and Kaverippa, K.M. 1996. Ethnobotany of Gowalis of Uttara Kannada district, Karnataka: 244-249. In Maheshwari, J.K. (*ed.*): *Ethnobotany in South Asia.* Sci. Publ., Jodhpur (India).

Bhatt, D.C., Mehta, M.R., Mehta, S.K. and Parmar, R.P. 2002. Studies on some ethnomedicinal plants from Talja taluka of Bhavnagar district, Gujarat: 295-310. In Trivedi, P.C. (*ed.*): *Ethnobotany.* Aavishkar Publ., Distributors, Jaipur (India).

Bhatt, D.C., Mehta, S.K. and Mitaliya, K.D. 1999. Ethnobotanical plants of Shetrunjaya hills of Palitana, Gujarat. *Ethnobotany* ***11***: 22-25.

Bhatt, D.C., Mitaliya, K.D. and Babaria, A.V. 2003. Observations on some ethnobotanical plants of Malnath hillocks of Gogha tehsil, dist. Bhavanagar, Gujarat state: 217-220. In Singh, V. and Jain, A.P. (*eds.*): *Ethnobotany and Medicinal Plants of India and Nepal*, Vol. I. Sci. Publ., Jodhpur (India).

Bhatt, R.P. and Sabnis, S.D. 1987. Contribution to the ethnobotany of Khedbrahma region of north Gujarat. *J. Econ. Tax. Bot.* ***9****(1)*: 139-145.

Bhatt, U.D. and Jain, B.K. 2009. Ethnomedicinal uses of some important plants recorded from forest areas of Shoolpaneshwar Wildlife Sanctury, Narmada district (Gujarat). *Ethnobotany* ***21***: 70-74.

Bhattacharjee, S.K, 2001. *Handbook of Medicinal Plants*. Pointer Publ., Jaipur (India).

Bhattacharya, A.K. and Patra, K. 2004. Medicinal plants for rural women and child health care-issues, option and strategy. *Ind. Fore.* ***130*** *(4)*: 385-397.

Bhattacharya, G. 1996. Medico-ethno-botanical value of Sourashtra weeds: 166-168. In Maheshwari, J.K. (*ed.*) *Ethnobotany in South Asia.* Sci. Publ., Jodhpur (India).

Bhogaonkar, P.Y. and Ahmad, S.A. 2007. Less known folklore Unani uses of plants in Amravati district, Maharashtra (India). *Ethnobotany* ***19***: 124-127.

Bhuyan, L.R. 2003. Some plants used as medicine by the Nishi tribe of Arunachal Pradesh: a preliminary study: 447-450. In Singh, V. and Jain, A.P. (*eds.*): *Ethnobotany and Medicinal Plants of India and Nepal*, Vol. I. Sci. Publ., Jodhpur (India).

Biswas, K. and Chopra, R.N. 1982. *Common Medicinal Plants of Darjeeling and Sikkim Himalayas*. Periodical Expert Book Agency, Delhi.

Biswas, S., Sahu, D.K., Dhal, N.K. and Brahmam, M. 2009. Indigenous phytotherapies among Koya tribe of Malkangiri district, Orissa. *Ethnobotany* ***21***: 99-102.

Black, R.E., Morris, S.S. and Bryce, J. 2003. Where and why are 10 million children dying every year? *Lancet* ***361***: 2226-2234.

Boissya, C.K. and Majumdar, R. 1980. Some folklore claims from the Brahmputra valley (Assam). *Ethnomedicine* ***6***: 139-144.

Bondya, S.L., Khanna, K.K. and Singh, K.P. 2006. Ethnomedicinal uses of leafy vegetables from the tribal folk-lore of Achanakmar-Amarkantak biosphere reserve (Madhya Pradesh and Chattisgarh). *Ethnobotany* ***18***: 145-148.

Borthakur, S.K. 1993. Native phytotherapy for child and women diseases from Assam. *Ethnobotany* ***5***: 87-91.

Borthakur, S.K., Choudhury, B.T. and Gogoi, R. 2004. Folklore hepato-protective herbal recipes from Assam in northeast India. *Ethnobotany* ***16***: 76-82.

Bose, K.C. 1932. *Pharmacopoeia Indica. Being A Collection of Vegetable, Mineral And Animal Drugs in Common Use in India.* Bishen Singh Mahendra Pal Singh, Dehra Dun.

Brahmam, M. and Saxena, H.O. 1990. Ethnobotany of Gandhamardan Hills-some noteworthy folk-medicinal uses. *Ethnobotany* ***2***: 71-79.

Caius, J.F. 2003 (*ed.*). *The Medicinal and Aromatic Poisonous Plants of India*. Sci. Publ., Jodhpur.

Chandra, S. 1990. *Foundations of Ethnobotany (Pre-1900 Ethnobotany). A Review and Bibliography.* Deep Publ., New Delhi.

Chandraker, R., Chakrabarty, S., Mitra, M. and Bharati, P. 2009. A study of reproductive and child health among the Dhur Gond tribal community of Mahasamund district, Chhattisgarh, India. *Stud Tribes Tribals* ***7*** *(2)*: 97-103.

Chatterjee, A. and Pakrashi, S.C. 1997. *The Treatise on Indian Medicinal Plants.* C.S.I.R., New Delhi (India).

Chauhan, D., Navneet and Prabhat. 2007. Medicinal plants of Rigveda: 281-298. In Chopra, A.K., Khanna, D.R., Prasad, G., Malik, D.S. and Bhutiani, R. (*eds.*): *Medicinal Plants: Conservation, Cultivation and Utilization.* Daya Publ. House, Delhi (India).

Chauhan, N.S. 1999. *Medicinal and Aromatic Plants of Himachal Pradesh*. Indus Publ. House, New Delhi.

Chauhan, N.S. 2001. Domestication of selected medicinal plants of Himachal Pradesh: 285-315. In Samant, S.S., Dhar, U. and Palni, M.S. (*eds.*): *Himalayan Medicinal Plants: Potential and Prospects*. Gyanodaya Prakashan, Nainital.

Chhetri, R.B. 1994. Further observations of ethnomedicobotany of Khasi hills in Meghalaya, India: *Ethnobotany* ***6***: 33-36.

Chopra, I.C. 1960. *Medicinal Plants of Arid Zone.* Today and Tomorrow's Printers and Publ., Delhi.

Chopra, R.N. 1933. *Chopra's Indigenous Drugs of India* (2nd *ed.* Revised and Enlarged by Chopra, R.N., Chopra, I.C., Handa, K.L. and Kapur, L.D. in 1958). Dhur and Sons Pvt. Ltd., Calcutta (India).

Chopra, R.N., Chopra, I.C. and Verma, B.S. 1974. *Supplement to Glossary of Indian Medicinal Plants.* Publications and Information Directorate Hillside Road, New Delhi.

Chopra, R.N., Nayar, S.L. and Chopra I.C. 1956. *Glossary of Indian Medicinal Plants.* C.S.I.R., New Delhi.

Chowdhury, J.U., Alam, M.K. and Hassan, M.A. 1996. Some traditional folk formularies against dysentery and diarrhoea in Bangladesh: 20-23. In Maheshwari, J.K. (*ed.*): *Ethnobotany in South Asia.* Sci. Publ., Jodhpur (India).

Dagar, H.S. 2003. Ethnobotanical evaluation of some endemic plants in Andaman and Nicobar Islands: 134-147. In Khan, S.S. (*ed.*): *Vistas in Ethnobotany*, Vol. I. Classic Printers and Publ., Bhopal (India).

Dagar, H.S. and Dagar, J.C. 1996. Ethnobotanical studies of the Nicobarese of Chowra Island of Nicobar group of Islands: 381-388. In Maheshwari, J.K. (*ed.*): *Ethnobotany in South Asia.* Sci. Publ., Jodhpur (India).

Dagar, H.S. and Dagar, J.C. 2000. Plant folk medicine for gynaecological, urino-genital and other related problems among aborigines of Andaman and Nicobar Islands: 561-567. In Maheshwari, J.K. (*ed.*): *Ethnobotany and Medicinal Plants of Indian Subcontinent*. Sci. Publ., Jodhpur (India).

Dam, D.P. and Hajra, P.K. 1981. Observation on ethnobotany of the Mompas of Kameng district, Arunachal Pradesh: 107-114. In Jain S.K. (*ed.*): *Glimpses of Indian Ethnobotany*. Oxford and IBH Publ. Co., New Delhi.

Das, A.K. and Sharma, G.D. 2003. Ethno-medicinal uses of plants by Manipur and Barman communities of Cachar district, Assam: 421-429. In Singh, V. and Jain, A.P. (*eds.*): *Ethnobotany and Medicinal Plants of India and Nepal*, Vol. I. Sci. Publ., Jodhpur (India).

Das, S. 2003. Usefulness of pteridophytes in India with special reference to medicine and conservation: 7-16. In Singh, V. and Jain, A.P. (*eds.*): *Ethnobotany and Medicinal Plants of India and Nepal*, Vol. I. Sci. Publ., Jodhpur (India).

Dash, S.S and Misra, M.K. 2000. Taxonomic survey and systematic census of economic plants of Narayanapatna hills of Koraput district, Orissa: 473-498. In Maheshwari, J.K. (*ed.*): *Ethnobotany and Medicinal Plants of Indian Subcontinent*. Sci. Publ., Jodhpur (India).

Dastur, J.F. 1970. *The Medicinal Plants of India and Pakistan*. D.B. Toporowala and Sons, Bombay.

D'Cruz, L. 2007. Ethnobotanical studies on Leguminous plants of Dediapada forests. *Ethnobotany* ***19***: 73-77.

Deshmukh, V.R., Muratkar, G.D. and Rothe, S.P. 2000. Preliminary observation on the medicinal and economically important leguminous plant species from Amravati Tehsil: 283-289. In Maheshwari, J.K. (*ed.*): *Ethnobotany and Medicinal Plants of Indian Subcontinent*. Sci. Publ., Jodhpur (India).

Devi, Y.N. and Singh, P.K. 2008. Ethnobotany of Verbenaceae in Manipur. *Ethnobotany* ***20***: 111-114.

Dey, A. and De, J.N. 2011. Traditional use of medicinal plants in paediatric and maternal care practiced by the ethnic groups of Purulia district, West Bengal, India. *Int. Med. Arom. Plants* ***1***: 189-194.

Dey, K.L. 1896. *The Indigenous Drugs of India.* Thacker Spink, Calcutta (India).

Dey, K.L. and Bahadur, R. 1973. *The Indigenous Drugs of India: Short Descriptive Notices of the Principal Medicinal Products Met with in British India*. Pama Primlane The Chronica Botanica, New Delhi, (India).

Dhiman, A.K. 2003. *Sacred Plants and Their Medicinal Uses*. Daya Publ. House, Delhi (India).

Dhiman, A.K. 2007. Conservation of some useful medicinal plants of Haridwar district in Uttaranchal state: 147-166. In Chopra, A.K., Khanna, D.R., Prasad, G., Malik, D.S. and Bhutiani, R. (*eds.*): *Medicinal Plants: Conservation, Cultivation and Utilization*. Daya Publ. House, Delhi (India).

Dixit, A.K. 2011. Some herbal remedies for common ailments in south India. *J. Biosci. and Tech.* **2** *(5)*: 379-384.

Dixit, R.D., Das, A. and Kar, B.D. 1978. Studies on ethnobotany III- on some less known edible, economic and medicinal ferns of Darjeeling district, West Bengal. *Nagarjun* **21** *(10)*: 1-4.

D'Rozario, A., Bera, S. and Mukherji, D. 1999. *A Handbook of Ethnobotany*. Kalyani Publ., New Delhi (India).

Dutt, N.B. 1928. *Commercial Drugs of India*. Bishen Singh Mahendra Pal Singh, Dehra Dun.

Dutta, A.C. 1985. *Dictionary of Economic and Medicinal Plants*. Assam Printing Works (P) Ltd. Jorhat, Assam.

Dutta, M. and Nath, S.C. 2000. Ethno-medico botany of the tai-ashoms of Assam, India: 591-598. In Maheshwari, J.K. (*ed.*): *Ethnobotany and Medicinal Plants of Indian Subcontinent*. Sci. Publ., Jodhpur (India).

Dutta, S.C. 1949. *Pharmacognosy of Indian Root and Rhizome Drugs.* Central Council for Research in Ayurveda and Siddha, New Delhi (India).

Dutta, U.C. 1877. *The Materia Medica of the Hindus.* Thacker Spink, Calcutta (India).

Dwarakan, P. and Ansari, A.A. 1996. Less known uses of plants of Kollimalai (Salem DT., Tamil Nadu) in south India: 284-286. In Maheshwari, J.K. (*ed.*): *Ethnobotany in South Asia*. Sci. Publ., Jodhpur (India).

Dwivedi, S.N. 2000. Traditional health care among tribals of Rewa district of Madhya Pradesh with special reference to conservation of endangered and vulnerable species: 315-320. In Maheshwari, J.K. (*ed.*): *Ethnobotany and Medicinal Plants of Indian Subcontinent.* Sci. Publ., Jodhpur (India).

Dymock, J.H., Warden, J.H. and Hooper, D. 1889-1893. *Pharmacographica Indica* **2**: 423. Trubner Co. Ltd., London (U.K.).

Dymock, W. 1883. *The Vegetable Materia Medica of Western India*. Edu. Society's Press, Bombay (India).

Farnsworth, N.R. and Soejarto, D.D. 1991. Global importance of medicinal plants: 29-35. In Akrele, O., Heywood, V. and Synge, H. (*eds.*): *Conservation of Medicinal Plants*. Cambridge Univ. Press, New York.

Farooq, S. 2005. *555 Medicinal Plants: Field and Laboratory Manual*. Intl. Book Distributors, Dehradun (India).

Flemming, J. 1810. A catalogue of Indian medicinal plants and drugs. *Asit. Res.* **11**: 153-189.

Fluckiger, F.A. and Handburg, D. 1874. *Pharmacographia*. McMillan and Co. London.

Ganesan, S. and Kesavan, L. 2003. Ethnomedicinal plants used by the ethnic group Valaiyans of Vellimalai hills (reserved forest), Tamil Nadu, India: 754-760. In Singh, V. and Jain, A.P. (*eds.*): *Ethnobotany and Medicinal Plants of India and Nepal*, Vol. II. Sci. Publ., Jodhpur (India).

Ghosh, S. and Sensarma, P. 1997. Ethnomedicine to modern medicine: an observational study in some villages of West Bengal. *Ethnobotany* ***9***: 80-84.

Girach, R.D. 1992. Medicinal plants used by Kondh tribe of district Phulbani (Orissa) in eastern India. *Ethnobotany* ***4***: 53-66.

Girach, R.D. and Aminuddin. 1995. Ethnomedicinal uses of plants among the tribals of Singhbum district, Bihar, India. *Ethnobotany* ***7***: 103-107.

Girach, R.D., Aminuddin, Brahmam, M. and Misra, M.K. 1997. Observations on ethnomedicinal plants of Bhadrak district, Orissa India. *Ethnobotany* ***9***: 44-46.

Goel, A.K. and Rajendran, A. 2000. Cross-cultural ethnobotanical studies of Santhal Pargana (eastern India) and Western Ghats (southern India): 147-150. In Maheshwari, J.K. (*ed.*): *Ethnobotany and Medicinal Plants of Indian Subcontinent*. Sci. Publ., Jodhpur (India).

Gogoi, J.C., Sharma, D.K. and Borthakur, S.K. 2003. Folklore hepatoprotective herbal recipes from Assam in northeastern India: 1-21. In Khan, S.S. (*ed.*): *Vistas In Ethnobotany*, Vol. I. Classic Printers and Publ., Bhopal (India).

Gogoi, R. and Borthakur, S.K. 2001. Notes on herbal recipes of Bodo tribe in Kamrup District, Assam. *Ethnobotany* ***13***: 15-23.

Gogoi, R. and Das, M.K. 2003. Observations on some weeds of medicinal importance in Brahmputra valley of Assam: 434-441. In Singh, V. and Jain, A.P. (*eds.*): *Ethnobotany and Medicinal Plants of India and Nepal*, Vol. I. Sci. Publ., Jodhpur (India).

Gopal, G.V. and Chandra, K.S.J. 2003. Ethnobotany of Soligas of Biligirirangana hills, Karnataka, India: 640-647. In Singh, V. and Jain, A.P. (*eds.*): *Ethnobotany and Medicinal Plants of India and Nepal*, Vol. II. Sci. Publ., Jodhpur (India).

Goyal, H.R. 1999. *Medicinal Plants of Kachchh (Gujarat)*. CCRAS, New Delhi.

Gupta, R. 1986. Prospects of cultivation of plants used in Ayurvedic medicine into agriculture and agro-forestry systems in India: 321-333. Proc. Silver Jubilee Celebrations, Nehru Ayurvedic Medicinal Plants Garden and Herbarium, Pune.

Hajra, P.K. and Chakraborty, P.C. 1981. A survey of wild plants sold in the Lal market of Gangtok. *Indian J. Fore.* ***4*** *(3)*: 217-220.

Haridasan, K., Shukla, G.P. and Deori, M.L. 2001 Cultivation prospects of medicinal plants of Arunanchal Pradesh: a review: 329-343. In Samant, S.S., Dhar, U. and Palni, L.M.S. (*eds.*): *Himalayan Medicinal Plants: Potential and Prospects*. Gyanodaya Prakashan, Nainital (India).

Hembrom, P.P. 1991. Tribal medicine in Chotanagpur and Santhal Parganas of Bihar, India. *Ethnobotany* ***3***: 97-99.

Hembrom, P.P. 1996. Contact therapy practised by Mundas of Chotanagpur (Bihar). *Ethnobotany* ***8***: 36-39.

Hooker, J.D. 1872-1897. *The Flora of British India*, Vol. I-VIII. Lalit Mohan Basu, Allahabad (India).

Hosagoudar, V.B. and Henry, A.N. 1996a. Ethnobotany of Kadars, Malasars and Muthuvans of the Anamalais in Coimbatore district, Tamil Nadu, India: 260-267. In Maheshwari, J.K. (*ed.*): *Ethnobotany in South Asia.* Sci. Publ., Jodhpur (India).

Hosagoudar, V.B. and Henry, A.N. 1996b. Ethnobotany of Soligas in Biligiri Rangana Betta, Karnataka, Southern India: 228-243. In Maheshwari, J.K. (*ed.*): *Ethnobotany in South Asia.* Sci. Publ., Jodhpur (India).

Hosagoudar, V.B. and Henry, A.N. 1996c. Ethnobotany of tribes Irular, Kurumban and Paniyan of Nilgiris in Tamil Nadu, Southern India: 272-283. In Maheshwari, J.K. (*ed.*): *Ethnobotany in South Asia.* Sci. Publ., Jodhpur (India).

Hussain, A. 1992. *Dictionary of Indian Medicinal Plants.* Central Institute of Medicinal and Aromatic Plants, Lucknow.

Islam, M. 1996. Ethnobotany of certain underground parts of plants of northeastern region, India: 338-343. In Maheshwari, J.K. (*ed.*): *Ethnobotany in South Asia.* Sci. Publ., Jodhpur (India).

Islam, M. A. and Jha, R.K. 2003. Ethnopharmacology of tribals of Ranchi district in Jharkhand: 300-310. In Singh, V. and Jain, A.P. (*eds.*): *Ethnobotany and Medicinal Plants of India and Nepal*, Vol. I. Sci. Publ., Jodhpur (India).

Iyenger, M.A. 1977. *Bibliography on Investigated Indian Plants (1950-75).* Kasturba Med. Coll., Manipur (India).

Jadhav, D. 2009. Ethnomedicinal plants used in leaf therapy in Ratlam District (Madhya Pradesh). *Ethnobotany* ***21***: 84-90.

Jain, A.K. and Patole, S.N. 2001. Less-known medicinal uses of plants among some tribal and rural communities of Pachmarhi forest (M.P.) *Ethnobotany* ***13***: 96-100.

Jain, S.K. 1987. *Manual of Ethnobotany.* Sci. Publ., Jodhpur.

Jain, S.K. 1989 (*ed.*). *Methods and approaches in Ethnobotany.* Deep Pub., New Delhi, India.

Jain, S.K. 1991. *Dictionary of Indian Folkmedicine and Ethnobotany.* Deep Publ., New Delhi (India).

Jain, S.K. 1994. *Medicinal Plants.* National Book Trust, New Delhi (India).

Jain, S.K. 1995. Ethnobotanical diversity in Zingibers of India. *Ethnobotany* ***7***: 83-88.

Jain, S.K. 2002. *Bibliography of Indian Ethnobotany.* Sci. Publ., Jodhpur.

Jain, S.K. and De Filipps, R. 1991. *Medicinal Plants of India,* Vols. I, II. Reference Publ. Algonac Mich. U.S.A.

Jain, S.K. and Mudgal, V. 1999. *A Handbook of Ethnobotany.* Bishen Singh Mahendra Pal Singh, Dehra Dun.

Jain, S.K. and Saklani, A. 1992. Cross-cultural ethnobotanical studies in Northeast India. *Ethnobotany* ***4***: 25-38.

Jain, S.K. and Sharma, P.P. 2000. A review of parallel or unique indigenous uses of twenty five plants in India and some African countries. *Ethnobotany* ***12***: 51-55.

Jain, S.K., Mudgal, V., Banerjee, D.K., Guha, A. and Das, D. 1984. *Bibliography of Ethnobotany*: 1-157. B.S.I., Calcutta.

Jain, S.K., Sinha, B.K. and Gupta, R.C. 1991. *Notable Plants in Ethnomedicine of India*. Deep Publ., New Delhi.

Jain, S.P., Singh, S.C., Singh, J. and Kumar, S. 2003. Ethno-medico-botanical survey of Raipur district, Chhattisgarh state: 266-271. In Singh, V. and Jain, A.P. (*eds.*): *Ethnobotany and Medicinal Plants of India and Nepal*, Vol. I. Sci. Publ., Jodhpur (India).

Jamir, N.S. and Rao, R.R. 1990. Fifty new or interesting medicinal plants used by the Zeliangs of Nagaland (India). *Ethnobotany* ***2***: 11-18.

Jamir, N.S., Lanusunep and Konyak, C.H.M. 2011. Some less known ethnomedicinal plants used by Angami-Naga tribes in Kohima district, Nagaland (India). *Ethnobotany* ***23***: 116-120.

Jha, A.K., Suman, N.R. and Rathore, D.K. 2003. Medicinal pteridophytes of Bastar, Chattisgarh: 993-996. In Singh, V. and Jain, A.P. (*eds.*): *Ethnobotany and Medicinal Plants of India and Nepal*, Vol. II. Sci. Publ., Jodhpur (India).

Jha, B.J. and Kumar, M. 2003. Herbal remedies for various diseases: 78-87. In Singh, V. and Jain, A.P. (*eds.*): *Ethnobotany and Medicinal Plants of India and Nepal*, Vol. I. Sci. Publ., Jodhpur (India).

Jha, R.R. and Varma, S.K. 1996. Ethnobotany of Sauria Paharias of Santhal Pargana, Bihar: 1. Medicinal plants. *Ethnobotany* ***8***: 31-35.

Jones, W. 1799. *Botanical Observations on Select Indian Plants*, London.

Joshi, S.G. 2000. *Medicinal Plants*. Oxford and IBH Publ., New Delhi.

Kala, C.P. 2003. *Medicinal Plants of Indian Trans-Himalaya*. (*Focus on Tibetan Use of Medicial Resources*). Bishen Singh Mahendra Pal Singh, Dehradun (India).

Kalpesh, I., Rao, V.R., Parabia, F.M., Kothari, I.L. and Parabia, M.H. 2004. Ethnobotany of Little Rann of Kachchh, Gujarat. *Ethnobotany* ***16***: 40-43.

Kapoor, S.L. and Mitra, R. 1979. *Herbal Drugs Used in Pharmaceutical Industry*. N.B.R.I., Lucknow.

Kapur, S.K. 1996a. Traditionally important medicinal plants of Bhaderwah hills-Jammu Province-III: 62-69. In Maheshwari, J.K. (*ed.*): *Ethnobotany in South Asia*. Sci. Publ., Jodhpur (India).

Kapur, S.K. 1996b. Traditionally important medicinal plants of Bhaderwah hills-Jammu Province-IV: 70-74. In Maheshwari, J.K. (*ed.*): *Ethnobotany in South Asia*. Sci. Publ., Jodhpur (India).

Kapur, S.K. and Nanda, S. 1996. Traditionally important medicinal plants of Bhaderwah hills (Jammu Province) Part II: 56-61. In Maheshwari, J.K. (*ed.*): *Ethnobotany in South Asia*. Sci. Publ., Jodhpur (India).

Kapur, S.K. and Singh, P. 1996. Traditionally important medicinal plants of Udhampur district (Jammu Province) Part-I: 75-81. In Maheshwari, J.K. (*ed.*): *Ethnobotany in South Asia*. Sci. Publ., Jodhpur (India).

Kapur, S.K. and Srivastava, T.N. 1996. Traditionally important medicinal plants of Udhampur district (Jammu Province) Part-II: 82-88. In Maheshwari, J.K. (*ed.*): *Ethnobotany in South Asia.* Sci. Publ., Jodhpur (India).

Kapur, S.K., Nanda, S. and Srivastava, T.N. 1996. Ethnobotanical uses of RRL-herbarium-III: 50-55. In Maheshwari, J.K. (*ed.*): *Ethnobotany in South Asia.* Sci. Publ., Jodhpur (India).

Karnick, C.R. 1994a. *Pharmacopoeial Standards of Herbal Plants.* Vol. I. Sri Satguru Publ., Delhi (India).

Karnick, C.R. 1994b. *Pharmacopoeial Standards of Herbal Plants.* Vol. II. Sri Satguru Publ., Delhi (India).

Karnick, C.R., Tiwari, K.C., Majumdar, R. and Bhattacharjee, S. 1981. Newer ethnobotanical and folk-lore studies of some medicinal plants of Gauhati and surrounding areas. *Nagarjun* ***24***: 240-245.

Katark, N.N. 1903. *Materia Medica of Indian and Their Therapeutics.* Neeraj Publ. House, Delhi.

Katewa, S.S. and Arora, A. 1997. Some plants in folk medicines of Udaipur district (Rajasthan). *Ethnobotany* ***9***: 48-51.

Katewa, S.S., Choudhary, B.L., Jain, A. and Takhar, H.K. 2001. Some plants in folk medicine of Rajsamand district (Rajasthan). *Ethnobotany* ***13***: 129-134.

Kaul, M.K. 1997. *Medicinal Plants of Kashmir and Ladakh Temperate and Cold Arid Himalaya.* Indus Publ. Co., New Delhi (India).

Keller, H., Prance, G.T., Tressens, S.G. and Duarte, R. 2010. Ethnobotanical resources of multiple-use Guarani reserve, misiones: their importance to indigenous communities of Caramelito and Taruma Poty-I. *Ethnobotany* ***22***: 38-54.

Keller, H., Prance, G.T., Tressens, S.G. and Duarte, R. 2011. Ethnobotanical resources of multiple-use Guarani reserve, misiones: their importance to indigenous communities of Caramelito and Taruma Poty-II. *Ethnobotany* ***23***: 33-50.

Kemp, R. 2003. Ethno-medicinal plants used by the Rengma tribe in Dimapur district, Nagaland (India): 485-488. In Singh V. and Jain, A.P. (*eds.*): *Ethnobotany and Medicinal Plants of India and Nepal,* Vol. I. Sci. Publ., Jodhpur (India).

Khan, A.A., Khan, S. and Singh, M.P. 2003. Weeds of soyabean [*Glycine max* (L.) Merrill] and their medicinal importance in Rewa division (M.P.) India: 245-248. In Singh, V. and Jain, A.P. (*eds.*): *Ethnobotany and Medicinal Plants of India and Nepal,* Vol. I. Sci. Publ., Jodhpur (India).

Khan, I.A. and Khanum, A. 2005. *Role of Biotechnology in Medicinal and Aromatic Plants.* Ukaaz Publ., Hyderabad, Andhra Pradesh.

Khanna, K.K. 2002a. Ethnobotany of five districts of Terai region, Uttar Pradesh: 128-145. In Trivedi, P.C. (*ed.*): *Ethnobotany.* Aavishkar Publ., Distributors, Jaipur (India).

Khanna, K.K. 2002b. Unreported ethnomedicinal uses of plants from the tribal and rural folklore of Gonda district, Uttar Pradesh. *Ethnobotany* ***14***: 52-56.

Khanna, K.K., Mudgal, V., Shukla, G. and Srivastava, P.K. 1996a. Unreported ethnomedicinal uses of plants from Mirzapur district, Uttar Pradesh: 112-117. In Maheshwari, J.K. (*ed.*): *Ethnobotany in South Asia*. Sci. Publ., Jodhpur (India).

Khanna, K.K., Shukla, G. and Mudgal, V. 1996b. New traditional medicinal uses of plants from Jalaun District, Uttar Pradesh: 108-111. In Maheshwari, J.K. (*ed.*): *Ethnobotany in South Asia*. Sci. Publ., Jodhpur (India).

Khanna, K.K., Srivastava, P.K., and Mudgal, V. 1996c. Noteworthy medicinal plant uses from rural folklore of Raebareli District, Uttar Pradesh: 118-122. In Maheshwari, J.K. (*ed.*): *Ethnobotany in South Asia*. Sci. Publ., Jodhpur (India).

Khare, C.P. (*ed.*) 2004. *Encyclopedia of Indian Medicinal Plants, Rational Western Therapy, Ayurvedic and other Traditional Usage, Botany.* Springer – Verlag Berlin Heidelberg, New York.

Kharkonger, P. and Joseph, J. 1981. Folklore medico-botany of rural Khasi and Jantia tribes in Meghalaya: 124-136. In Jain, S.K. (*ed.*): *Glimpses of Indian Ethnobotany*. Oxford and IBH Publ. Co., New Delhi.

Khomdram, S.D., Devi, Y.N. and Singh, P.K. 2011. Ethnobotanical uses of Lamiaceae in Manipur, India. *Ethnobotany* ***23***: 64-69.

Khory, R.N. 1887. *The Bombay Materia Medica and Their Therapeutics*. Periodical Expert Book Agency, Delhi (India).

Khullar, S.P. 2001. Ferns in medicine and how to identify them: 65-72. In Samant, S.S., Dhar, U. and Palni, L.M.S. (*eds.*): *Himalayan Medicinal Plants: Potential and Prospects.* Gyanodaya Prakashan, Nainital (India).

Kirtikar, K.R. and Basu, B.D. 1935. *Indian Medicinal Plants*, Vol. I-IV. Lalit Mohan Basu, Allahabad, India.

Kirtikar, K.R. and Basu, B.D. 1984. *Indian Medicinal Plants.* Vol. I-IV. Bishen Singh Mahendra Pal Singh, New Delhi (India).

Krishnamurthy, S. 1953. *Horticultural and Economical Plants of the Nilgiris.* Govt. Bot. Gdn. Centenary Committee, Udhagamandalam.

Kshirsagar, R.D. and Singh, N.P. 2000a. Less-known ethnomedicinal uses of plants in Coorg district of Karnataka state, Southern India. *Ethnobotany* ***12***: 12-16.

Kshirsagar, R.D. and Singh, N.P. 2000b. Less-known ethnomedicinal uses of plants reported by Jenukuruba tribe of Mysore district, Southern India. *Ethnobotany* ***12***: 118-122.

Kshirsagar, R.D. and Singh, N.P. 2003. Less-known herbal remedies from Soligas of Mysore, southern India: 656-661. In Singh, V. and Jain, A.P. (*eds.*): *Ethnobotany and Medicinal Plants of India and Nepal*, Vol. II. Sci. Publ., Jodhpur (India).

Kshirsagar, R.D. and Singh, N.P. 2007. *Ethnobotany of Mysore and Coorg, Karnataka State*. Bishen Singh Mahendra Pal Singh, Dehradun (India).

Kothari, M.J. and Londhe, A.N. 2000. Ethnobotany in human health care of Chikhaldara, Amaravati district in Maharashtra state, India: 273-281. In Maheshwari, J.K. (*ed.*): *Ethnobotany and Medicinal Plants of Indian Subcontinent.* Sci. Publ., Jodhpur (India).

Kothari, M.J. and Rao, K.M. 2000. Ethnobotanical studies of Thane district Maharashtra: 265-272. In Maheshwari, J.K. (*ed.*): *Ethnobotany and Medicinal Plants of Indian Subcontinent.* Sci. Publ., Jodhpur (India).

Kothari, M.J., Jadhav, C.R. and Singh, N.P. 2003. Ethnobotanical wealth of two sacred grooves in Junnar Taluka, Pune, Maharashtra: 585-591. In Singh, V. and Jain, A.P. (*eds.*): *Ethnobotany and Medicinal Plants of India and Nepal*, Vol. II. Sci. Publ., Jodhpur (India).

Kumar, A. 1996. Some ethnomedicinal plants of the Murias of the Indravati tiger reserve, Bastar (Madhya Pradesh): 201-205. In Maheshwari, J.K. (*ed.*): *Ethnobotany in South Asia.* Sci. Publ., Jodhpur (India).

Kumar, A. and Lal, J. 2003. Medicinal plant resources of the Kanha tiger reserve, Madhya Pradesh: 255-265. In Singh, V. and Jain, A.P. (*eds.*): *Ethnobotany and Medicinal Plants of India and Nepal*, Vol. I. Sci. Publ., Jodhpur (India).

Kumar, K.K., Sasidharan, N. and Swarupanandan, K. 2000. Ethnobotanical studies on the hill tribes in the Shola forests of high ranges, Kerala, South India: 451-466. In Maheshwari, J.K. (*ed.*): *Ethnobotany and Medicinal Plants of Indian Subcontinent.* Sci. Publ., Jodhpur (India).

Kumar, M.S.B. and Balakishnan, N.P. 1996. Ethnobotanical studies of the genus *Euphorbia* L. (Euphorbiaceae): 46-49. In Maheshwari, J.K. (*ed.*): *Ethnobotany in South Asia.* Sci. Publ., Jodhpur (India).

Kumar, R.V. and Pullaiah, T. 1998. Medicinal plants used by tribals of Prakasam district, Andhra Pradesh. *Ethnobotany* ***10***: 97-102.

Kumar, S. 1999. *The Economic Plants of North-East India*. Sci. Publ., Jodhpur.

Kumar, S. 2002. *The Medicinal Plants of North-East India*. Sci. Publ., Jodhpur.

Kumar, S. and Goel, A.K. 2008. A preview of odoriferous plants found around Itchagarh hills in Ormanjhi block of Jharkhand. *Ethnobotany* ***20***: 135-137.

Kumar, S. and Kumari, B. 2007. Some less-known 'Sags' (leafy vegetables) utilized by the tribals and others of Panch Pargana area in Jharkhand. *Ethnobotany* ***19***: 62-66.

Kumar, S. and Mathur, R. 2000. Some important plants used in homeopathic system of medicine: 85-88. In Maheshwari, J.K. (*ed.*): *Ethnobotany and Medicinal Plants of Indian Subcontinent.* Sci. Publ., Jodhpur (India).

Kumar, T.D.C. and Pullaiah, T. 2000. Ethno-medicinal uses of some plants of Mahabubnagar district, Andhra Pradesh, India: 341-345. In Maheshwari, J.K. (*ed.*): *Ethnobotany and Medicinal Plants of Indian Subcontinent.* Sci. Publ., Jodhpur (India).

Kumar, V. and Jain, S.K. 1998. A contribution to ethnobotany of Surguja District in Madhya Pradesh, India. *Ethnobotany* ***10***: 89-96.

Kumar, V. and Rao, R.R. 2001. Some plant beverages used in traditional medicine. *Ethnobotany* ***13***: 36-39.

Kumar, Y., Haridasan, K. and Rao, R.R. 1980. Ethnobotanical notes on certain medicinal plants among some Garo people around Balphakram Sanctuary in Meghalaya. *Bull. Bot. Surv. India* ***22***: 161-165.

Lalramnghinglova, H. 2003. *Ethno-Medicinal Plants of Mizoram*. Bishen Singh Mahendra Pal Singh, Dehradun (India).

Lalramnghinglova, H. and Jha, L.K. 1999. New records of ethnomedicinal plants from Mizoram. *Ethnobotany* ***11***: 57-64.

Lalramnghinglova, J.H. 1996. Ethnobotany of Mizoram – a preliminary survey: 439-459. In Maheshwari, J.K. (*ed.*): *Ethnobotany in South Asia*. Sci. Publ., Jodhpur (India).

Lanjewar, S., Luka, R., Parveen, S. and Dubey, S. 2011. Traditional medicinal practices among the rural people of Tilda, Raipur, India. *Int. Multidisc. Res. J.* ***1*** *(9)*: 4-5.

Lawn, J.E., Cousens, S. and Zupan, J. 2005. 4 million neonatal deaths when? Where? Why? *Lancet* ***365*** *(9462)*: 891-900.

Mahato, A.K. and Mahato, P. 1996. Ethnobotanical wealth of Chottanagpur plateau-IV. Some medicinal palnts used against intestinal worms: 389-391. In Maheshwari, J.K. (*ed.*): *Ethnobotany in South Asia*. Sci. Publ., Jodhpur (India).

Maheshwari, J.K. 1987a. Interdisciplinary approaches in ethnobotany: 19-34. In Jain, S.K. (*ed.*) *A Manual of Ethnobotany*, Sci. Publ., Jodhpur, India.

Maheshwari, J.K. 1987b. Tribal ecosystem- an overview. *Tribal Res. Bull.* ***9***: 1-14.

Maheshwari, J.K. (*ed.*) 1996. *Ethnobotany in South Asia*. Sci. Publ., Jodhpur.

Maheshwari, J.K. (*ed.*) 2000. *Ethnobotany and Medicinal Plants of Indian Subcontinent*. Sci. Publ., Jodhpur.

Maheshwari, P. and Singh, U. 1965. *Dictionary of Economic Plants in India*. I.C.A.R., New Delhi.

Maikhuri, R.K., Nautiyal, S. and Rao, K.S. 2001. Medicinal plant cultivation practices of Bhotiyas in Nanda Devi biosphere reserve villages of Garhwal Himalaya: 317-328. In Samant, S.S., Dhar, U. and Palni, M.S. (*eds.*): *Himalayan Medicinal Plants: Potential and Prospects*. Gyanodaya Prakashan, Nainital.

Maliya, S.D. 2004. Some new or less-known folkmedicines of district Bahraich, Uttar Pradesh, India-II. *Ethnobotany* ***16***: 113-115.

Maliya, S.D. 2007. Traditional fruit and leafy therapy among Tharus and indigenous people of district Bahraich, India. *Ethnobotany* ***19***: 131-133.

Mandal, S.K. and Basu, S.K. 1996. Ethnobotanical studies among some tribals of Nilgiri district, Tamil Nadu: 268-271. In Maheshwari, J.K. (*ed.*): *Ethnobotany in South Asia*. Sci. Publ., Jodhpur (India).

Mehra, P.N., Bhatnagar, J.K. and Handa, S.S. 1969. Researches in pharmacognosy in India. *Res. Bull. Punjab Univ.* ***20***: 261-337.

Mehrotra, S., Rawat, A.K.S., Singh, H.K. and Shome, U. 1995. Standardization of popular ayurvedic adaptogenic preparation 'Chyavanprash' and ethnobotany of its ingredients. *Ethnobotany 7*: 1-15.

Mishra, S. 2008. Ethnomedicinal studies of Kokru tribe, with Gotra names derived from plant names, from Khandwa district in Madhya Pradesh. *Ethnobotany* ***20***: 122-127.

Mitaliya, K.D. and Bhatt, D.C. 2003. Rare and endangered medicinal plants of Gujarat: 845-850. In Singh, V. and Jain, A.P. (*eds.*): *Ethnobotany and Medicinal Plants of India and Nepal*, Vol. II. Sci. Publ., Jodhpur (India).

Mitra, R. 1985. *Bibliography on Pharmacognosy of Medicinal Plants*. N.B.R.I. Lucknow (India).

Mitra, R. and Jain, S.K. 1991. Medicinal plant research in India, an overview. *Ethnobotany* ***3***: 65-77.

Mohanty, R.B., Padhy, S.N. and Dash, S.K. 1996. Traditional phytotherapy for diarrhoeal diseases in Ganjam and Phulbani districts of south Orissa, India. *Ethnobotany* ***8***: 60-65.

Molla, H.A. and Roy, B. 1996. Some ethnobotanical claims from the Jalpaiguri district of West Bengal: 322-324. In Mahehwari, J.K. (*ed.*): *Ethnobotany in South Asia.* Sci. Publ., Jodhpur (India).

Molla, M.A. and Roy, B. 1985. Traditional uses of some medicinal plants by the Rabha tribals in Jalpaiguri district, West Bengal. *J. Econ. Tax. Bot.* ***7*** *(3)*: 578-580.

Mudgal, V., Pal, D.C., Kayal, R.N. and Saha, S. 1999. *Ethnobotany of Totopara*. Bishen Singh Mahendra Pal Singh, Dehradun (India).

Mukherjee, A., Mukherjee, S. and Mukherjee, M. 1986. Less known medicinal plants of Rubiaceae in Darjeeling Hills, West Bengal (India). *Indian J. Applied and Pure Biol.* ***1***: 86-91.

Mulay, J.R. and Sharma, P.P. 2012. Plants in child care in Ahmednagar district, Maharashtra (India). *Recent Research in Science and Technology* ***4*** *(10)*: 11-15.

Murugesan, M., Balasubramaniam, V. and Arumugasamy, K. 2011. Ethnomedicinal diversity of Malasars in Velliangiri hills, Western Ghats, Tamil Nadu. *Ethnobotany* ***23***: 89-99.

Muthukumarsamy, S., Mohan, V.R., Kumaresan, S. and Chelladurai, V. 2003. Herbal medicinal plants used by Palliyars to obtain relief from gastro-intestinal complaints: 711-714. In Singh, V. and Jain, A.P. (*eds.*): *Ethnobotany and Medicinal Plants of India and Nepal,* Vol. II. Sci. Publ., Jodhpur (India).

Muthukumarsamy, S., Mohan, V.R., Kumaresan, S. and Chelladurai, V. 2004. Traditional medicinal practices of Palliyar tribe of Srivilliputhur in antenatal and post-natal care of mother and child. *Natural Product Radiance* ***3*** *(6)*: 422-426.

Nadanakunjidam, S. 2003. Ethnomedicinal observation from Attapadi hills of Western Ghats: 732-740. In Singh, V. and Jain, A.P. (*eds.*): *Ethnobotany and Medicinal Plants of India and Nepal*, Vol. II. Sci. Publ., Jodhpur (India).

Nadeem, M., Kumar, A., Nandi, S.K. and Palni, L.M.S. 2001. Tissue culture of medicinal plants with particular reference to Kumaon Himalaya: 231-268. In Samant, S.S., Dhar, U. and Palni, L.M.S. (*eds.*): *Himalayan Medicinal Plants: Potential and Prospects*. Gyanodaya Prakashan, Nainital (India).

Nadkarni, K.M. 1926. *Indian Materia Medica*. Popular Book Depot, Bombay (India).

Nambiar, V.P. 1985. *Studies on the Medicinal Plants of Kerala Forest*. Kerala Forest Res. Inst., Peechi, Kerala.

Nanjan, M.J. 2010. Quality control and standardization of traditional medicine/ herbal drugs: 173-175. An International Conference on Challenging and Emerging Dimensions in Medicinal/Herbal Plants and their Products: A Global Perspective: 26-28 November 2010; Chennai Trade Centre, Chennai, India.

Nath, A. and Maiti, G.G. 2003. Ethnobotany of Barak Valley (Southern Assam) with special reference to folk medicine: 965-971. In Singh, V. and Jain, A.P. (*eds.*): *Ethnobotany and Medicinal Plants of India and Nepal*, Vol. II. Sci. Publ., Jodhpur (India).

Nath, K.K., Deka, P. and Borthakur, S.K. 2007. Ethnomedicinal aspects of some weeds from Darrang district of Assam. *Ethnobotany* ***19***: 82-87.

Nath, S.C. and Bordoloi, D.N. 1989. Ethnobotanical observation on some medicinal folklores of Tirap district, Arunachal Pradesh. *J. Econ. Tax. Bot.* ***13***: 321-325.

Nautiyal, S., Maikhuri, R.R., Rao, K.S. and Saxena, K.G. 2003. Ethnobotany of Tolccha Bhotiya tribe of the buffer zone villages in Nanda Devi Biosphere reserve, India: 119-142. In Jain, V. and Jain, A.P. (*eds.*): *Ethnobotany and Medicinal Plants of India and Nepal*, Vol. I. Sci. Publ., Jodhpur (India).

Nayak, P.K., Nayak, R.K. and Choudhury, B.P. 2003a. A check list of medicinal plants of Kalahandi District in Orissa: 519-532. In Singh, V. and Jain, A.P. (*eds.*): *Ethnobotany and Medicinal plants of India and Nepal*, Vol. II. Sci. Publ., Jodhpur (India).

Nayak, R.K., Nayak, P.K. and Choudhury, B.P. 2003b. Enumeration of some potential economic plants of Paradeep in Mahanadi Delta: 539-545. In Singh, V. and Jain, A.P. (*eds.*): *Ethnobotany and Medicinal plants of India and Nepal*, Vol. II. Sci. Publ., Jodhpur (India).

Nayak, R.K., Nayak, P.K. and Choudhury, B.P. 2003c. Some medicinal weeds of Mahanadi Delta: 533-538. In Singh, V. and Jain, A.P. (*eds.*): *Ethnobotany and Medicinal plants of India and Nepal*, Vol. II. Sci. Publ., Jodhpur (India).

Nayar, M.P., Ramamurthy, K. and Agarwal, V.S. 1989. *Economic Plants of India*, Vols. I and II. Bot. Survey of India, Calcutta.

Nisha, V.M. and Sivadasan, M. 2007. Ethnodermatologically significant plants used by traditional healers of Wayanad district, Kerala. *Ethnobotany* ***19***: 55-61.

O' Shaugnessy, W.B. 1841. *The Bengal Dispensatory and Pharmacopoeia*. Thacker, Spink and Co., Calcutta (India).

Padhy, S.N. and Dash, S.K. 2003. Ethnobotanical survey of insect repellent plant species in folklore and literature of Orissa state, India: 505-512. In Singh, V. and Jain, A.P. (*eds.*): *Ethnobotany and Medicinal Plants of India and Nepal*, Vol. II. Sci. Publ., Jodhpur (India).

Painuli, R.M. and Maheshwari, J.K. 1996. Some interesting ethnomedicinal plants used by Saharia tribe of Madhya Pradesh: 179-185. In Maheshwari, J.K. (*ed.*): *Ethnobotany in South Asia*. Sci. Publ., Jodhpur (India).

Pal, D.C. 2000. Some plants in traditional and homeopathic systems of medicine in India: 79-83. In Maheshwari, J.K. (*ed.*): *Ethnobotany and Medicinal Plants of Indian Subcontinent*. Sci. Publ., Jodhpur (India).

Pal, D.C., Guha, A. and Sen, R. 2000. Medicinal plants and plant product used in children diseases among aboriginals in India: 121-130. In Maheshwari, J.K. (*ed.*): *Ethnobotany and Medicinal Plants of Indian Subcontinent.* Sci. Publ., Jodhpur (India).

Panda, B.K. 2007. Some ethnomedicinal plants of Karlapat reserve forest, District Kalahandi, Orissa. *Ethnobotany* ***19***: 134-136.

Panda, P.C. and Das, P. 2000. Medicinal plant-lore of the tribals of Baliguda sub-division, Phulbani district, Orissa: 515-521. In Maheshwari, J.K. (*ed.*): *Ethnobotany and Medicinal Plants of Indian Subcontinent*. Sci. Publ., Jodhpur (India).

Panda, S. 1996. Plant use- a recent perspective from eastern Himalayas: 332-337. In Maheshwari, J.K. (*ed.*) *Ethnobotany in South Asia.* Sci. Publ., Jodhpur (India).

Pande, P.C. and Joshi, G.C. 2001. Indigenous medicinal uses of cultivated plants in Kumaon Himalaya: 117-124. In Samant, S.S., Dhar, U. and Palni, L.M.S. (*eds.*) *Himalayan Medicinal Plants: Potential and Prospects*. Gyanodaya Prakashan, Nainital (India).

Pande, P.C., Tiwari, L. and Pande, H.C. 2006. *Folk Medicine and Aromatic Plants of Uttranchal*. Bishen Singh Mahendra Pal Singh, Dehra Dun (India).

Pandey, A.K. and Rout, S.D. 2006. Ethnobotanical uses of plants of Similipal Biosphere Reserve (Orissa). *Ethnobotany* ***18***: 102-106.

Pandey, G. 1995. *Medicinal Plants of Himalaya*. Sri Satguru Pub., Delhi.

Pandey, H.P. and Verma, B.K. 2002. Plants in oral healthcare among the aborigins of Gonda and Balrampur regions, U.P., India. *Ethnobotany* ***14***: 81-86.

Pandey, I.B. 2003. Some traditional herbal home remedies used in around Kanpur city (Uttar Pradesh), India. *Ethnobotany* ***15***: 129-131.

Pandey, P.H. and Verma, B.K. 2003. Magico-therapeutic plants of Uttar Pradesh, India: 178-187. In Khan, S.S. (*ed.*): *Vistas In Ethnobotany*, Vol. I. Classic Printers and Publ., Bhopal (India).

Parmar, C. and Kaushal, M.K. 1982. *Wild Fruits of the Sub-Himalayan Region*. Kalyani Publ., New Delhi-Ludhiana (India).

Parrotta, J.A. 2001. *Healing Plants of Peninsular India.* CABI Publ., Wallingford (U.K.).

Pathak, S.K. and Pathak, S. 2007. Ethnomedicinal flora of West Nimar (Khargone) district, M.P., India: 207-211. In Chopra, A.K., Khanna, D.R., Prasad, G., Malik, D.S. and Bhutiani, R. (*eds.*): *Medicinal Plants: Conservation, Cultivation and Utilization.* Daya Publ. House, Delhi (India).

Patil, D.A. 2004. Origin of medicines *Vis-à-vis* doctrine of signatures. *Ethnobotany* ***16***: 52-58.

Patil, U.S. 2007. Ethnomedicinal plants in Melghat of Amravati district: a need for conservation: 53-63. In Chopra, A.K., Khanna, D.R., Prasad, G., Malik, D.S. and Bhutiani, R. (*eds.*) *Medicinal Plants: Conservation, Cultivation and Utilization.* Daya Publ. House, Delhi (India).

Pawar, S. and Patil. D.A. 2010. Indigenous phytotherapy for paediatric diseases in Jalgaon district (Maharashtra). *J. Nat. Remedies* ***10***/*1*: 61-63.

Prajapati, N.D., Purohit, S.S., Sharma, A.K. and Kumar, T. 2006. *A Handbook of Medicinal Plants (A Complete Source Book).* Agrobios, Jodhpur (India).

Prasad, P.N., Singh, A.J.A.R., Narayanan, L.M. and Natarajan, C.R. 1996. Ethnobotany of the Kanikkars of South Tamil Nadu-I: 292-298. In Maheshwari, J.K. (*ed.*): *Ethnobotany in South Asia.* Sci. Publ., Jodhpur (India).

Prasad, V.K., Rajagopal, T. and Badarinath, K.V.S. 2003. Notes on economic importance of wild plants of Rampa agency – East Godavari district, Andhra Pradesh, India: 603-612. In Singh, V. and Jain, A.P. (*eds.*): *Ethnobotany and Medicinal plants of India and Nepal,* Vol. II. Sci. Publ., Jodhpur (India).

Prusti, A.B. 2007. Plants used as ethnomedicine by Bondo tribe of Malkangiri district, Orissa. *Ethnobotany* ***19***: 105-110.

Punjani, B.L. 2002. Ethnobotanical aspects of some weeds from Gujarat. *Ethnobotany* ***14***: 78-80.

Pushpangadan, P. 1994. *Ethnobiology in India. A Staus Report.* Ministry of Environment and Forests, GOI, New Delhi.

Pushpangadan, P. and George, V. 2010. Ethnomedicinal practices of rural and tribal populations of India with special reference to the mother and childcare. *Indian J. Trad. Know* ***9****(1)*: 9-17.

Radhika, M. and Balasubramanian, V. 1990. Mother and childcare in traditional medicine. Parts I and II LSPSS Monographs Nos. 3 and 4 (LSPSS: Madras).

Raghunathan, K. 1976. *Medicinal Plants Growing in the Districts of Andhra Pradesh Through Which River Krishna Flows.* Central Council for Research in Indian Medicine and Homeopathy, Delhi.

Rahman, M.A. 2000. Ethno-medico-botanical knowledge among tribals of Bangladesh: 89-93. In Maheshwari, J.K. (*ed.*): *Ethnobotany and Medicinal Plants of Indian Subcontinent.* Sci. Publ., Jodhpur (India).

Rahman, M.A. and Yusuf, M. 1996. Diversity, ecology and ethnobotany of the *Zingiberaceae* of Bangladesh: 13-19. In Maheshwari, J.K. (*ed.*): *Ethnobotany in South Asia*. Sci. Publ., Jodhpur (India).

Rai, L.K. 1994. *Medicinal Plants of Sikkim Himalaya: Status, Usage and Potential*. Bishen Singh Mahendra Pal Singh, Dehra Dun.

Rai, L.K. and Sharma, E. 2001. Diversity and indigenous uses of medicinal plants of Sikkim: 157-163. In Samant, S.S., Dhar, U. and Palni, L.M.S. (*eds.*): *Himalayan Medicinal Plants: Potential and Prospects*. Gyanodaya Prakashan, Nainital (India).

Rajan, S., Baburaj, D.S., Sethuraman, M. and Parimala, S. 2003. Folk medicine among Nilgiri Irulas: an emerging trend in primary health care practice: 791-795. In Singh, V. and Jain, A.P. (*eds.*): *Ethnobotany and Medicinal Plants of India and Nepal*, Vol. II. Sci. Publ., Jodhpur (India).

Rajan, S., Gupta, H.C. and Kumar, S. 2000. Exotic medicinal plants of Lucknow: 205-222. In Maheshwari, J.K. (*ed.*): *Ethnobotany and Medicinal Plants of Indian Subcontinent*. Sci. Publ., Jodhpur (India).

Rajendran, A. and Henry, A.N. 1994. Plants used by the tribe Kadar in Anamalai hills of Tamil Nadu. *Ethnobotany* ***6***: 19-24.

Rajendran, K., Balakishnan, R. and Chandrasekaran, S. 2003. Common medicinal plants and their utilization by villagers in east coast districts of Tamil Nadu: 727-731. In Singh, V. and Jain, A.P. (*eds.*): *Ethnobotany and Medicinal Plants of India and Nepal*, Vol. II. Sci. Publ., Jodhpur (India).

Rajendran, S.M. 2003. West Sikkim: a treasure house of medicinal plants and their utility by local tribals: 311-316. In Singh, V. and Jain, A.P. (*eds.*): *Ethnobotany and Medicinal Plants of India and Nepal*, Vol. I. Sci. Publ., Jodhpur (India).

Ram, A.J., Raja, K., Reddy, K.E. and Raju, R.R.V. 2002. Medicinal plant lore of Sugalis of Gooty forests, Andhra Pradesh. *Ethnobotany* ***14***: 37-42.

Rana, T.S., Datt, B. and Rao, R.R. 1996. Strategies for sustainable utilization of plant resources by the tribals of the Tons Valley, Western Himalaya. *Ethnobotany* ***8***: 96-104.

Ranjan, P. 2000. A contribution to some of the medicinal plants of Indo-Nepal border area adjoining the districts of Madhubani and Sitamarhi: 651-659. In Maheshwari, J.K. (*ed.*): *Ethnobotany and Medicinal Plants of Indian Subcontinent*. Sci. Publ. Jodhpur (India).

Rao, M.K.V. and Shanpru, R. 1981. Some plants in the life of Garos of Meghalaya: 153-160. In Jain, S.K. (*ed.*): *Glimpses of Indian Ethnobotany*. Oxford and IBH Publ. Co., New Delhi.

Rao, N.R. and Henry, A.N. 1995. *The Ethnobotany of Eastern Ghats in Andhra Pradesh, India. 1996*. Botanical Survey of India, Calcutta.

Rao, N.R. and Henry, A.N. 1996. *The Ethnobotatany of Eastern Ghats in Andhra Pradesh, India*. Botanical Survey of India, Calcutta.

Rastogi, R.P. and Mehrotra, B.N. 1998. *Compendium of Medicinal Plants.* Vol. V. (1990-1994). Central Drugs Research Institute, Lucknow and National Institute of Science Communication, New Delhi (India).

Rastogi, R.P. and Mehrotra, B.N. 1999a. *Compendium of Medicinal Plants.* Vol. I. (1960-1969). Central Drug Research Institute, Lucknow and National Institute of Science Communication, New Delhi (India).

Rastogi, R.P. and Mehrotra, B.N. 1999b. *Compendium of Medicinal Plants.* Vol. II. (1970-79). Central Drugs Research Institute, Lucknow and National Institute of Science Communication, New Delhi (India).

Rastogi, R.P. and Mehrotra, B.N. 2001. *Compendium of Medicinal Plants.* Vol. III. (1980-84). Central Drugs Research Institute, Lucknow and National Institute of Science Communication, New Delhi (India).

Rastogi, R.P. and Mehrotra, B.N. 2002. *Compendium of Medicinal Plants.* Vol. V. (1985-89). Central Drugs Research Institute, Lucknow and National Institute of Science Communication, New Delhi (India).

Ravikumar, K. and Sankar, R.V. 2003. Ethnobotany of Malayali tribes in Melpattu village, Javvadhu hills of Eastern Ghats, Tiruvannamalai districts of Tamil Nadu : 715-726. In Singh, V. and Jain, A.P. (*eds.*): *Ethnobotany and Medicinal Plants of India and Nepal*, Vol. II. Sci. Publ., Jodhpur (India).

Rawat, D. and Kharwal, A. 2013. Studies on traditional herbal pediatrics practices in Jaisinghpur, district Kangra (Himachal Pradesh, India). *Global J. Res. Med. Plants and Indigen. Med.* ***2***: 219-230.

Rawat, M.S. and Chaudhary, S. 1998. *Ethno Medico Botany of Arunachal Pradesh (Nishi and Atapani Tribe).* Bishen Singh Mahendra Pal Singh, Dehra Dun.

Ray, S. 2003. Ethnomedicinal plants of Dewas district (M.P.) India: In Khan, S.S. (*ed.*): *Vistas In Ethnobotany*, Vol. I. Classic Printers and Publ., Bhopal (India).

Reddy, P.R., Rao, P.P and Prabhakar, M. 2003. Ethnomedicinal practices amongst Chenchus of Nagarjunasagar Srisailam tiger reserve (NSTR), Andhra Pradesh-plant remedies for cuts, wounds and boils. *Ethnobotany* ***15***: 67-70.

Retnam, K.R. and Martin, P. 2005. *Ethnomedicinal Plants.* Agrobios (India), Jodhpur.

Robert, K. 1932. *Vegetable Drugs of India.* New Delhi.

Rosakutty, P.J., Roslin, A.S. and Ignacimuthu, S. 2000. Some traditional folklore medicinal plants of Kanyakumari district (Tamil Nadu): 369-375. In Maheshwari, J.K. (*ed.*): *Ethnobotany and Medicinal Plants of Indian Subcontinent.* Sci. Publ., Jodhpur (India).

Roxburgh, W. 1832. *Flora Indica.* Thakur, Spink and Co., Calcutta.

Royle, J.F. 1837. *An Essay on Antiquity of Hindu Medicine.* London.

Sahoo, A.K. and Bahali, D.D. 2003. Plants used in ethnomedicines and medico-religious beliefs in Phulbani district of Orissa: 500-504. In Singh, V. and Jain, A.P. (*eds.*): *Ethnobotany and Medicinal Plants of India and Nepal*, Vol. I. Sci. Publ., Jodhpur (India).

Sahoo, A.K. and Mudgal, V. 1995. Less known ethnobotanical uses of plants of Phulbani District, Orissa, India. *Ethnobotany* **7**: 63-67.

Saikia, N. and Nath, S.C. 2003a. Ethnomedicinal observations of some species of the genus *Curcuma* Linn. growing in Assam: 430-433. In Singh, V. and Jain, A.P. (*eds.*): *Ethnobotany and Medicinal Plants of India and Nepal*, Vol. I. Sci. Publ., Jodhpur (India).

Saikia, N. and Nath, S.C. 2003b. Traditional use of some plants in religio-cultural purposes by 'Koch' community in Assam, India: 997-1001. In Singh, V. and Jain, A.P. (*eds.*): *Ethnobotany and Medicinal Plants of India and Nepal*, Vol. II. Sci. Publ., Jodhpur (India).

Saini, D.C. 1996. Ethnobotany of Tharus of Basti district, Uttar Pradesh: 138-153. In Maheshwari, J.K. (*ed.*): *Ethnobotany in South Asia*. Sci. Publ., Jodhpur (India).

Saklani, A. and Jain, S.K. 1994. *Cross-Cultural Ethnobotany of Northeast India*. Deep Publ., New Delhi.

Salam, S., Jamir, N.S. and Singh, P.K. 2011. Ethnomedicinal studies on Tangkhul-Naga tribe in Ukhrul District, Manipur. *Ethnobotany* **23**: 129-134.

Samant, S.S., Dhar, U. and Rawal, R.S. 2001a. Diversity and distribution of wild edible plants of the Indian Himalaya: 421-482. In Pande, P.C. and Samant, S.S. (*eds.*): *Plant Diversity of the Himalaya.* Gyanodaya Prakashan, Nainital (India).

Samant, S.S., Dhar, U. and Rawal, R.S. 2001b. Diversity, distribution and indigenous uses of threatened medicinal plants of Askot wildlife sanctuary in West Himalaya; conservation and management perspectives: 167-184. In Samant, S.S., Dhar, U. and Palni, L.M.S. (*eds.*): *Himalayan Medicinal Plants: Potential and Prospects*. Gyanodaya Prakashan, Nainital (India).

Sandhya, B., Rao, J.K. and Reddi, T.V.V.S. 2011. Plants used in magico-religious beliefs by the tribals of Visakhapatnam district, Andhra Pradesh. *Ethnobotany* **23**: 100-105.

Sanjay, Navneet, Manohar, M. and Prabhat 2007. Antimicrobial properties of herbal tooth powders: 303-317. In Chopra, A.K., Khanna, D.R., Prasad, G., Malik, D.S. and Bhutiani, R. (*eds.*): *Medicinal Plants: Conservation, Cultivation and Utilization.* Daya Publ. House, Delhi (India).

Sanyal, D. 1984. *Vegetable Drugs of India*. Bishen Singh Mahendra Pal Singh, Dehra Dun.

Saren, A.M., Sen, R. and Pal, D.C. 2000. A contribution to the ethnobotany of Bankura district, West Bengal: 545-555. In Maheshwari, J.K. (*ed.*): *Ethnobotany and Medicinal Plants of Indian Subcontinent.* Sci. Publ., Jodhpur (India).

Sarma, S.K., Bhattacharjya, D.K. and Devi, B. 2002. Traditional use of herbal medicines by Madahi tribe of Nalbari district of Assam. *Ethnobotany* **14**: 103-111.

Satapathy, K.B. 2008. Interesting ethnobotanical uses from Juang, Kolha and Munda tribes of Keonjhar District, Orissa. *Ethnobotany* **20**: 99-105.

Satapathy, K.B. and Brahmam, M. 2000. Some interesting phytotherapeutic claims of tribals of Jajpur Dist. (Orissa), India: 467-472. In Maheshwari, J.K. (*ed.*): *Ethnobotany and Medicinal Plants of Indian Subcontinent*. Sci. Publ., Jodhpur (India).

Satapathy, K.B. and Chand, P.K. 2003. Plants used in healthcare of tribal women and children of Sundargarh district of Orissa. *Plant Sci. Res.* ***25****(1 and 2)*: 52-57.

Saxena, H.O. 1971. *Medicinal Plants of Patalkot*. SFRI, Jabalpur.

Saxena, H.O., Brahmam, M. and Dutta, P.K. 1981. Ethnobotanical studies in Orissa: 232-244. In Jain S.K. (*ed.*): *Glimpses of Indian Ethnobotany*. Oxford and IBH Publ. Co., New Delhi.

Schultes, R.E. 1960. Tapping our heritage of ethnobotanical lore. *Eco. Bot.* ***14***: 257-262.

Seetharam, Y.N., Chalageri, G., Haleshi, C. and Vijay. 1999. Folk medicine and ethnomedicine of northeastern Karnataka. *Ethnobotany* ***11***: 32-37.

Sen, S. and Batra, A. 1997. Ethno-medico-botany of household remedies of Phagi tehsil of Jaipur district (Rajasthan). *Ethnobotany* ***9***: 122-128.

Sen, S.K. and Behera, L.M. 2003. Ethnomedicinal plants used against skin diseases in Bargarh district in Orissa. *Ethnobotany* ***15***: 90-96.

Sen, S.K. and Behera, L.M. 2007. Ethnomedicinal plants used in touch therapy at Bargarh district of Orissa. *Ethnobotany* ***19***: 100-104.

Shah, N.C. and Virmani, O.P. 1982. *Directory of Crude Drugs and Aromatic Plants, Dealers, Producers and Exporters in India*: 1-98. Centr. Inst. Med. Arom. Pl., Lucknow.

Sharma, B.D. and Rana, J.C. 2000. Traditional medicinal uses of plants of Himachal hills: 173-176. In Maheshwari, J.K. (*ed.*): *Ethnobotany and Medicinal Plants of Indian Subcontinent*. Sci. Publ., Jodhpur (India).

Sharma, H.M., Devi, A.R. and Sharma, B.M. 2003a. Ethnomedicinal uses of monocotyledonous plants by the Meiteis of Manipur: 473-480. In Singh, V. and Jain, A.P. (*eds.*): *Ethnobotany and Medicinal Plants of India and Nepal*, Vol. I. Sci. Publ., Jodhpur (India).

Sharma, K.R. and Sood, M. 1997. *Important Medicinal Plants of Himachal Pradesh*. Dr. Y.S. Parmar Univ. of Hort. and Forestry, Nauni, H.P. (India).

Sharma, N.K. 2002. Ethno-medico-religious plants of Hadoti plateau (S.E. Rajasthan)-a preliminary survey: 394-411. In Trivedi, P.C. (*ed.*): *Ethnobotany*. Aavishkar Publ., Distributors, Jaipur (India).

Sharma, P.K., Chauhan, N.S. and Lal, B. 2003b. Commercially important medicinal and aromatic plants of Parvati Valley, Himachal Pradesh: 937-942. In Singh, V. and Jain, A.P. (*eds.*): *Ethnobotany and Medicinal Plants of India and Nepal*, Vol. II. Sci. Publ., Jodhpur (India).

Sharma, P.P. and Singh, N.P. 1999. Less-known ethnobotanical uses of plants in Dadra and Nagar Haveli (U.T.): Part I (A-K). *Ethnobotany* ***11***: 109-114.

Sharma, P.P. and Singh, N.P. 2000. Less-known ethnobotanical uses of plants in Dadra and Nagar Haveli (U.T.): Part II (L-Z). *Ethnobotany* ***12***: 81-85.

Sharma, P.P. and Singh, N.P. 2001. Ethnomedicinal uses of some edible plants in Dadra Nagar Haveli and Daman (U.T.). *Ethnobotany* ***13***: 121-125.

Sharma, R. 2003. *Medicinal Plants of India – An Encyclopaedia*. Daya Publ. House, Delhi (India).

Sharma, U.K. 2004. *Medicinal Plants of Assam*. Bishen Singh Mahendra Pal Singh, Dehra Dun (India).

Shivaprasad, B. and Chandrashekar, K.R. 2003. Plants used in the treatment of jaundice in Dakshina Kannada district: 864-868. In Singh, V. and Jain, A.P. (*eds.*): *Ethnobotany and Medicinal Plants of India and Nepal*, Vol. II. Sci. Publ., Jodhpur (India).

Sikarwar, R.L.S. 2002. Ethnogynaecological uses of plants new to India. *Ethnobotany* ***14***: 112-115.

Sikarwar, R.L.S. and Kaushik, J.P.1992. Some less known medicinal uses of trees among the Sahariyas of Morena district, M.P. India. *Ethnobotany* ***4***:71-74.

Singh, A.K. 2000. A contribution to the ethnobotany of sub-Himalayan region of eastern Uttar Pradesh: 237-246. In Maheshwari, J.K. (*ed.*): *Ethnobotany and Medicinal Plants of Indian Subcontinent*. Sci. Publ., Jodhpur (India).

Singh, B., Chaurasia, O.P. and Jadhav, C.K.L. 1996. An ethnobotany study of Indus valley (Ladakh): 92-101. In Maheshwari, J.K. (*ed.*): *Ethnobotany and Medicinal Plants of Indian Subcontinent*. Sci. Publ., Jodhpur (India).

Singh, G.S. 2002. Minor forest products of Sariska national park: An ethnobotanical profile: 221-238. In Trivedi, P.C. (*ed.*): *Ethnobotany*. Aavishkar Publ., Distributors, Jodhpur, India.

Singh, G.S. 2003. Ethnobotanical inventory of Sariska National Park in the Aravalli hills of eastern Rajasthan, India: 181-195. In Singh, V. and Jain, A.P. (*eds.*): *Ethnobotany and Medicinal Plants of India and Nepal*, Vol. I. Sci. Publ., Jodhpur (India).

Singh, H.B. 2000. Potential medicinal pteridophytes of India and their chemical constituents: 63-78. In Maheshwari, J.K. (*ed.*): *Ethnobotany and Medicinal Plants of Indian Subcontinent*. Sci. Publ., Jodhpur (India).

Singh, H.B. and Viswanathan, M.V. 1996. Useful pteridophytes of India- a gift of nature to human beings: 24-36. In Maheshwari, J.K. (*ed.*): *Ethnobotany in South Asia*. Sci. Publ., Jodhpur (India).

Singh, J., Bhuyan, T.C. and Ahmed, A. 1996. Ethnobotanical studies on the Mishing tribes of Assam with special reference to food and medicinal plants-I: 350-356. In Maheshwari, J.K. (*ed.*): *Ethnobotany in South Asia*. Sci. Publ., Jodhpur (India).

Singh, K.K. and Kumar, K. 2000. *Ethnobotanical Wisdom of Gaddi Tribe in Western Himalaya*. Bishen Singh Mahendra Pal Singh, Dehra Dun (India).

Singh, K.K. and Prakash, A. 1996. Observation on ethnobotany of the Kol tribe of Varanasi District, Uttar Pradesh, India: 133-137. In Maheshwari, J.K. (*ed.*): *Ethnobotany in South Asia*. Sci. Publ., Jodhpur (India).

Singh, P., Sisodia, S. and Shah, J. 2007. Angiospermous seeds of medicinal importance in Gujarat state: 17-24. In Chopra, A.K., Khanna, D.R., Prasad, G., Malik, D.S. and Bhutiani, R. (*eds.*): *Medicinal Plants: Conservation, Cultivation and Utilization.* Daya Publ. House, Delhi (India).

Singh, P.B. 1999. *Illustrated Field Guide to Commercially Important Medicinal and Aromatic Plants of Himachal Pradesh.* Soc. for Herbal Med. and Himalayan Biodiversity, H.P. (India).

Singh, P.K. and Singh, K.I. 2003a. First-aid remedies: an ethno-medico-botanical study of the Meitei community of Manipur: 466-472. In Singh, V. and Jain, A.P. (*eds.*): *Ethnobotany and Medicinal Plants of India and Nepal*, Vol. I. Sci. Publ., Jodhpur (India).

Singh, P.K. and Singh, K.I. 2003b. Mother and Child health: 1 – An ethnobotanical study of the Meitei community of Manipur state, India: 457-465. In Singh, V. and Jain, A.P. (*eds.*): *Ethnobotany and Medicinal Plants of India and Nepal*, Vol. I. Sci. Publ., Jodhpur (India).

Singh, S.C. and Srivastava, G.N. 2000. Exotic medicinal plants of Lucknow district, U.P., India: 223-235. In Maheshwari, J.K. (*ed.*): *Ethnobotany and Medicinal Plants of Indian Subcontinent.* Sci. Publ., Jodhpur (India).

Singh, S.P., Tripathi, S. and Shukla, R.S. 2003. Ethnomedicinal heritage of bioprospecting and drug development in North-Eastern states of India: 384-395. In Singh, V. and Jain, A.P. (*eds.*): *Ethnobotany and Medicinal Plants of India and Nepal*, Vol. I. Sci. Publ., Jodhpur (India).

Singh, U. 1996. *Dictionary of Economic Plants in India.* I.C.A.R., New Delhi.

Singh, U. and Narain, S. 2008. A short note on traditional remedy through plant wreath. *Ethnobotany* ***20***: 152-153.

Singh, U., Wadhwani, A.M. and John, B.M. 1983. *Dictionary of Economic Plants of India.* I.C.A.R., New Delhi.

Singh, V. and Pandey, R.P. 1998. *Ethnobotany of Rajasthan, India.* Sci. Publ., Jodhpur (India).

Singh, V.K. 1984. *Medicinal Plants of Gwalior Forest Division, Madhya Pradesh.* Central Council for Research in Unani Medicine, Delhi.

Sinha, B.K., Maina, V. and Padhye, P.M. 1996. Ethno-medicinal plants of Bay Islands for skin care: 375-380. In Maheshwari, J.K. (*ed.*): *Ethnobotany in South Asia.* Sci. Publ., Jodhpur (India).

Sinha, R.K. 1996. Conservation of cultural diversity of indigenous people essential for protection of biological diversity: 280-283. In Jain, S.K. (*ed.*): *Ethnobiology in Human Welfare.* Deep Publ., New Delhi.

Sinha, S.C. 1986. *Ethnobotanical Study of Manipur.* Ph.D. Thesis, Manipur University, Manipur.

Sinha, S.C. 1987. Ethnobotany of Manipur-medicinal plants. *Frontr. Bot.* ***1***: 123 – 152.

Sinha, S.C. 1987. Ethnobotany-medicinal plants. *Frontr. Bot.**1***: 123-152.

Sood, S.K. and Sudershana. 2008. *Root Drug Plants of Ethnic India.* Satish Serial Publ. House, Azadpur, Delhi.

Sood, S.K. and Thakur, S. 2004. *Ethnobotany of Rewalsar Himalaya* (*Dist. Mandi, Himachal Pradesh, India*). Deep Publ., New Delhi (India).

Sood, S.K., Bhardwaj, R. and Lakhanpal, T.N. 2005a. *Ethnic Indian Plants in Cure of Diabetes.* Sci. Publ., Jodhpur.

Sood, S.K., Kumar, S. and Lakhanpal, T.N. 2013. *Sacred Plants for Sustenance of Mankind.* Daya Publ., House, Delhi (India).

Sood, S.K., Kumar, S., Bassi, S.K. and Rana, J.C. 2012. *Ethnobotany of the Heritage Region of Shiwalik Himalaya.* Anamika Publ. and Distributors (P) LTD. New Delhi (India).

Sood, S.K., Nath, R. and Kalia, D.C. 2001. *Ethnobotany of Cold Desert Tribes of Lahaul-Spiti (N.W. Himalaya).* Deep Publ., New Delhi (India).

Sood, S.K., Parmar, S. and Lakhanpal, T.N. 2005b. *Ethnic Plants of India Used in Cancer Cure (A Compendium).* Bishen Singh Mahendra Pal Singh, Dehradun.

Sood, S.K., Rana, S. and Lakhanpal, T.N. 2005c. *Ethnic Aphrodisiac Plants.* Sci. Publ., Jodhpur.

Sood, S.K., Sharma, S.K., Kumar, N. and Kumar, H. 2009. *Ethnobotanical Studies on Trees, Shrubs and Climbers of Himalaya.* Satish Serial Publ. House, Delhi (India).

Srivastava, R.C., Singh, V.P. and Singh, M.K. 2003a. Medicinal plants of Jaunpur district (U.P.): 148-159. In Singh, V. and Jain, A.P. (*eds.*): *Ethnobotany and Medicinal Plants of India and Nepal*, Vol. I. Sci. Publ., Jodhpur (India).

Srivastava, S.C. and Rout, N. 1994. Some plants of ethnopaediatric importance in district Koraput, Orissa. *Bull. Bot. Surv. India* ***36***: 166-168.

Srivastava, S.C., Mitra, S. and Bandyopadhyay, S. 2003b. Some plants of ethnomedicinal importance from Darjeeling district, West Bengal: 972-977. In Singh, V. and Jain, A.P. (*eds.*): *Ethnobotany and Medicinal Plants of India and Nepal*, Vol. II. Sci. Publ., Jodhpur (India).

Srivastava, S.K., Mehrotra, S., Srivastava, G.K. and Chauhan, D.K. 2003c. Folk uses of herbs from Mahableshwar region of Sahyadri hills: 857-863. In Singh, V. and Jain, A.P. (*eds.*): *Ethnobotany and Medicinal Plants of India and Nepal*, Vol. II. Sci. Publ., Jodhpur (India).

Srivastava, T.N., Kapahi, B.K. and Atal, C.K. 1987. Ethnomedico-botanical investigation in Sikkim. *J. Econ. Tax. Bot.* ***11***: 413-421.

Subramani, S.P. and Goraya, G.S. 2003. Some folklore medicinal plants of Kolli hills: Record of a Natti Vaidyas Sammelan: 665-678. In Singh, V. and Jain, A.P. (*eds.*): *Ethnobotany and Medicinal Plants of India and Nepal*, Vol. II. Sci. Publ., Jodhpur (India).

Subramaniam, A. 2000. A survey of medicinal plants from Chitheri hills in Dharmapuri district, Tamil Nadu: 395-416. In Maheshwari, J.K. (*ed.*): *Ethnobotany and Medicinal Plants of Indian Subcontinent*. Sci. Publ., Jodhpur (India).

Subramanian, A., Mohan, V.R., Kumaresan, S. and Chelladurai, V. 2003. Medicinal plants used by the Valaiyans of Madurai District, Tamil Nadu: 785-787. In Singh, V. and Jain, A.P. (*eds.*): *Ethnobotany and Medicinal Plants of India and Nepal*, Vol. II. Sci. Publ., Jodhpur (India).

Sundararaj, D. 1959. *Guide to Economic Plants of South India*. Periodical Expert Book Agency, Delhi.

Sundaravelu, S., Selvaraju, M. and Sundaramoorthy, P. 2003. Medicinal weed flora of south Arcot Cuddalore district in Tamil Nadu: 87-94. In Khan, S.S. (*ed.*): *Vistas in Ethnobotany*, Vol. I. Classic Printers and Publ., Bhopal (India).

Sur, P.R. and Halder, A.C. 2002. Observation on the ethnobotany of Purulia District, West Bengal, India: 146-168. In Trivedi, P.C. (*ed.*): *Ethnobotany*. Aavishkar Publ., Distributors, Jaipur (India).

Swami, A. and Gupta, B.K. 1996. A note on some commonly occurring medicinal weeds of Udhampur district (J &K): 89-91. In Maheshwari, J.K. (*ed.*): *Ethnobotany in South Asia*. Sci. Publ., Jodhpur (India).

Tewari, K.C., Joshi, G.C. and Misra, S.S. 2001. Plants and allied material used in folk medicines of Assam, Meghalaya and Arunachal Pradesh (India): 523-546. In Pande, P.C. and Samant, S.S. (*eds.*): *Plant Diversity of the Himalaya.* Gyanodaya Prakashan, Nainital (India).

Thakor, A.B. 2009. Ethnomedicinal plants of Vaarli tribe in Valsad district (Gujarat). *Ethnobotany* ***21***:116-120.

Thakur, R.S., Puri, H.S., and Husain, A. 1989. *Major Medicinal Plants of India*. Central Institute of Medicinal and Aromatic Plants, Lucknow, India.

Thomas, J. and Britto, A.J.D. 2003. Ethnobotanical study in Wynad District in Kerala: 815-824. In Singh, V. and Jain, A.P. (*eds.*): *Ethnobotany and Medicinal Plants of India and Nepal*, Vol. II. Sci. Publ., Jodhpur (India).

Tirkey, A. 2004. Some ethnomedicinal plant species of Chhattisgarh state. *Ethnobotany* ***16***: 118-124.

Tiwari, J.K., Ballabha, R. and Tiwari, P. 2010. Ethnopaediatrics in Garhwal Himalaya, Uttarakhand, India (psychomedicine and medicine). *New York Sci. J.* ***3*** *(4)*: 123-126.

Tiwari, K.C. and Tiwari, V.P. 1996. Some important medicinal plants of the tropical, sub-tropical and temperate regions of Siang, Subansiri and Tirap districts of Arunanchal Pradesh: 359-363. In Maheshwari, J.K. (*ed.*): *Ethnobotany in South Asia*. Sci. Publ., Jodhpur (India).

Tiwari, K.P., 1998. *Medicinal Plants of Madhya Pradesh: Distribution, Cultivation and Trade*. SFRI Jabalpur.

Topno, S. and Ghosh, T.K. 2000. Correlation of uses of medicinal plants by tribals of Chotanagpur with other tribals of India:148-146. In Maheshwari, J.K. (*ed.*): *Ethnobotany and Medicinal Plants of Indian Subcontinent.* Sci. Publ., Jodhpur (India).

Tribedi, G.N., Kayal, R.N. and Chaudhuri Rai, H.N. 1982. Some medicinal plants of Mayurbhanj (Orissa). *Bull. Bot. Surv. India.* ***24***: 117-120.

Tripathi, G. 2001. Indigenous knowledge and traditional practices of some Himalayan medicinal plants: 151-156. In Samant, S.S., Dhar, U. and Palni, L.M.S. (*eds.*): *Himalayan Medicinal Plants: Potential and Prospects.* Gyanodaya Prakashan, Nainital (India).

Tripathi, S. and Goel, A.K. 2001. Ethnobotanical diversity of Zingiberaceae in North-Eastern India. *Ethnobotany* ***13***: 67-79.

Tyler, V.E. 1986. Plant Drugs in the twenty first century. *Econ. Bot.* ***40***: 279.

Uniyal, M.R. 1989. *Medicinal Flora of Garhwal Himalayas.* Baidyanath Ayurvedic Research Centre, Patna.

Uniyal, M.R. 1997. *Medicinal Plants and Minerals of Uttarakhand Himalaya.* Baidyanath Ayurvedic Research Centre, Patna.

Upadhyay, R. and Chauhan, S.V.S. 2000. Ethnobotanical observations on Koya tribe of Gundala Mandal of Khammam district, Andhra Pradesh. *Ethnobotany* ***12***: 93-99.

Upadhyay, R. and Chauhan, S.V.S. 2003. Ethnobotanical uses of plant gums by the tribals: 601-602. In Singh, V. and Jain, A.P. (*eds.*): *Ethnobotany and Medicinal Plants of India and Nepal*, Vol. II. Sci. Publ., Jodhpur (India).

Upadhyay, R. and Singh, J. 2005. Ethno-medicinal uses of plants from Tikri forest of Gonda district (U.P.). *Ethnobotany* ***17***: 167-170.

Upadhye, A.S., Vartak, V.D. and Kumbhojkar, M.S. 1994. Ethno-medico-botanical studies in Western Maharastra, India. *Ethnobotany* ***6***: 25-31.

Uphof, J.C.T. 1986. *Dictionary of Economic Plants* (2nd *ed.*). Stechert-Hafner Service Agency, Inc., New York.

Varghese, E. 1996. *Applied Ethnobotany - A Case Study Among the Kharias of Central India.* Deep Publ., New Delhi (India).

Vedavathy, S., Mrudula, V. and Sudhakar, A. 1997. *Tribal Medicine of Chittore District, A.P.* Intl. Devlt. Res. Ctre., Canada and Sri Venkateshwara Arts College.

Venkataratnam, K. and Raju, R.R.V. 2004. Folk medicine from Gundlabrahmeswaram wild-life sanctuary, Andhra Pradesh. *Ethnobotany 16*: 33-39.

Venugopal, S.N. 1992. *Medicinal Plants Useful to Mother and Child Health.* Department of Family Welfare, Ministry of Health and Family Welfare, Govt. of India, New Delhi.

Verma, P., Khan, A.A. and Singh, K.K. 1995. Traditional phytotherapy among the Baiga tribe of Shahdol district of Madhya Pradesh, India. *Ethnobotany* ***7***: 69-73.

Verma, S. and Chauhan, N.S. 2006. Studies on ethno-medico-botany of Kunihar forest division, district Solan (H.P.). *Ethnobotany* ***18***: 160-165.

Vihari, V. 1995. Ethnobotany of cosmetics of Indo-Nepal border. *Ethnobotany* ***7***: 89-94.

Vishwanathan, M.B. 1997. Ethnobotany of the Malayalis in north Arcot district, Tamil Nadu, India. *Ethnobotany* ***9***: 77-79.

Vishwanathan, M.B., Kumar, E.H.P. and Ramesh, N. 2001. Ethnomedicines of Kanis in Kalakkad – Mundanthurai tiger reserve, Tamil Nadu. *Ethnobotany* ***13***: 60-66.

Vishwanathan, M.B., Kumar, E.H.P. and Ramesh, N. 2006. *Ethnobotany of the Kanis (Kalakkad-Mundanthurai tiger reserve in Tirunelveli district, Tamil Nadu, India).* Bishen Singh Mahendra Pal Singh, Dehradun (India).

Vishwanathan, M.V. 2000. Edible and medicinal plants of Ladakh (J and K): 151-154. In Maheshwari, J.K. (*eds.*): *Ethnobotany and Medicinal Plants of Indian Subcontinent.* Sci. Publ., Jodhpur (India).

Wambebe, C.O.N. 1990. Natural products in developing economy. In Igbocchi, A.C. and Osisigu, I.U.W. (*eds.*): *National Workshop on Natural Products.* Univ. of Benin Press, Nigeria.

Warrier, P.K., Nambiar, V.P.K. and Raman, K. 1994-1996. *Indian Medicinal Plants – A Compendium of 500 Species.* Vol. I-V. Orient Longman Limited, Hyderabad.

Watt, G. 1889-96. *A Dictionary of the Economic Products of India*, Vol. I-VI. Periodical Experts, New Delhi (India).

WHO, 2002. WHO Traditional Medicine Strategy 2002-2005. World Health Organization, Geneva, 2002.

Yasodamma, N., Mehar, S.K. and Paramageetham, C. 2009. Threat assessment (IUCN categorization) for ethnomedicinal plants used by Chenchu tribe of Gundlabramheswaram in Nallamalai hills in Andhra Pradesh. *Ethnobotany* ***21***: 51-60.

Appendices

Appendix I
Index to Families, Genera and Species

Division	*No. of Families*	*Total No. of Genera/ Species*	*Total No. of Species in a Genera*
Fungi	**1**	**1/1**	*Calvatia* (1).
Lichens	**1**	**1/1**	*Parmelia* (1).
Pteridophytes	**14**	**17/19**	*Adiantum* (2), *Asplenium* (1), *Athyrium* (1), *Blechnum* (1), *Christella* (1), *Diplazium* (1), *Equisetum* (2), *Hemionitis* (1), *Huperzia* (1), *Lycopodium* (1), *Lygodium* (1), *Marsilea* (1), *Osmunda* (1), *Psilotum* (1), *Pteris* (1), *Selaginella* (1), *Tectaria* (1).
Gymnosperms	**3**	**5/5**	*Abies* (1), *Cedrus* (1), *Ephedra* (1), *Pinus* (1), *Thuja* (1).
Dicots	**116**	**483/683**	*Abrus* (1), *Abutilon* (2), *Acacia* (5), *Acalypha* (1), *Achillea* (1), *Achyranthes* (1), *Acmella* (1), *Aconitum* (4), *Acranthera* (1), *Acronychia* (1), *Adansonia* (1), *Aeginetia* (1), *Aegle* (1), *Aerva* (1), *Aeschynanthus* (1), *Aethusa* (1), *Ageratum* (1), *Agrimonia* (1), *Ailanthus* (3), *Ajuga* (1), *Alangium* (1), *Albizia* (1), *Alcea* (1), *Alhagi* (1), *Allophylus* (1), *Alstonia* (1), *Alternanthera* (1), *Althaea* (1), *Alysicarpus* (1), *Amaranthus* (2), *Ampelocissus* (1), *Amphineurion* (1), *Anacardium* (1), *Anacolosa* (1), *Anacyclus* (1), *Anaphalis* (1), *Andrographis* (3), *Anemone* (2), *Anethum* (1), *Angelica* (1), *Anisochilus* (1), *Anisomeles* (2), *Annona* (1), *Anthemis* (1), *Antiaris* (1), *Antidesma* (1), *Apocynum* (1), *Aporosa* (1), *Aquilaria* (1), *Arachis* (1), *Archidendron* (1), *Ardisia* (1), *Argemone* (1), *Argyreia* (1), *Aristolochia* (4), *Artemisia* (9), *Artocarpus* (2), *Asclepias* (2), *Asystasia* (1), *Atropa* (1), *Atylosia* (1), *Averrhoa* (2), *Avicennia* (1), *Azadirachta* (1), *Azima* (1), *Baccaurea* (1), *Bacopa* (1), *Balanites* (1), *Baliospermum* (1), *Baphicacanthus* (1), *Baptisia* (1), *Barleria* (1), *Barringtonia* (1), *Basella* (1), *Basilicum* (1), *Bauhinia* (3), *Benincasa* (1), *Berberis* (3), *Berchemia* (1), *Bergenia* (2), *Bidens* (1), *Bischofia* (1), *Bistorta* (1), *Bixa* (1), *Blepharispermum* (1), *Blumea* (3), *Blumeopsis* (1),

Contd..

Appendix–I–*Contd...*

Division	*No. of Families*	*Total No. of Genera/ Species*	*Total No. of Species in a Genera*
			Boerhavia (1), *Bombax* (2), *Breonia* (1), *Breynia* (1), *Bridelia* (2), *Bryophyllum* (1), *Buchanania* (1), *Butea* (2), *Cadaba* (1), *Caesalpinia* (1), *Cajanus* (3), *Calendula* (1), *Callicarpa* (2), *Calotropis* (2), *Cananga* (1), *Cannabis* (1), *Capparis* (2), *Capsicum* (1), *Caralluma* (1), *Cardiospermum* (1), *Careya* (1), *Carica* (1), *Carissa* (1), *Carthamus* (1), *Carum* (1), *Cascabela* (1), *Casearia* (1), *Cassia* (2), *Cassine* (1), *Cassytha* (1), *Catharanthus* (1), *Catunaregam* (1), *Ceiba* (1), *Celastrus* (1), *Celosia* (1), *Centaurea* (1), *Centella* (1), *Centratherum* (1), *Cerasus* (1), *Ceriscoides* (1), *Ceropegia* (1), *Chenopodium* (1), *Chrozophora* (2), *Cicer* (1), *Cinchona* (1), *Cinnamomum* (5), *Cirsium* (1), *Cissampelos* (1), *Cissus* (1), *Citrullus* (1), *Citrus* (2), *Cleistanthus* (1), *Clematis* (2), *Cleome* (2), *Clerodendrum* (3), *Clitoria* (1), *Coccinia* (1), *Cocculus* (1), *Codiaeum* (1), *Codiocarpus* (1), *Coffea* (1), *Coleus* (1), *Combretum* (3), *Commiphora* (2), *Coptis* (1), *Corallocarpus* (1), *Corchorus* (1), *Coriandrum* (1), *Coscinium* (1), *Cotula* (1), *Crossandra* (1), *Crotalaria* (4), *Croton* (2), *Cryptolepis* (1), *Cucumis* (1), *Cucurbita* (1), *Cuscuta* (1), *Cyamopsis* (1), *Cyclea* (1), *Cydonia* (1), *Cynoglossum* (1), *Cyphostemma* (1), *Dactylicapnos* (1), *Dalbergia* (2), *Datura* (2), *Decaspermum* (1), *Dendrocnide* (1), *Dendrophthoe* (1), *Descurainia* (1), *Desmodium* (4), *Dichrostachys* (1), *Dicliptera* (1), *Dillenia* (1), *Diospyros* (2), *Diplocyclos* (1), *Diplopenta* (1), *Dipteracanthus* (1), *Dischidia* (1), *Drosera* (1), *Drymaria* (1), *Dysphania* (1), *Echinacea* (1), *Echinops* (1), *Eclipta* (1), *Elaeocarpus* (1), *Elatostema* (1), *Elephantopus* (1), *Elsholtzia* (3), *Elytraria* (1), *Embelia* (3), *Emilia* (1), *Enicostemma* (1), *Entada* (1), *Enydra* (1), *Eranthemum* (1), *Erythrina* (3), *Eschscholzia* (1), *Eucalyptus* (2), *Euodia* (1), *Eupatorium* (1), *Euphorbia* (9), *Evolvulus* (1), *Fagonia* (1), *Ferula* (2), *Ficus* (6), *Flacourita* (2), *Flemingia* (2), *Flueggea* (1), *Foeniculum* (1), *Forsythia* (1), *Fragaria* (1),

Contd..

Appendix–I–*Contd...*

Division	*No. of Families*	*Total No. of Genera/ Species*	*Total No. of Species in a Genera*
			Fraxinus (1), *Galium* (1), *Garcinia* (2), *Gardenia* (3), *Gelsomium* (1), *Geranium* (1), *Givotia* (1), *Glinus* (1), *Glochidion* (1), *Glossocardia* (1), *Glycine* (1), *Glycosmis* (1), *Gmelina* (1), *Gnaphalium* (1), *Gossypium* (2), *Grewia* (2), *Guaiacum* (1), *Gymnema* (1), *Gymnosporia* (1), *Gynocardia* (1), *Haematoxylum* (1), *Haldina* (1), *Hedera* (2), *Hedyotis* (1), *Helicteres* (1), *Heliotropium* (1), *Hemidesmus* (1), *Hibiscus* (1), *Holarrhena* (1), *Holoptelea* (1), *Houttuynia* (1), *Hybanthus* (1), *Hydrastis* (1), *Hydrocotyle* (2), *Hygrophila* (1), *Hyptis* (1), *Ichnocarpus* (1), *Ilex* (1), *Incarvillea* (1), *Indigofera* (4), *Inula* (1), *Ipomoea* (3), *Ixora* (1), *Jasminum* (4), *Jatropha* (4), *Juglans* (1), *Justicia* (2), *Kalanchoe* (1), *Kallstroemia* (1), *Kickxia* (1), *Lactuca* (2), *Lagenaria* (1), *Lannea* (1), *Lantana* (2), *Lasianthus* (1), *Launaea* (3), *Lawsonia* (1), *Lens* (1), *Leonotis* (1), *Lepidagathis* (2), *Lepidium* (1), *Lepionurus* (1), *Leucas* (2), *Limonia* (1), *Litchi* (1), *Lithospermum* (1), *Lobelia* (1), *Lonicera* (1), *Ludwigia* (1), *Luffa* (2), *Macrotyloma* (1), *Madhuca* (1), *Mallotus* (1), *Malus* (1), *Mangifera* (1), *Manilkara* (1), *Martynia* (1), *Matricaria* (1), *Melanthera* (1), *Melastoma* (1), *Melilotus* (1), *Melochia* (1), *Mentha* (4), *Mesua* (1), *Meyna* (2), *Microglossa* (1), *Micromelum* (1), *Mikania* (1), *Mimosa* (1), *Mimusops* (1), *Momordica* (2), *Morinda* (2), *Moringa* (1), *Morus* (1), *Mucuna* (1), *Murraya* (2), *Mussaenda* (2), *Myrica* (1), *Myristica* (1), *Myrtus* (1), *Myxopyrum* (1), *Naravelia* (1), *Nauclea* (1), *Nelsonia* (1), *Nelumbo* (1), *Nepeta* (1), *Nicotiana* (2), *Nyctanthes* (1), *Nymphaea* (1), *Ocimum* (4), *Olea* (1), *Onosma* (1), *Ophiorrhiza* (1), *Opuntia* (1), *Origanum* (1), *Oroxylum* (1), *Orthosiphon* (2), *Osbeckia* (2), *Oxalis* (1), *Oxystelma* (1), *Padus* (1), *Paederia* (1), *Paeonia* (1), *Pandanus* (1), *Papaver* (1), *Parochetus* (1), *Passiflora* (1), *Pavetta* (2), *Pedalium* (1), *Pegolettia* (1), *Pentanema* (1), *Pentapetes* (1), *Pepromia* (1), *Pergularia* (1), *Persicaria* (1), *Peucedanum* (1),

Contd..

Appendix–I–*Contd...*

Division	*No. of Families*	*Total No. of Genera/ Species*	*Total No. of Species in a Genera*
			Phaseolus (1), *Phlogacanthus* (1), *Phyla* (1), *Phyllanthus* (8), *Physalis* (1), *Phytolacca* (2), *Pimpinella* (1), *Piper* (3), *Pisonia* (1), *Pistacia* (1), *Plantago* (2), *Platycodon* (1), *Plectranthus* (2), *Pleurospermum* (1), *Pogostemon* (1), *Polycarpon* (1), *Polygonum* (2), *Polyscias* (1), *Pometia* (1), *Pongamia* (1), *Populus* (1), *Portulaca* (1), *Pouzolzia* (1), *Premna* (1), *Primula* (1), *Prinsepia* (1), *Prosopis* (3), *Prunus* (2), *Psidium* (1), *Pterolobium* (1), *Pterospermum* (2), *Pueraria* (1), *Punica* (1), *Quercus* (3), *Radermachera* (1), *Ranunculus* (1), *Rauvolfia* (1), *Rhazya* (1), *Rheum* (3), *Rhododendron* (1), *Rhus* (3), *Ricinus* (1), *Rorippa* (1), *Rosa* (2), *Roylea* (1), *Rubia* (1), *Rubus* (3), *Rungia* (2), *Ruta* (2), *Salomonia* (1), *Salvadora* (1), *Salvia* (1), *Sambucus* (1), *Santalum* (1), *Sapindus* (2), *Sauropus* (2), *Schefflera* (1), *Schleichera* (1), *Schrebera* (1), *Scoparia* (1), *Scrophularia* (1), *Semecarpus* (1), *Senna* (4), *Seriphidium* (2), *Sesamum* (1), *Sesbania* (2), *Shorea* (1), *Sida* (2), *Skimmia* (1), *Solanum* (9), *Solidago* (1), *Sphaeranthus* (1), *Spondias* (1), *Stachytarpheta* (1), *Stephania* (1), *Stereospermum* (1), *Streblus* (1), *Strychnos* (2), *Stylosanthes* (1), *Swertia* (1), *Syzygium* (5), *Tabernaemontana* (1), *Tadehagi* (1), *Tagetes* (1), *Tamarindus* (1), *Tamilnadia* (1), *Tanacetum* (1), *Taraxacum* (1), *Tarenna* (1), *Tephrosia* (1), *Terminalia* (5), *Thunbergia* (2), *Tinospora* (1), *Toddalia* (1), *Toona* (1), *Tournefortia* (1), *Trachyspermum* (2), *Tragia* (1), *Tribulus* (1), *Trichaurus* (1), *Trichodesma* (1), *Trichosanthes* (1), *Trifolium* (2), *Trigonella* (1), *Tylophora* (2), *Uncaria* (2), *Uraria* (2), *Urtica* (1), *Vaccinium* (1), *Valeriana* (1), *Ventilago* (1), *Verbascum* (1), *Vernonia* (3), *Vicia* (1), *Vigna* (2), *Viola* (3), *Vitex* (3), *Vitis* (1), *Volutarella* (1), *Withania* (1), *Woodfordia* (1), *Wrightia* (1), *Xanthium* (1), *Ximenia* (1), *Xylocarpus* (1), *Xylosma* (1), *Zanthoxylum* (1), *Ziziphus* (2), *Zornia* (2).

Contd..

Appendix–I–*Contd...*

Division	*No. of Families*	*Total No. of Genera/ Species*	*Total No. of Species in a Genera*
Monocots	**28**	**61/91**	*Acorus* (1), *Allium* (5), *Alocasia* (1), *Aloe* (2), *Alpinia* (2), *Amomum* (1), *Amorphophallus* (1), *Ananas* (1), *Aneilema* (1), *Areca* (1), *Asparagus* (5), *Bambusa* (2), *Bulbophyllum* (1), *Calamus* (2), *Canna* (1), *Carex* (1), *Cocos* (1), *Coix* (1), *Colchicum* (1), *Costus* (1), *Crocus* (1), *Cryptocoryne* (1), *Curculigo* (1), *Curcuma* (6), *Cymbopogon* (5), *Cynodon* (1), *Cyperus* (2), *Dactyloctenium* (1), *Dactylorhiza* (1), *Dioscorea* (1), *Eleusine* (1), *Flagellaria* (1), *Gloriosa* (1), *Hedychium* (2), *Heteropogon* (1), *Hordeum* (2), *Imperata* (1), *Iris* (2), *Kaempferia* (2), *Lodoicea* (1), *Melocalamus* (1), *Monochoria* (2), *Musa* (1), *Orchis* (1), *Oryza* (2), *Panicum* (1), *Pinellia* (1), *Polianthes* (1), *Pothos* (1), *Remusatia* (1), *Saccharum* (1), *Sansevieria* (1), *Scleria* (1), *Setaria* (1), *Smilax* (2), *Syagrus* (1), *Tacca* (1), *Typha* (2), *Typhonium* (1), *Veratrum* (1), *Zingiber* (1).

Appendix II
Index to Families of Plants

Name of Family	*Genus/Species*	*Total No. of Species under a Genus*
Acanthaceae	16/22	*Andrographis* (3), *Asystasia* (1), *Baphicacanthus* (1), *Barleria* (1), *Crossandra* (1), *Dicliptera* (1), *Dipteracanthus* (1), *Elytraria* (1), *Eranthemum* (1), *Hygrophila* (1), *Justicia* (2), *Lepidagathis* (2), *Nelsonia* (1), *Phlogacanthus* (1), *Rungia* (2), *Thunbergia* (2).
Acoraceae	1/1	*Acorus* (1).
Adiantaceae	2/3	*Adiantum* (2), *Hemionitis* (1).
Agavaceae	1/1	*Polianthes* (1).
Alangiaceae	1/1	*Alangium* (1).
Aloaceae	1/2	*Aloe* (2).
Amaranthaceae	5/6	*Achyranthes* (1), *Aerva* (1), *Alternanthera* (1), *Amaranthus* (2), *Celosia* (1).
Anacardiaceae	8/10	*Anacardium* (1), *Buchanania* (1), *Lannea* (1), *Mangifera* (1), *Pistacia* (1), *Rhus* (3), *Semecarpus* (1), *Spondias* (1).
Annonaceae	2/2	*Annona* (1), *Cananga* (1).
Apiaceae	13/16	*Aethusa* (1), *Anethum* (1), *Angelica* (1), *Carum* (1), *Centella* (1), *Coriandrum* (1), *Ferula* (2), *Foeniculum* (1), *Hydrocotyle* (2), *Peucedanum* (1), *Pimpinella* (1), *Pleurospermum* (1), *Trachyspermum* (2).
Apocynaceae	12/13	*Alstonia* (1), *Amphineurion* (1), *Apocynum* (1), *Carissa* (1), *Cascabela* (1), *Catharanthus* (1), *Holarrhena* (1), *Ichnocarpus* (1), *Rauvolfia* (1), *Rhazya* (1), *Tabernaemontana* (1), *Wrightia* (2).
Aquifoliaceae	1/1	*Ilex* (1).
Araceae	7/7	*Alocasia* (1), *Amorphophallus* (1), *Cryptocoryne* (1), *Pinellia* (1), *Pothos* (1), *Remusatia* (1), *Typhonium* (1).
Araliaceae	3/4	*Hedera* (2), *Polyscias* (1), *Schefflera* (1).
Arecaceae	5/6	*Areca* (1), *Calamus* (2), *Cocos* (1), *Lodoicea* (1), *Syagrus* (1).
Aristolochiaceae	1/4	*Aristolochia* (4)

Contd...

Appendix–II–*Contd...*

Name of Family	*Genus/Species*	*Total No. of Species under a Genus*
Asclepiadaceae	11/14	*Asclepias* (2), *Calotropis* (2), *Caralluma* (1), *Ceropegia* (1), *Cryptolepis* (1), *Dischidia* (1), *Gymnema* (1), *Hemidesmus* (1), *Oxystelma* (1), *Pergularia* (1), *Tylophora* (2).
Asparagaceae	1/5	*Asparagus* (5).
Aspleniaceae	1/1	*Asplenium* (1).
Asteraceae	44/60	*Achillea* (1), *Acmella* (1), *Ageratum* (1), *Anacyclus* (1), *Anaphalis* (1), *Anthemis* (1), *Artemisia* (9), *Bidens* (1), *Blepharispermum* (1), *Blumea* (3), *Blumeopsis* (1), *Calendula* (1), *Carthamus* (1), *Centaurea* (1), *Centratherum* (1), *Cirsium* (1), *Cotula* (1), *Echinacea* (1), *Echinops* (1), *Eclipta* (1), *Elephantopus* (1), *Emilia* (1), *Enydra* (1), *Eupatorium* (1), *Glossocardia* (1), *Gnaphalium* (1), *Inula* (1), *Lactuca* (2), *Launaea* (3), *Matricaria* (1), *Melanthera* (1), *Microglossa* (1), *Mikania* (1), *Pegolettia* (1), *Pentanema* (1), *Seriphidium* (2), *Solidago* (1), *Sphaeranthus* (1), *Tagetes* (1), *Tanacetum* (1), *Taraxacum* (1), *Vernonia* (3), *Volutarella* (1), *Xanthium* (1).
Balanitaceae	1/1	*Balanites* (1).
Bambusaceae	1/2	*Bambusa* (2).
Basellaceae	1/1	*Basella* (1).
Berberidaceae	1/3	*Berberis* (3).
Bignoniaceae	4/4	*Incarvillea* (1), *Oroxylum* (1), *Radermachera* (1), *Stereospermum* (1).
Bixaceae	1/1	*Bixa* (1).
Blechnaceae	1/1	*Blechnum* (1).
Bombacaceae	2/3	*Adansonia* (1), *Bombax* (2).
Boraginaceae	6/6	*Cynoglossum* (1), *Heliotropium* (1), *Lithospermum* (1), *Onosma* (1), *Tournefortia* (1), *Trichodesma* (1).
Brassicaceae	3/3	*Descurainia* (1), *Lepidium* (1), *Rorippa* (1).
Bromeliaceae	1/1	*Ananas* (1).
Burseraceae	1/2	*Commiphora* (2).
Cactaceae	1/1	*Opuntia* (1).
Campanulaceae	2/2	*Lobelia* (1), *Platycodon* (1).

Contd...

Appendix–II–*Contd...*

Name of Family	*Genus/Species*	*Total No. of Species under a Genus*
Cannabinaceae	1/1	*Cannabis* (1).
Cannaceae	1/1	*Canna* (1).
Capparaceae	3/5	*Cadaba* (1), *Capparis* (2), *Cleome* (2).
Caprifoliaceae	2/2	*Lonicera* (1), *Sambucus* (1).
Caricaceae	1/1	*Carica* (1).
Caryophyllaceae	2/2	*Drymaria* (1), *Polycarpon* (1).
Celastraceae	3/3	*Cassine* (1), *Celastrus* (1), *Gymnosporia* (1).
Chenopodiaceae	2/2	*Chenopodium* (1), *Dysphania* (1).
Clusiaceae	2/3	*Garcinia* (2), *Mesua* (1).
Colchicaceae	1/1	*Colchicum* (1).
Combretaceae	2/8	*Combretum* (3), *Terminalia* (5).
Commelinaceae	1/1	*Aneilema* (1).
Convolvulaceae	4/6	*Argyreia* (1), *Cuscuta* (1), *Evolvulus* (1), *Ipomoea* (3).
Costaceae	1/1	*Costus* (1).
Crassulaceae	2/2	*Bryophyllum* (1), *Kalanchoe* (1).
Cucurbitaceae	11/13	*Benincasa* (1), *Citrullus* (1), *Coccinia* (1), *Corallocarpus* (1), *Cucumis* (1), *Cucurbita* (1), *Diplocyclos* (1), *Lagenaria* (1), *Luffa* (2), *Momordica* (2), *Trichosanthes* (1).
Cupressaceae	1/1	*Thuja* (1).
Cyperaceae	3/4	*Carex* (1), *Cyperus* (2), *Scleria* (1).
Dilleniaceae	1/1	*Dillenia* (1).
Dioscoreaceae	1/1	*Dioscorea* (1).
Dipterocarpaceae	1/1	*Shorea* (1).
Dracaenaceae	1/1	*Sansevieria* (1).
Droseraceae	1/1	*Drosera* (1).
Dryopteridaceae	1/1	*Tectaria* (1).
Ebenaceae	1/2	*Diospyros* (2).
Elaeocarpaceae	1/1	*Elaeocarpus* (1).
Ephedraceae	1/1	*Ephedra* (1).
Equisetaceae	1/2	*Equisetum* (2).

Contd...

Appendix–II–*Contd...*

Name of Family	*Genus/Species*	*Total No. of Species under a Genus*
Ericaceae	2/2	*Rhododendron* (1), *Vaccinium* (1).
Euphorbiaceae	22/44	*Acalypha* (1), *Antidesma* (1), *Aporosa* (1), *Baccaurea* (1), *Baliospermum* (1), *Bischofia* (1), *Breynia* (1), *Bridelia* (2), *Chrozophora* (2), *Cleistanthus* (1), *Codiaeum* (1), *Croton* (2), *Euphorbia* (9), *Flueggea* (1), *Givotia* (1), *Glochidion* (1), *Jatropha* (4), *Mallotus* (1), *Phyllanthus* (8), *Ricinus* (1), *Sauropus* (2), *Tragia* (1).
Fabaceae	50/82	*Abrus* (1), *Acacia* (5), *Albizia* (1), *Alhagi* (1), *Alysicarpus* (1), *Arachis* (1), *Archidendron* (1), *Atylosia* (1), *Baptisia* (1), *Bauhinia* (3), *Butea* (2), *Caesalpinia* (1), *Cajanus* (3), *Cassia* (2), *Cicer* (1), *Clitoria* (1), *Crotalaria* (4), *Cyamopsis* (1), *Dalbergia* (2), *Desmodium* (4), *Dichrostachys* (1), *Entada* (1), *Erythrina* (3), *Flemingia* (2), *Glycine* (1), *Haematoxylum* (1), *Indigofera* (4), *Lens* (1), *Macrotyloma* (1), *Melilotus* (1), *Mimosa* (1), *Mucuna* (1), *Parochetus* (1), *Phaseolus* (1), *Pongamia* (1), *Prosopis* (3), *Pterolobium* (1), *Pueraria* (1), *Senna* (4), *Sesbania* (2), *Stylosanthes* (1), *Tadehagi* (1), *Tamarindus* (1), *Tephrosia* (1), *Trifolium* (2), *Trigonella* (1), *Uraria* (1), *Vicia* (1), *Vigna* (2), *Zornia* (2).
Fagaceae	1/3	*Quercus* (3).
Flacourtiaceae	3/4	*Flacourita* (2), *Gynocardia* (1), *Xylosma* (1).
Flagellariaceae	1/1	*Flagellaria* (1).
Gentianaceae	2/2	*Enicostemma* (1), *Swertia* (1).
Geraniaceae	1/1	*Geranium* (1).
Gesneriaceae	1/1	*Aeschynanthus* (1).
Hypoxidaceae	1/1	*Curculigo* (1).
Icacinaceae	1/1	*Codiocarpus* (1).
Iridaceae	2/3	*Crocus* (1), *Iris* (2).
Juglandaceae	1/1	*Juglans* (1).
Lamiaceae	22/39	*Ajuga* (1), *Anisochilus* (1), *Anisomeles* (2), *Basilicum* (1), *Callicarpa* (2), *Clerodendrum* (3), *Coleus* (1), *Elsholtzia* (3), *Hyptis* (1), *Leonotis* (1), *Leucas* (2), *Mentha* (4), *Nepeta* (1), *Ocimum* (4), *Origanum* (1), *Orthosiphon* (2), *Plectranthus* (2), *Pogostemon* (1), *Premna* (1), *Roylea* (1), *Salvia* (1), *Vitex* (3).

Contd...

Appendix–II–*Contd...*

Name of Family	Genus/Species	Total No. of Species under a Genus
Lauraceae	2/6	*Cassytha* (1), *Cinnamomum* (5).
Lecythidaceae	1/1	*Barringtonia* (1).
Liliaceae	2/6	*Allium* (5), *Gloriosa* (1).
Loganiaceae	2/3	*Gelsomium* (1), *Strychnos* (2).
Loranthaceae	1/1	*Dendrophthoe* (1).
Lycoperdaceae	1/1	*Calvatia* (1).
Lycopodiaceae	2/2	*Huperzia* (1), *Lycopodium* (1).
Lythraceae	2/2	*Lawsonia* (1), *Woodfordia* (1).
Malvaceae	9/12	*Abutilon* (2), *Alcea* (1), *Althaea* (1), *Ceiba* (1), *Diplopenta* (1), *Gossypium* (2), *Hibiscus* (1), *Pterospermum* (1), *Sida* (2).
Marsileaceae	1/1	*Marsilea* (1).
Martyniaceae	1/1	*Martynia* (1).
Melanthiaceae	1/1	*Veratrum* (1).
Melastomataceae	2/3	*Melastoma* (1), *Osbeckia* (2).
Meliaceae	3/3	*Azadirachta* (1), *Toona* (1), *Xylocarpus* (1).
Menispermaceae	6/6	*Cissampelos* (1), *Cocculus* (1), *Coscinium* (1), *Cyclea* (1), *Stephania* (1), *Tinospora* (1).
Molluginaceae	1/1	*Glinus* (1).
Moraceae	5/11	*Antiaris* (1), *Artocarpus* (2), *Ficus* (6), *Morus* (1), *Streblus* (1).
Moringaceae	1/1	*Moringa* (1).
Musaceae	1/1	*Musa* (1).
Myricaceae	1/1	*Myrica* (1).
Myristicaceae	1/1	*Myristica* (1).
Myrsinaceae	2/4	*Ardisia* (1), *Embelia* (3).
Myrtaceae	6/11	*Careya* (1), *Decaspermum* (1), *Eucalyptus* (2), *Myrtus* (1), *Psidium* (1), *Syzygium* (5).
Nelumbonaceae	1/1	*Nelumbo* (1).
Nyctaginaceae	2/2	*Boerhavia* (1), *Pisonia* (1).
Nymphaeaceae	1/1	*Nymphaea* (1).
Olacaceae	2/2	*Anacolosa* (1), *Ximenia* (1).

Contd...

Appendix–II–*Contd...*

Name of Family	*Genus/Species*	*Total No. of Species under a Genus*
Oleaceae	7/10	*Forsythia* (1), *Fraxinus* (1), *Jasminum* (4), *Myxopyrum* (1), *Nyctanthes* (1), *Olea* (1), *Schrebera* (1).
Onagraceae	1/1	*Ludwigia* (1).
Opiliaceae	1/1	*Lepionurus* (1).
Orchidaceae	3/3	*Bulbophyllum* (1), *Dactylorhiza* (1), *Orchis* (1).
Orobanchaceae	1/1	*Aeginetia* (1).
Osmundaceae	1/1	*Osmunda* (1).
Oxalidaceae	2/3	*Averrhoa* (2), *Oxalis* (1).
Paeoniaceae	1/1	*Paeonia* (1).
Pandanaceae	1/1	*Pandanus* (1).
Papaveraceae	4/4	*Argemone* (1), *Dactylicapnos* (1), *Eschscholzia* (1), *Papaver* (1).
Parmeliaceae	1/1	*Parmelia* (1).
Passifloraceae	1/1	*Passiflora* (1).
Pedaliaceae	2/2	*Pedalium* (1), *Sesamum* (1).
Phytolaccaceae	1/2	*Phytolacca* (2).
Pinaceae	3/3	*Abies* (1), *Cedrus* (1), *Pinus* (1).
Piperaceae	2/4	*Pepromia* (1), *Piper* (3).
Plantaginaceae	1/2	*Plantago* (2).
Poaceae	13/19	*Coix* (1), *Cymbopogon* (5), *Cynodon* (1), *Dactyloctenium* (1), *Eleusine* (1), *Heteropogon* (1), *Hordeum* (2), *Imperata* (1), *Melocalamus* (1), *Oryza* (2), *Panicum* (1), *Saccharum* (1), *Setaria* (1).
Polygonaceae	5/8	*Bistorta* (1), *Persicaria* (1), *Polygonum* (2), *Rheum* (3), *Salomonia* (1).
Pontederiaceae	1/2	*Monochoria* (2).
Portulacaceae	1/1	*Portulaca* (1).
Primulaceae	1/1	*Primula* (1).
Psilotaceae	1/1	*Psilotum* (1).
Pteridaceae	1/1	*Pteris* (1).
Punicaceae	1/1	*Punica* (1).

Contd...

Appendix–II–*Contd...*

Name of Family	*Genus/Species*	*Total No. of Species under a Genus*
Ranunculaceae	7/12	*Aconitum* (4), *Anemone* (2), *Clematis* (2), *Coptis* (1), *Hydrastis* (1), *Naravelia* (1), *Ranunculus* (1).
Rhamnaceae	3/4	*Berchemia* (1), *Ventilago* (1), *Ziziphus* (2).
Rosaceae	10/14	*Agrimonia* (1), *Cerasus* (1), *Cydonia* (1), *Fragaria* (1), *Malus* (1), *Padus* (1), *Prinsepia* (1), *Prunus* (2), *Rosa* (2), *Rubus* (3).
Rubiaceae	23/30	*Acranthera* (1), *Breonia* (1), *Catunaregam* (1), *Ceriscoides* (1), *Cinchona* (1), *Coffea* (1), *Galium* (1), *Gardenia* (3), *Haldina* (1), *Hedyotis* (1), *Ixora* (1), *Lasianthus* (1), *Meyna* (2), *Morinda* (2), *Mussaenda* (2), *Nauclea* (1), *Ophiorrhiza* (1), *Paederia* (1), *Pavetta* (2), *Rubia* (1), *Tamilnadia* (1), *Tarenna* (1), *Uncaria* (2).
Rutaceae	12/15	*Acronychia* (1), *Aegle* (1), *Citrus* (2), *Euodia* (1), *Glycosmis* (1), *Limonia* (1), *Micromelum* (1), *Murraya* (2), *Ruta* (2), *Skimmia* (1), *Toddalia* (1), *Zanthoxylum* (1).
Salicaceae	1/1	*Populus* (1).
Salvadoraceae	2/2	*Azima* (1), *Salvadora* (1).
Samydaceae	1/1	*Casearia* (1).
Santalaceae	1/1	*Santalum* (1).
Sapindaceae	6/7	*Allophylus* (1), *Cardiospermum* (1), *Litchi* (1), *Pometia* (1), *Sapindus* (2), *Schleichera* (1).
Sapotaceae	3/3	*Madhuca* (1), *Manilkara* (1), *Mimusops* (1).
Saururaceae	1/1	*Houttuynia* (1).
Saxifragaceae	1/2	*Bergenia* (2).
Schizaeaceae	1/1	*Lygodium* (1).
Scrophulariaceae	5/5	*Bacopa* (1), *Kickxia* (1), *Scoparia* (1), *Scrophularia* (1), *Verbascum* (1).
Selaginellaceae	1/1	*Selaginella* (1).
Simaroubaceae	1/3	*Ailanthus* (3).
Smilacaceae	1/2	*Smilax* (2).
Solanaceae	7/17	*Atropa* (1), *Capsicum* (1), *Datura* (2), *Nicotiana* (2), *Physalis* (1), *Solanum* (9), *Withania* (1).

Contd...

Appendix–II–*Contd...*

Name of Family	*Genus/Species*	*Total No. of Species under a Genus*
Sterculiaceae	4/4	*Helicteres* (1), *Melochia* (1), *Pentapetes* (1), *Pterospermum* (1).
Taccaceae	1/1	*Tacca* (1).
Tamaricaceae	1/1	*Trichaurus* (1).
Thelypteridaceae	1/1	*Christella* (1).
Thymelaeaceae	1/1	*Aquilaria* (1).
Tiliaceae	2/3	*Corchorus* (1), *Grewia* (2).
Typhaceae	1/2	*Typha* (2).
Ulmaceae	1/1	*Holoptelea* (1).
Urticaceae	4/4	*Dendrocnide* (1), *Elatostema* (1), *Pouzolzia* (1), *Urtica* (1).
Valerianaceae	1/1	*Valeriana* (1).
Verbenaceae	5/6	*Avicennia* (1), *Gmelina* (1), *Lantana* (2), *Phyla* (1), *Stachytarpheta* (1).
Violaceae	2/4	*Hybanthus* (1), *Viola* (3).
Vitaceae	4/4	*Ampelocissus* (1), *Cissus* (1), *Cyphostemma* (1), *Vitis* (1).
Woodsiaceae	2/2	*Athyrium* (1), *Diplazium* (1).
Zingiberaceae	6/14	*Alpinia* (2), *Amomum* (1), *Curcuma* (6), *Hedychium* (2), *Kaempferia* (2), *Zingiber* (1).
Zygophyllaceae	4/4	*Fagonia* (1), *Guaiacum* (1), *Kallstroemia* (1), *Tribulus* (1).

Appendix III
Index to Plant Part/s Used

Part/s Used	*Plants*
Aerial	
Aerial Parts	*Agrimonia eupatoria, Artemisia abrotanum, Dysphania ambrosioides, Elsholtzia blanda, Eschscholzia californica, Gnaphalium uliginosum, Nepeta cataria, Origanum vulgare, Scleria lithosperma, Solidago virgaurea, Urtica dioica.*
Branch	*Euphorbia neriifolia, Uncaria rhynchophylla.*
Fronds	*Diplazium esculentum, Pteris multifida.*
Shoot	*Acalypha indica, Andrographis paniculata, Averrhoa carambola, Boerhavia diffusa, Cajanus cajan, Flacourtia jangomas, Gardenia gummifera, Gymnosporia montana, Hedera helix, Justicia adhatoda, Mussaenda frondosa, Psidium guajava, Sansevieria roxburghiana, Uncaria sessilifructus.*
Twigs	*Achyranthes aspera, Asparagus filicinus, Calotropis procera, Crotalaria spectabilis, Eupatorium cannabinum, Murraya paniculata.*
Bark	
Inner Bark	*Cinnamomum zeylanicum, Morus rubra.*
Resin	*Commiphora mukul.*
Root	*Aegle marmelos, Aneilema scapiflorum, Casearia tomentosa, Chrozophora plicata, Chrozophora rottleri, Clitoria ternatea, Dichrostachys cinerea, Erythrina stricta, Glycosmis pentaphylla, Hemidesmus indicus, Limonia acidissima, Oroxylum indicum, Psidium guajava.*
Stem	*Acacia caesia, Acacia pennata, Ailanthus altissima, Alstonia scholaris, Aristolochia indica, Bauhinia purpurea, Bauhinia variegata, Bombax ceiba, Bombax insigne, Breonia chinensis, Bridelia stipularis, Buchanania cochinchinensis, Caesalpinia crista, Careya arborea, Cassia fistula, Celastrus paniculatus, Cinchona calisaya, Cinnamomum tamala, Clitoria ternatea, Crossandra infundibuliformis, Dalbergia volubilis, Embelia tsjeriam-cottam, Erythrina variegata, Ficus religiosa, Flacourita indica, Glochidion ellipticum, Jatropha curcas, Lantana camara, Madhuca indica, Mangifera indica, Manilkara kauki, Morinda citrifolia, Mussaenda raiateensis, Oroxylum indicum, Phyla nodiflora, Pongamia pinnata, Prunus persica, Prunus serotina, Psidium guajava, Pterolobium hexapetalum, Pterospermum acerifolium, Pterospermum canescens, Quercus incana, Quercus robur, Saccharum bengalense, Schefflera octophylla, Sesbania grandiflora, Spondias pinnata, Syzygium*

Contd..

Appendix–III–*Contd...*

Part/s Used	*Plants*
	corynocarpum, Syzygium cumini, Syzygium malaccense, Tarenna sambucina, Terminalia bellirica, Terminalia catappa, Toona ciliata, Wrightia tinctoria, Ximenia americana, Zanthoxylum armatum, Ziziphus abyssinica, Ziziphus jujuba.
Fruit	**Berry** *Jasminum grandiflorum, Vitis vinifera.*
Endosperm	*Givotia moluccana.*
Epicarp	*Sapindus mukorossi, Trichosanthes tricuspidata.*
Fruit	*Aegle marmelos, Amomum aromaticum, Anethum graveolens, Argemone mexicana, Averrhoa carambola, Benincasa hispida, Berberis aristata, Berberis asiatica, Breonia chinensis, Bridelia retusa, Capsicum annum, Carica papaya, Carum carvi, Citrus aurantium* var. *bergamia, Coix lacryma-jobi, Cydonia oblonga, Datura innoxia, Decaspermum fructicosum, Dendrophthoe falcata, Diospyros malabarica, Elaeocarpus sphaericus, Embelia ribes, Ferula asafoetida, Ficus hispida, Ficus racemosa, Flagellaria indica, Foeniculum vulgare, Forsythia suspensa, Fragaria nubicola, Garcinia mangostana, Helicteres isora, Indigofera cassioides, Juglans regia, Limonia acidissima, Litchi chinensis, Lobelia angulata, Malus pumila, Mangifera indica, Martynia annua, Momordica charantia, Moringa oleifera, Musa balbisiana, Mussaenda frondosa, Myristica fragrans, Nelumbo nucifera, Oxystelma esculentum, Papaver somniferum, Pavetta crassicaulis, Phyllanthus emblica, Piper longum, Pongamia pinnata, Punica granatum, Quercus incana, Rubus idaeus, Rubus moluccanus, Rubus niveus, Sapindus mukorossi, Sapindus laurifolius, Senna alexandrina, Solanum anguivi, Solanum erianthum, Solanum surattense, Solanum torvum, Spondias pinnata, Syzygium aromaticum, Tabernaemontana alternifolia, Terminalia bellirica, Terminalia chebula, Trachyspermum roxburghianum, Uraria picta, Vaccinium myrtillus, Wrightia tinctoria, Xanthium strumarium, Ziziphus jujuba.*
Grain	*Eleusine coracana, Hordeum distichon, Hordeum vulgare, Oryza sativa, Setaria italica.*
Latex	*Argemone mexicana, Carica papaya.*
Mesocarp	*Sapindus mukorossi.*
Nut	*Cocos nucifera.*
Pod	*Acacia nilotica* subsp. *indica, Cyamopsis tetragonoloba.*
Pulp	*Balanites aegyptiaca, Cassia fistula, Ficus carica, Limonia acidissima, Senna occidentalis, Tamarindus indica.*
Rind	*Punica granatum.*

Contd...

Appendix–III–*Contd...*

Part/s Used	*Plants*
Inflorescence, Bud, Flower, Pollen Grains, Spores	
Calyx	*Mussaenda frondosa.*
Capitulum	*Bidens pilosa.*
Flower	*Acmella paniculata, Aconitum violaceum, Acranthera grandiflora, Aeschynanthus parviflorus, Anisomeles indica, Bauhinia purpurea, Butea monosperma, Cannabis sativa, Carthamus tinctorius, Catharanthus roseus, Dendrophthoe falcata, Diospyros malabarica, Gardenia taitensis, Grewia rothii, Hibiscus rosa-sinensis, Jasminum grandiflorum, Lantana camara, Leonotis nepetifolia, Lepidagathis hamiltoniana, Leucas aspera, Leucas helianthemifolia, Litchi chinensis, Madhuca indica, Matricaria chamomilla, Mimusops elngi, Moringa oleifera, Murraya paniculata, Nelumbo nucifera, Ocimum gratissimum, Osbeckia nepalensis, Pavetta indica* var. *tomentosum, Primula veris, Prunus persica, Psidium guajava, Pterospermum acerifolium, Pterospermum canescens, Ranunculus muricatus, Rhododendron arboreum, Rosa damascena, Sambucus nigra, Seriphidium thomsonianum, Solanum surattense, Sphaeranthus indicus, Tanacetum gracile, Viola pilosa, Woodfordia fruticosa.*
Bud	*Bauhinia variegata, Punica granatum.*
Flower Oil	*Verbascum thapsus.*
Flowering Tops	*Artemisia indica, Nepeta cataria.*
Inflorescence	*Artemisia maritima, Elsholtzia blanda, Elsholtzia communis, Mangifera indica, Nelsonia canescens, Phlogacanthus thyrsiflorus, Trifolium pratense.*
Petals	*Gossypium arboreum, Rosa centifolia, Viola odorata.*
Spores	*Calvatia cyathiformis, Lycopodium clavatum, Psilotum nudum.*
Stigma, Styles	*Crocus sativus.*
Leaf	
Bud	*Azadirachta indica, Gardenia gummifera, Ilex paraguariensis, Populus alba, Solanum americanum, Toddalia asiatica.*
Gall	*Ficus racemosa.*
Leaf	*Abies spectabilis, Acacia nilotica* subsp. *indica, Acacia pennata, Acalypha indica, Achyranthes aspera, Acmella paniculata, Acorus calamus, Acranthera grandiflora, Acronychia pedunculata, Ageratum conyzoides, Ailanthus excelsa, Ajuga integrifolia, Albizia amara, Allium tuberosum, Allophylus cobbe, Aloe vera, Amaranthus spinosus, Anacolosa crassipes, Ananas comosus, Anaphalis margaritacea, Andrographis*

Contd..

Appendix–III–*Contd...*

Part/s Used	*Plants*
	laxiflora, Andrographis paniculata, Anemone vitifolia, Anisochilus carnosus, Anisomeles indica, Anisomeles malabarica, Annona squamosa, Archidendron clypearia, Ardisia solanacea, Aristolochia bracteolata, Aristolochia indica, Artemisia absinthium, Artemisia indica, Artocarpus heterophyllus, Asparagus racemosus, Atropa belladonna, Averrhoa bilimbi, Averrhoa carambola, Azadirachta indica, Azima tetracantha, Bacopa monnieri, Bambusa tulda, Baphicacanthus cusia, Baptisia tinctoria, Barleria prionitis, Barringtonia acutangula, Basella alba, Basilicum polystachyon, Blumea balsamifera, Blumea lacera, Blumea membranacea, Blumeopsis flava, Breonia chinensis, Breynia vitis-idaea, Bridelia stipularis, Bryophyllum pinnatum, Butea monosperma, Butea superba, Caesalpinia crista, Calotropis gigantea, Calotropis procera, Cananga odorata, Cannabis sativa, Cascabela thevetia, Cassia fistula, Cassia italica, Catharanthus roseus, Celastrus paniculatus, Centella asiatica, Cerasus vulgaris, Chenopodium album, Cissampelos pareira, Cissus quadrangularis, Citrus aurantifolia, Clerodendrum infortunatum, Cocculus hirsutus, Codiaeum variegatum, Codiocarpus andamanicus, Coffea benghalensis, Coix lacryma-jobi, Coleus forskohlii, Coriandrum sativum, Costus speciosus, Cryptolepis buchananii, Curcuma angustifolia, Cymbopogon schoenanthus, Cynodon dactylon, Cyphostemma auriculatum, Dactyloctenium aegyptium, Datura metel, Dendrocnide harveyi, Desmodium triflorum, Dicliptera bupleuroides, Diospyros lanceifolia, Diplopenta odorata, Dischidia benghalensis, Drymaria cordata, Eclipta prostrata, Elatostema platyphyllum, Embelia subcoriacea, Embelia tsjeriam-cottam, Emilia sonchifolia, Enicostemma littorale, Enydra fluctuans, Erythrina variegata, Eucalyptus globules, Eucalyptus tereticornis, Eupatorium cannabinum, Euphorbia hirta, Euphorbia neriifolia, Euphorbia nivulia, Euphorbia plumerioides, Euphorbia thymifolia, Ficus hispida, Ficus religiosa, Flacourtia jangomas, Flemingia strobilifera, Flueggea virosa, Glycine max, Glycosmis pentaphylla, Gymnosporia montana, Hedera helix, Hedera nepalensis, Hedychium coronarium, Holoptelea integrifolia, Houttuynia cordata, Hydrocotyle javanica, Hydrocotyle sibthorpioides, Hygrophila auriculata, Hyptis suaveolens, Indigofera cordifolia, Ipomoea mauritiana, Ixora coccinea, Jasminum arborescens, Jasminum multiflorum, Jasminum rottlerianum, Jatropha curcas, Jatropha gossypifolia, Justicia adhatoda, Justicia gendarussa, Kalanchoe lanciniata, Lactuca serriola, Lantana camara, Lantana indica, Lasianthus longicaudus, Launaea nudicaulis, Launaea procumbens, Lawsonia inermis, Leonotis nepetifolia, Leucas aspera, Leucas helianthemifolia, Limonia acidissima, Luffa acutangula, Madhuca indica, Mangifera indica, Marsilea minuta, Martynia annua,

Contd..

Appendix–III–*Contd...*

Part/s Used	*Plants*
	Melanthera biflora, Melastoma malabathricum, Melochia corchorifolia, Mentha arvensis, Mentha piperita, Mentha spicata, Mesua ferrea, Meyna laxiflora, Micromelum minutum, Mimosa pudica, Momordica charantia, Morinda citrifolia, Morinda pubescens, Murraya koenigii, Murraya paniculata, Musa balbisiana, Mussaenda frondosa, Mussaenda raiateensis, Myrtus communis, Nelsonia canescens, Nepeta cataria, Nicotiana rustica, Nicotiana tabacum, Nyctanthes arbor-tristis, Ocimum americanum, Ocimum basilicum, Ocimum gratissimum, Ocimum sanctum, Onosma echioides, Orthosiphon thymiflorus, Osbeckia stellata, Oxalis corniculata, Paederia foetida, Pandanus odoratissimus, Parochetus communis, Passiflora edulis, Pergularia daemia, Phaseolus lunatus, Phlogacanthus thyrsiflorus, Phyla nodiflora, Phyllanthus acidus, Phyllanthus airy-shawii, Phyllanthus reticulatus, Phyllanthus simplex, Phyllanthus urinaria, Physalis angulata, Piper betle, Piper hongkongense, Pisonia aculeata, Plantago major, Plectranthus amboinicus, Plectranthus barbatus, Polycarpon prostratum, Polygonum chinensis, Polyscias fruticosa, Pometia pinnata, Portulaca oleracea, Pothos scandens, Pouzolzia zeylanica, Premna tomentosa, Prosopis farcta, Prunus persica, Psidium guajava, Punica granatum, Radermachera xylocarpa, Rhazya stricta, Rhus coriaria, Ricinus communis, Roylea cinerea, Rubus idaeus, Rungia pectinata, Rungia repens, Ruta chalepensis, Ruta graveolens, Salvia officinalis, Sambucus nigra, Sauropus androgynus, Sauropus quadrangularis, Scoparia dulcis, Senna alexandrina, Senna auriculata, Senna occidentalis, Senna tora, Seriphidium brevifolium, Seriphidium thomsonianum, Sesamum indicum, Sesbania grandiflora, Sida acuta, Skimmia laureola, Smilax perfoliata, Solanum nigrum, Sphaeranthus indicus, Stephania japonica, Stylosanthes fruticosa, Syagrus romanzoffiana, Syzygium corynocarpum, Syzygium cumini, Syzygium samarangense, Tagetus minuta, Tamarindus indica, Tanacetum gracile, Terminalia bellirica, Terminalia catappa, Thuja occidentalis, Thunbergia fragrans, Toddalia asiatica, Trichaurus ericoides, Uncaria sessilifructus, Urtica dioica, Valeriana jatamansi, Vernonia divergens, Vigna marina, Vitex leucoxylon, Vitex negundo, Vitex trifolia, Xylosma longifolia, Ziziphus jujuba.
Leaf Base	*Ananas comosus.*
Oil	*Cymbopogon citratus.*
Petiole	*Alocasia macrorrhizos, Mangifera indica.*
Root, its Modification and Exudates	
Aerial Roots	*Mentha Canadensis.*
Exudates	*Ferula asafoetida, Ferula narthex.*

Contd...

Appendix–III–*Contd...*

Part/s Used	*Plants*
Root	*Acacia caesia, Acacia pennata, Achyranthes aspera, Aconitum ferox, Aconitum lethale, Aconitum violaceum, Acorus calamus, Adiantum lunulatum, Aegle marmelos, Aerva lanata, Aeschynanthus parviflorus, Aethusa cynapium, Ageratum conyzoides, Alysicarpus vaginalis, Amaranthus spinosus, Anacyclus pyrethrum, Andrographis paniculata, Angelica glauca, Apocynum cannabinum, Aristolochia fontanesii, Aristolochia indica, Aristolochia saccata, Asclepias speciosa, Asparagus gonoclados, Asparagus racemosus, Asparagus sarmentosus, Atropa belladonna, Baliospermum solanifolium, Baphicacanthus cusia, Baptisia tinctoria, Berberis aristata, Bombax ceiba, Caesalpinia crista, Cajanus crassus, Calotropis gigantea, Canna edulis, Capparis sepiaria, Carex baccans, Cassia fistula, Celastrus paniculatus, Ceriscoides turgida, Chrozophora plicata, Cissampelos pareira, Citrullus colocynthis, Clematis hexapetala, Clerodendrum indicum, Clerodendrum infortunatum, Clerodendrum multiflorum, Cocculus hirsutus, Coleus forskohlii, Combretum nanum, Coriandrum sativum, Crotalaria juncea, Crotalaria linifolia, Crotalaria prostrata, Cyclea arnottii, Cynoglossum wallichii, Cyperus rotundus, Dalbergia lanceolaria* subsp. *paniculata, Datura metel, Desmodium laxiflorum, Desmodium velutinum, Dioscorea hispida, Echinacea angustifolia, Echinops cornigerus, Elephantopus scaber, Elytraria acaulis, Equisetum giganteum, Eranthemum roseum, Erythrina stricta, Euphorbia antiquorum, Evolvulus alsinoides, Ficus benghalensis, Flueggea virosa, Geranium walliachinum, Gloriosa superba, Glycosmis pentaphylla, Hedyotis scandens, Hemidesmus indicus, Hemionitis arifolia, Holarrhena antidysenterica, Hybanthus enneaspermus, Indigofera tinctoria, Inula helenium, Jatropha glandulifera, Lawsonia inermis, Lepionurus sylvestris, Lithospermum officinale, Ludwigia octovalvis, Manilkara kauki, Martynia annua, Melastoma malabathricum, Mimosa pudica, Mucuna pruriens, Myxopyrum serratulum, Naravelia zeylanica, Nelumbo nucifera, Ocimum basilicum, Origanum vulgare, Osmunda regalis, Padus cornuta, Paeonia emodi, Pavetta indica* var. *tomentosum, Pentanema indicum, Phyllanthus fraternus, Phyllanthus reticulatus, Phyllanthus urinaria, Phytolacca acinosa, Phytolacca americana, Pleurospermum angelicoides, Pogostemon benghalense, Pongamia pinnata, Prosopis glandulosa, Psidium guajava, Punica granatum, Radermachera xylocarpa, Rauvolfia serpentina, Rheum australe, Rheum spiciforme, Rhus semialata, Rhus succedanea, Rubia cordifolia, Scoparia dulcis, Scrophularia ningpoensis, Solanum anguivi, Solanum giganteum, Solanum stramoniifolium, Solanum surattense, Stachytarpheta jamaicensis, Tadehagi triquetrum, Tamilnadia uliginosa, Tephrosia purpurea, Toddalia asiatica,*

Contd..

Appendix–III–*Contd...*

Part/s Used	*Plants*
	Tournefortia montana, Tragia involucrata, Trichodesma indicum, Typhonium trilobatum, Uraria picta, Vitex negundo, Withania somnifera, Wrightia arborea, Wrightia tinctoria, Xanthium strumarium, Zornia diphylla, Zornia gibbosa.
Root Tuber	*Aconitum heterophyllum, Ipomoea mauritiana, Tacca plantaginea.*
Seed	**Cotyledon** *Entada phaseoloides.*
Kernel	*Entada phaseoloides, Ricinus communis.*
Nutmeg	*Myristica fragrans.*
Oil	*Alangium salviifolium, Anethum graveolens, Arachis hypogaea, Celastrus paniculatus, Celosia argentea, Gynocardia odorata, Juglans regia, Madhuca indica, Olea europaea, Prinsepia utilis, Prunus persica, Ricinus communis, Salvadora persica, Semecarpus anacardium, Sesamum indicum.*
Seed	*Abutilon indicum, Alstonia scholaris, Anacardium occidentale, Anethum graveolens, Aristolochia indica, Artocarpus lakoocha, Avicennia officinalis, Balanites aegyptiaca, Barringtonia acutangula, Bombax ceiba, Butea monosperma, Caesalpinia crista, Cajanus scarabaeoides, Capsicum annum, Cardiospermum helicacabum, Centaurea cyanus, Cicer arietinum, Cinnamomum impressinervium, Cinnamomum iners, Cleome gynandra, Cleome viscosa, Clitoria ternatea, Combretum indicum, Coriandrum sativum, Cucumis melo, Cucurbita pepo, Datura metel, Descurainia sophia, Entada phaseoloides, Euphorbia thymifolia, Foeniculum vulgare, Gossypium hirsutum, Helicteres isora, Holarrhena antidysenterica, Jatropha curcas, Lens culinaris, Leonotis nepetifolia, Lepidagathis hamiltoniana, Madhuca indica, Melilotus indicus, Meyna spinosa, Mimusops elngi, Nelumbo nucifera, Ocimum basilicum, Ocimum gratissimum, Ocimum sanctum, Oroxylum indicum, Pergularia daemia, Phaseolus lunatus, Phyllanthus urinaria, Pimpinella anisum, Plantago ovata, Pongamia pinnata, Quercus incana, Ricinus communis, Senna occidentalis, Sesbania sesban, Smilax zeylanica, Stereospermum cholonoides, Strychnos nux-vomica, Syagrus romanzoffiana, Tephrosia purpurea, Terminalia bellirica, Terminalia chebula, Trigonella foenum-graceum, Vernonia anthelmintica, Vicia sativa* subsp. *nigra, Vigna umbellata, Xanthium strumarium, Xylocarpus granatum.*
Seed Memb.	*Momordica cochinchinesis.*
Stem, its Modification and Inclusions	
Bulb	*Allium cepa, Allium sativum, Polianthes tuberosa.*
Fibre	*Careya arborea.*

Contd...

Appendix–III–*Contd...*

Part/s Used	*Plants*
Gall	*Pistacia chinensis* ssp. *integerrima, Quercus infectoria, Rhus succedanea, Terminalia chebula.*
Gum	*Azadirachta indica, Bombax ceiba, Ceiba pentandra, Ferula narthex, Gardenia resinifera, Pistacia chinensis* ssp. *integerrima.*
Latex	*Jatropha curcas, Ficus benghalensis, Ficus hispida, Ficus racemosa, Streblus asper.*
Resin	*Centratherum anthelminticum, Pinus roxburghii, Shorea robusta.*
Rhizome	*Acorus calamus, Aeginetia indica, Alpinia galangal, Bistorta macrophylla, Blechnum orientale, Coptis teeta, Costus speciosus, Cryptocoryne spiralis, Curculigo orchioides, Curcuma angustifolia, Curcuma aromatica, Curcuma caesia, Curcuma longa, Curcuma montana, Curcuma zedoaria, Cyperus rotundus, Gelsomium elegans, Hedychium coronarium, Hedychium spicatum, Hydrastis canadensis, Inula helenium, Iris domestica, Iris germanica, Kaempferia galangal, Kaempferia rotunda, Lygodium flexuosum, Nelumbo nucifera, Nymphaea alba, Persicaria bistorta, Platycodon grandiflorum, Polygonum aviculare, Pteris multifida, Rheum officinalis, Tectaria coadunata, Typha australis, Typha elephantina, Veratrum viride, Zingiber officinale.*
Sap	*Melocalamus compactiflorus.*
Stem	*Anisochilus carnosus, Asparagus racemosus, Basella alba, Bischofia javanica, Cassytha filiformis, Cissus quadrangularis, Costus speciosus, Dactyloctenium aegyptium, Ephedra gerardiana, Equisetum arvense, Euodia hortensis, Ichnocarpus frutescens, Leonotis nepetifolia, Leucas aspera, Lonicera japonica, Piper hongkongense, Santalum album, Sphaeranthus indicus, Urtica dioica.*
Stolon	*Imperata cylindrica.*
Tuber	*Ceropegia candelabrum, Corallocarpus epigaeus, Cyperus longus, Gloriosa superba, Orchis mascula, Pinellia pedatisecta, Remusatia vivipara, Withania somnifera.*
Whole Plant	
Latex	*Cascabela thevetia, Euphorbia hirta, Euphorbia nivulia.*
Pulp	*Salomonia cantoniensis.*
Whole Plant	*Abrus precatorius, Abutilon indicum, Abutilon theophrasti, Acacia farnesiana, Acacia pennata, Acalypha indica, Achillea millefolium, Achyranthes aspera, Aconitum ferox, Acorus calamus, Adansonia digitata, Adiantum incisum, Adiantum lunulatum, Aegle marmelos, Aerva lanata, Ailanthus excelsa, Ailanthus triphysa, Alangium*

Contd..

Appendix–III–*Contd...*

Part/s Used	*Plants*
	salviifolium, Alcea rosea, Alhagi maurorum, Allium ascalonicum, Allium stracheyi, Aloe officinalis, Alpinia galangal, Alpinia officinarum, Alstonia scholaris, Alternanthera sessilis, Althaea officinalis, Amaranthus spinosus, Amaranthus viridis, Amorphophallus sylvaticus, Ampelocissus barbata, Amphineurion marginatum, Andrographis echioides, Andrographis paniculata, Anemone pulsatilla, Anethum graveolens, Angelica glauca, Anisochilus carnosus, Anthemis odontostephana, Antiaris toxicaria, Antidesma acidum, Aporosa cardiosperma, Areca catechu, Argemone mexicana, Argyreia nervosa, Aristolochia indica, Artemisia absinthium, Artemisia capillaris, Artemisia dracunculus, Artemisia indica, Artemisia japonica, Artemisia maritima, Artemisia persica, Artemisia sievarsiana, Artocarpus heterophyllus, Asclepias curassavica, Asparagus adscenden, Asparagus racemosus, Asplenium ruta-muraria, Asystasia gangetica, Athyrium pectinatum, Atylosia scarabaeoides, Averrhoa carambola, Azadirachta indica, Baccaurea ramiflora, Bacopa monnieri, Baliospermum solanifolium, Bambusa arundinacea, Bambusa tulda, Barleria prionitis, Bauhinia racemosa, Bauhinia variegata, Berberis lycium, Berchemia lineata, Bergenia ciliata, Bergenia pacumbis, Bidens pilosa, Bixa orellana, Blepharispermum sub-sessile, Boerhavia diffusa, Breonia chinensis, Bulbophyllum fuscopurpureum, Butea monosperma, Butea superba, Cadaba trifoliata, Caesalpinia crista, Calamus rotang, Calamus travancoricus, Calendula officinalis, Callicarpa longifolia, Callicarpa tomentosa, Calotropis gigantea, Calotropis procera, Cannabis sativa, Capparis deciduas, Caralluma aucheriana, Cardiospermum helicacabum, Carissa carandas, Carthamus tinctorius, Cascabela thevetia, Cassia fistula, Cassia italica, Cassytha filiformis, Catharanthus roseus, Catunaregam spinosa, Cedrus deodara, Celastrus paniculatus, Centella asiatica, Christella dentata, Chrozophora plicata, Chrozophora rottleri, Cinnamomum camphora, Cirsium wallichii, Cissampelos pareira, Cleistanthus collinus, Clematis gouriana, Clitoria ternatea, Coccinia grandis, Cocculus hirsutus, Coix lacryma-jobi, Colchicum luteum, Combretum album, Commiphora myrrha, Corchorus capsularis, Cotula anthemoides, Crocus sativus, Croton persimilis, Croton tiglium, Cryptolepis buchananii, Curculigo orchioides, Cuscuta reflexa, Cymbopogon jwarancusa, Cymbopogon jwarancusa subsp. *olivieri, Cymbopogon martini, Cynoglossum wallichii, Dactylicapnos macrocapnos, Dactylorhiza hatagirea, Dendrophthoe falcata, Descurainia Sophia, Desmodium gangeticum, Desmodium triflorum, Dillenia indica, Diplocyclos palmatus, Dipteracanthus prostratus, Drosera burmanni, Drymaria cordata, Eclipta prostrata, Elephantopus scaber, Elsholtzia ciliata, Elsholtzia communis, Elytraria acaulis, Embelia*

Contd..

Appendix–III–*Contd...*

Part/s Used	*Plants*
	ribes, Emilia sonchifolia, Entada phaseoloides, Erythrina suberosa, Erythrina variegata, Euphorbia heyneana, Euphorbia hirta, Euphorbia hypericifolia, Euphorbia serpens, Euphorbia thymifolia, Evolvulus alsinoides, Fagonia cretica, Ficus benjamina, Ficus racemosa, Ficus religiosa, Flacourita indica, Flemingia chappar, Foeniculum vulgare, Fraxinus ornus, Galium rotundifolium, Garcinia xanthochymus, Gardenia gummifera, Gardenia taitensis, Glinus lotoides, Glossocardia bidens, Glycosmis pentaphylla, Gmelina arborea, Grewia microcos, Gymnema sylvestre, Haldina cordifolia, Hedyotis scandens, Helicteres isora, Heliotropium indicum, Hemidesmus indicus, Heteropogon contortus, Holarrhena antidysenterica, Holoptelea integrifolia, Houttuynia cordata, Huperzia phlegmaria, Hydrocotyle sibthorpioides, Ichnocarpus frutescens, Incarvillea emodi, Indigofera linifolia, Indigofera tinctoria, Ipomoea coptica, Ipomoea turbinata, Ixora coccinea, Jatropha multifida, Justicia adhatoda, Kallstroemia pubescens, Kickxia ramosissima, Lactuca virosa, Lannea coromandelica, Lantana camara, Launaea sarmentosa, Lens culinaris, Leonotis nepetifolia, Lepidagathis cristata, Lepidium didymum, Leucas aspera, Lodoicea maldivica, Luffa acutangula, Luffa cylindrica, Macrotyloma uniflorum, Mallotus philippensis, Melastoma malabathricum, Melilotus indicus, Mentha arvensis, Mentha canadensis, Mentha piperita, Mesua ferrea, Microglossa pyrifolia, Micromelum minutum, Mikania micrantha, Mimusops elngi, Monochoria hastata, Monochoria vaginalis, Moringa oleifera, Mucuna pruriens, Myrica esculenta, Myristica fragrans, Nauclea orientalis, Ocimum americanum, Ocimum gratissimum, Ocimum sanctum, Onosma echioides, Ophiorrhiza rugosa var. *argentea, Opuntia dillenii, Oroxylum indicum, Orthosiphon rubicundus, Oryza rufipogon, Osmunda regalis, Oxalis corniculata, Paederia foetida, Panicum antidotale, Parmelia tinctorium, Pedalium murex, Pegolettia senegalensis, Pentapetes phoenicea, Pepromia pellucida, Peucedanum dhana, Phyla nodiflora, Phyllanthus amarus, Phyllanthus fraternus, Phyllanthus reticulatus, Phyllanthus urinaria, Polygonum aviculare, Pongamia pinnata, Pouzolzia zeylanica, Prosopis cineraria, Psilotum nudum, Pterospermum acerifolium, Pueraria tuberosa, Ricinus communis, Rorippa sarmentosa, Rungia pectinata, Ruta graveolens, Salomonia cantoniensis, Schleichera oleosa, Schrebera swieteniodes, Scoparia dulcis, Selaginella bryopteris, Semecarpus anacardium, Senna occidentalis, Senna tora, Seriphidium thomsonianum, Shorea robusta, Sida rhombifolia, Smilax perfoliata, Smilax zeylanica, Solanum dulcamara, Solanum nigrum, Strychnos potatorum, Swertia chirayita, Tamarindus indica, Taraxacum mongolicum, Tephrosia purpurea, Terminalia alata, Terminalia arjuna, Terminalia catappa, Thunbergia fragrans* var. *laevis,*

Contd..

Appendix–III–*Contd...*

Part/s Used	*Plants*
	Tinospora cordifolia, Trachyspermum ammi, Tribulus terrestris, Trifolium repens, Tylophora asthmatica, Tylophora flexuosa, Ventilago madraspatana, Vernonia cinerea, Viola biflora, Vitex negundo, Volutarella divaricata, Withania somnifera, Xanthium strumarium, Zanthoxylum armatum.
Wood	
Heart Wood	*Acacia catechu, Haematoxylum campechianum, Santalum album.*
Gum	*Guaiacum officinale.*
Wood	*Coscinium fenestratum, Toona ciliata.*
Wood Oil	*Aquilaria malaccensis.*

Appendix IV
Index to Ailments and Botanicals Employed

Diseases	*No. of Species*	*Name of Plant*
Diarrhoea/Dysentery	145	*Acacia catechu, Acacia nilotica subsp. indica, Acorus calamus, Aegle marmelos, Aerva lanata, Agrimonia eupatoria, Ailanthus altissima, Allium cepa, Allium sativum, Andrographis paniculata, Antidesma acidum, Argyreia nervosa, Asparagus racemosus, Azadirachta indica, Bacopa monnieri, Baliospermum solanifolium, Barleria prionitis, Berberis lyceum, Bombax ceiba, Buchanania cochinchinensis, Bulbophyllum fuscopurpureum, Butea monosperma, Butea superba, Cananga odorata, Cannabis sativa, Careya arborea, Cassia fistula, Cassine glauca, Catharanthus roseus, Centella asiatica, Cinnamomum impressinervium, Cinnamomum iners, Cinnamomum zeylanicum, Cissampelos pareira, Cleome gynandra, Coix lacryma-jobi, Crotalaria prostrata, Curcuma longa, Curcuma zedoaria, Cyclea arnottii, Cydonia oblonga, Cyperus rotundus, Dactyloctenium aegyptium, Dalbergia lanceolaria subsp. paniculata, Dalbergia volubilis, Desmodium triflorum, Diplopenta odorata, Dipteracanthus prostratus, Drosera burmanni, Drymaria cordata, Eclipta prostrata, Elephantopus scaber, Eleusine coracana, Elytraria acaulis, Embelia ribes, Euphorbia hirta, Euphorbia nivulia, Euphorbia thymifolia, Ficus benghalensis, Ficus hispida, Garcinia mangostana, Garcinia xanthochymus, Gardenia gummifera, Gardenia resinifera, Glossocardia bidens, Gymnosporia montana, Haematoxylum campechianum, Hedyotis scandens, Helicteres isora, Hemidesmus indicus, Hemidesmus indicus, Holarrhena antidysenterica, Hordeum distichon, Hydrocotyle sibthorpioides, Hygrophila auriculata, Indigofera cordifolia, Iris germanica, Lagenaria siceraria, Lawsonia inermis, Ludwigia octovalvis, Luffa cylindrica, Malus pumila, Mangifera indica, Manilkara kauki, Melastoma malabathricum, Melilotus indicus, Melochia corchorifolia, Meyna*

Contd..

Appendix–IV–*Contd...*

Diseases	*No. of Species*	*Name of Plant*
		spinosa, Mikania micrantha, Morinda citrifolia, Morinda pubescens, Myristica fragrans, Nauclea orientalis, Nelsonia canescens, Nelumbo nucifera, Ocimum basilicum, Orchis mascula, Paederia foetida, Pergularia daemia, Persicaria bistorta, Phyla nodiflora, Phyllanthus fraternus, Phyllanthus reticulatus, Plantago ovata, Plectranthus barbatus, Pleurospermum angelicoides, Polygonum aviculare, Prosopis cineraria, Psidium guajava, Psilotum nudum, Pterospermum acerifolium, Punica granatum, Quercus incana, Rauvolfia serpentina, Rhododendron arboretum, Rhus coriaria, Rhus succedanea, Ricinus communis, Ruta graveolens, Schrebera swieteniodes, Sesamum indicum, Shorea robusta, Smilax perfoliata, Smilax zeylanica, Solanum giganteum, Solanum surattense, Sphaeranthus indicus, Spondias pinnata, Streblus asper, Strychnos potatorum, Stylosanthes fruticosa, Syzygium cumini, Tarenna sambucina, Tectaria coadunata, Tephrosia purpurea, Terminalia alata, Terminalia bellirica, Terminalia chebula, Toona ciliata, Trichodesma indicum, Vaccinium myrtillus, Valeriana jatamansi, Woodfordia fruticosa, Xanthium strumarium, Zanthoxylum armatum.
Fever	103	*Abies spectabilis, Abutilon theophrasti, Achillea millefolium, Aconitum ferox, Aconitum heterophyllum, Adansonia digitata, Adiantum incisum, Adiantum lunulatum, Aegle marmelos, Aerva lanata, Ailanthus excelsa, Ailanthus triphysa, Ajuga integrifolia, Alangium salviifolium, Albizia amara, Alcea rosea, Allium sativum, Allium tuberosum, Allophylus cobbe, Alstonia scholaris, Alternanthera sessilis, Amorphophallus sylvaticus, Amphineurion marginatum, Andrographis echioides, Andrographis paniculata, Anthemis odontostephana, Antiaris toxicaria, Aporosa cardiosperma, Argyreia nervosa, Aristolochia indica, Artemisia absinthium, Artemisia dracunculus, Artemisia maritima, Artemisia persica,*

Contd..

Appendix–IV–*Contd...*

Diseases	*No. of Species*	*Name of Plant*
		Artemisia sievarsiana, Asparagus gonoclados, Averrhoa carambola, Azadirachta indica, Barleria prionitis, Berchemia lineata, Bergenia ciliata, Bergenia pacumbis, Bixa orellana, Breonia chinensis, Bridelia stipularis, Butea monosperma, Caesalpinia crista, Calamus rotang, Callicarpa longifolia, Callicarpa tomentosa, Calotropis gigantea, Caralluma aucheriana, Careya arborea, Carissa carandas, Carthamus tinctorius, Cassia fistula, Cedrus deodara, Cinnamomum iners, Cissus quadrangularis, Coffea benghalensis, Cyclea arnottii, Dactylicapnos macrocapnos, Dicliptera bupleuroides, Elephantopus scaber, Elsholtzia blanda, Emilia sonchifolia, Equisetum giganteum, Euphorbia neriifolia, Evolvulus alsinoides, Ficus benghalensis, Flacourtia jangomas, Haldina cordifolia, Helicteres isora, Hemidesmus indicus, Incarvillea emodi, Indigofera cassioides, Ipomoea mauritiana, Ipomoea turbinata, Jatropha curcas, Lantana camara, Launaea nudicaulis, Launaea procumbens, Microglossa pyrifolia, Mucuna pruriens, Nepeta cataria, Nyctanthes arbor-tristis, Ocimum sanctum, Pentanema indicum, Pistacia chinensis Bunge ssp. *integerrima, Pogostemon benghalense, Rubia cordifolia, Senna occidentalis, Senna tora, Seriphidium thomsonianum, Sesbania grandiflora, Sida rhombifolia, Stachytarpheta jamaicensis, Tarenna sambucina, Trichosanthes tricuspidata, Vigna marina, Volutarella divaricata, Wrightia arborea, Ziziphus abyssinica.*
Smallpox	103	*Acacia caesia, Acronychia pedunculata, Alhagi maurorum, Amomum aromaticum, Ampelocissus barbata, Anacardium occidentale, Anemone vitifolia, Archidendron clypearia, Areca catechu, Argyreia nervosa, Aristolochia saccata, Artocarpus heterophyllus, Asclepias curassavica, Asparagus filicinus, Asparagus racemosus, Atylosia scarabaeoides, Avicennia officinalis, Azadirachta indica, Azima tetracantha, Bambusa tulda, Bombax ceiba, Bryophyllum pinnatum, Caesalpinia crista, Calotropis gigantea, Calotropis procera, Careya*

Contd..

Appendix–IV–*Contd...*

Diseases	*No. of Species*	*Name of Plant*
		arborea, Centella asiatica, Cinchona calisaya, Cinnamomum camphora, Citrus aurantium var. *bergamia, Coix lacryma-jobi, Corallocarpus epigaeus, Curcuma angustifolia, Curcuma aromatica, Curcuma longa, Dactyloctenium aegyptium, Dendrophthoe falcata, Desmodium laxiflorum, Diplocyclos palmatus, Drymaria cordata, Elaeocarpus sphaericus, Elephantopus scaber, Eleusine coracana, Embelia subcoriacea, Enydra fluctuans, Eranthemum roseum, Eupatorium cannabinum, Fagonia cretica, Ficus racemosa, Ficus religiosa, Flagellaria indica, Flueggea virosa, Gloriosa superba, Gmelina arborea, Grewia microcos, Hedychium spicatum, Hydrastis canadensis, Justicia adhatoda, Kalanchoe lanciniata, Lantana indica, Lawsonia inermis, Lens culinaris, Leucas aspera, Limonia acidissima, Litchi chinensis, Lithospermum officinale, Macrotyloma uniflorum, Melastoma malabathricum, Mimosa pudica, Mimusops elngi, Momordica charantia, Morinda citrifolia, Moringa oleifera, Nelumbo nucifera, Oroxylum indicum, Oryza sativa, Pandanus odoratissimus, Panicum antidotale, Phyllanthus reticulatus, Polygonum chinensis, Pothos scandens, Pterospermum acerifolium, Pterospermum canescens, Ricinus communis, Rungia pectinata, Santalum album, Sesbania grandiflora, Sesbania sesban, Skimmia laureola, Solanum nigrum, Solanum surattense, Sphaeranthus indicus, Terminalia bellirica, Terminalia chebula, Tragia involucrata, Trigonella foenum-graceum, Tylophora flexuosa, Vicia sativa* subsp. *nigra, Withania somnifera, Woodfordia fruticosa, Xanthium strumarium, Zanthoxylum armatum, Ziziphus jujuba.*
Tonsillitis	76	*Acacia catechu, Aconitum ferox, Aconitum lethale, Aeschynanthus parviflorus, Allium sativum, Anacyclus pyrethrum, Andrographis laxiflora, Baphicacanthus cusia, Baptisia tinctoria, Bombax ceiba, Bombax insigne, Breynia vitis-idaea, Capsicum annum, Carica papaya, Cassia italica,*

Contd..

Appendix–IV–*Contd...*

Diseases	*No. of Species*	*Name of Plant*
		Cinchona calisaya, Clematis hexapetala, Commiphora mukul, Commiphora myrrha, Crocus sativus, Drymaria cordata, Echinacea angustifolia, Elephantopus scaber, Elsholtzia blanda, Elsholtzia communis, Elytraria acaulis, Embelia tsjeriam-cottam, Emilia sonchifolia, Flacourita indica, Forsythia suspensa, Galium rotundifolium, Gnaphalium uliginosum, Guaiacum officinale, Hedychium coronarium, Hemidesmus indicus, Hydrocotyle javanica, Iris domestica, Justicia adhatoda, Kaempferia galanga, Leucas aspera, Lobelia angulata, Luffa acutangula, Madhuca indica, Martynia annua, Naravelia zeylanica, Origanum vulgare, Oroxylum indicum, Phytolacca acinosa, Phytolacca americana, Polyscias fruticosa, Quercus incana, Quercus robur, Ranunculus muricatus, Rheum australe, Rheum spiciforme, Rosa damascene, Roylea cinerea, Rubus idaeus, Saccharum bengalense, Salvia officinalis, Sapindus mukorossi, Sauropus quadrangularis, Scrophularia ningpoensis, Semecarpus anacardium, Solanum torvum, Syzygium aromaticum, Taraxacum mongolicum, Tephrosia purpurea, Terminalia chebula, Toona ciliata, Trachyspermum roxburghianum, Typhonium trilobatum, Uncaria sessilifructus, Veratrum viride, Vitex negundo, Zingiber officinale.
Skin Diseases (Eczema, Ringworm, Scabies, Itching, Prurigo)	72	*Acmella paniculata, Acranthera grandiflora, Alpinia galanga, Alpinia officinarum, Anacolosa crassipes, Argemone mexicana, Argyreia nervosa, Azadirachta indica, Baliospermum solanifolium, Bischofia javanica, Butea monosperma, Cannabis sativa, Cascabela thevetia, Catharanthus roseus, Celastrus paniculatus, Centella asiatica, Christella dentata, Chrozophora plicata, Chrozophora rottleri, Cleistanthus collinus, Cocos nucifera, Combretum album, Crossandra infundibuliformis, Curcuma longa, Cyperus rotundus, Dillenia indica, Diplazium esculentum, Elatostema platyphyllum, Elephantopus scaber, Elsholtzia blanda, Erythrina variegata, Ficus benjamina, Flueggea virosa,*

Contd..

Appendix–IV–*Contd...*

Diseases	*No. of Species*	*Name of Plant*
		Helicteres isora, Heliotropium indicum, Hibiscus rosa-sinensis, Holoptelea integrifolia, Huperzia phlegmaria, Ixora coccinea, Jasminum rottlerianum, Kickxia ramosissima, Leonotis nepetifolia, Leucas aspera, Leucas helianthemifolia, Lithospermum officinale, Lycopodium clavatum, Mikania micrantha, Myxopyrum serratulum, Nicotiana tabacum, Ophiorrhiza rugosa Wall. var. *argentea, Oroxylum indicum, Orthosiphon rubicundus, Phyllanthus simplex, Phyllanthus urinaria, Pongamia pinnata, Pteris multifida, Rhazya stricta, Schleichera oleosa, Scoparia dulcis, Semecarpus anacardium, Senna occidentalis, Senna tora, Sesamum indicum, Shorea robusta, Smilax zeylanica, Terminalia catappa, Urtica dioica, Ventilago madraspatana, Vernonia anthelmintica, Vitex negundo, Vitex trifolia, Xylosma longifolia.*
Stomach Troubles	64	*Acacia farnesiana, Acalypha indica, Aconitum violaceum, Ajuga integrifolia, Alternanthera sessilis, Anethum graveolens, Angelica glauca, Asparagus racemosus, Bacopa monnieri, Balanites aegyptiaca, Bauhinia racemosa, Bauhinia variegata, Bistorta macrophylla, Calotropis procera, Cassia fistula, Cissampelos pareira, Cocculus hirsutus, Curcuma angustifolia, Cymbopogon citratus, Cyperus longus, Dicliptera bupleuroides, Embelia ribes, Erythrina variegata, Euphorbia nivulia, Geranium walliachinum, Glochidion ellipticum, Glycosmis pentaphylla, Haldina cordifolia, Helicteres isora, Heteropogon contortus, Holarrhena antidysenterica, Hydrocotyle javanica, Indigofera linifolia, Indigofera tinctoria, Kallstroemia pubescens, Limonia acidissima, Mallotus philippensis, Matricaria chamomilla, Melastoma malabathricum, Mentha spicata, Micromelum minutum, Monochoria vaginalis, Morinda pubescens, Murraya paniculata, Myristica fragrans, Ocimum basilicum, Ocimum sanctum, Oxalis corniculata, Parochetus communis, Phaseolus lunatus, Prosopis cineraria, Prunus persica, Rauvolfia serpentina, Rheum australe, Ricinus*

Contd..

Appendix–IV–*Contd...*

Diseases	*No. of Species*	*Name of Plant*
		communis, Schrebera swieteniodes, Scleria lithosperma, Senna occidentalis, Seriphidium brevifolium, Smilax perfoliata, Smilax zeylanica, Solanum nigrum, Terminalia chebula.
Measles	54	*Abies spectabilis, Acacia caesia, Acacia pennata, Anacolosa crassipes, Anemone pulsatilla, Artemisia indica, Asclepias speciosa, Averrhoa carambola, Azadirachta indica, Centella asiatica, Cinnamomum camphora, Citrus aurantium* var. *bergamia, Clerodendrum infortunatum, Coccinia grandis, Coriandrum sativum, Crocus sativus, Crotalaria juncea, Descurainia sophia, Eleusine coracana, Elsholtzia ciliata, Flueggea virosa, Glycosmis pentaphylla, Houttuynia cordata, Ichnocarpus frutescens, Ilex paraguariensis, Jatropha curcas, Lepidagathis cristata, Lithospermum officinale, Lonicera japonica, Mentha canadensis, Mesua ferrea, Momordica charantia, Moringa oleifera, Ocimum sanctum, Oryza sativa, Phyllanthus acidus, Phyllanthus airy-shawii, Platycodon grandiflorum, Polycarpon prostratum, Prunus serotina, Santalum album, Senna tora, Setaria italica, Sida rhombifolia, Solanum americanum, Solanum dulcamara, Solanum torvum, Streblus asper, Typha australis, Typha elephantina, Vicia sativa* subsp. *nigra, Vigna umbellata, Ximenia americana, Ziziphus jujuba.*
Cough	53	*Acalypha indica, Aconitum heterophyllum, Aconitum violaceum, Acorus calamus, Ailanthus excelsa, Albizia amara, Alhagi maurorum, Alocasia macrorrhizos, Anisochilus carnosus, Aristolochia indica, Bacopa monnieri, Barleria prionitis, Bridelia stipularis, Canna edulis, Chrozophora plicata, Chrozophora rottleri, Cinnamomum camphora, Cinnamomum impressinervium, Cinnamomum iners, Coleus forskohlii, Cryptocoryne spiralis, Desmodium triflorum, Eclipta prostrata, Elytraria acaulis, Euphorbia heyneana, Euphorbia hirta, Ficus benghalensis, Glycosmis pentaphylla, Hemidesmus indicus, Hyptis suaveolens, Inula helenium, Justicia adhatoda, Lasianthus longicaudus, Leucas aspera,*

Contd..

Appendix–IV–*Contd...*

Diseases	*No. of Species*	*Name of Plant*
		Melocalamus compactiflorus, Mimosa pudica, Mimusops elngi, Mussaenda frondosa, Ocimum sanctum, Phlogacanthus thyrsiflorus, Phyllanthus reticulatus, Piper betle, Plectranthus amboinicus, Plectranthus barbatus, Psidium guajava, Ruta graveolens, Solanum torvum, Tabernaemontana alternifolia, Terminalia bellirica, Terminalia chebula, Toddalia asiatica, Trichaurus ericoides, Trichosanthes tricuspidata.
Helminthiasis (Tape-worm, Roundworm)	52	*Achyranthes aspera, Acorus calamus, Aerva lanata, Ailanthus excelsa, Allium stracheyi, Aloe vera, Alstonia scholaris, Alternanthera sessilis, Amaranthus spinosus, Ananas comosus, Andrographis paniculata, Athyrium pectinatum, Baliospermum solanifolium, Carica papaya, Chenopodium album, Cleome viscosa, Clerodendrum infortunatum, Clitoria ternatea, Combretum indicum, Cucurbita pepo, Elatostema platyphyllum, Erythrina variegata, Flemingia strobilifera, Garcinia xanthochymus, Gardenia gummifera, Glycosmis pentaphylla, Gymnosporia montana, Helicteres isora, Kallstroemia pubescens, Mallotus philippensis, Mikania micrantha, Momordica charantia, Musa balbisiana, Nelumbo nucifera, Nymphaea alba, Pavetta crassicaulis, Pavetta indica* L.var. *tomentosum, Pergularia daemia, Peucedanum dhana, Piper longum, Pongamia pinnata, Pterospermum acerifolium, Punica granatum, Ruta chalepensis, Ruta graveolens, Senna occidentalis, Smilax perfoliata, Smilax zeylanica, Tagetes minuta, Tephrosia purpurea, Xanthium strumarium, Zanthoxylum armatum.*
Mumps	48	*Abies spectabilis, Aeginetia indica, Ardisia solanacea, Averrhoa bilimbi, Bauhinia variegata, Breynia vitis-idaea, Cajanus crassus, Calotropis procera, Capparis sepiaria, Cascabela thevetia, Cinnamomum tamala, Costus speciosus, Cymbopogon jwarancusa, Datura innoxia, Datura metel, Dendrophthoe falcata, Dischidia benghalensis, Entada phaseoloides, Erythrina*

Contd..

Appendix–IV–*Contd...*

Diseases	*No. of Species*	*Name of Plant*
		stricta, Erythrina suberosa, Euphorbia nivulia, Ficus racemosa, Forsythia suspensa, Hedera helix, Hedera nepalensis, Hibiscus rosa-sinensis, Kaempferia rotunda, Lantana camara, Leucas aspera, Meyna laxiflora, Oroxylum indicum, Papaver somniferum, Phytolacca acinosa, Phytolacca americana, Piper betle, Radermachera xylocarpa, Remusatia vivipara, Rheum australe, Rheum spiciforme, Santalum album, Sapindus laurifolius, Senna occidentalis, Sesbania grandiflora, Terminalia chebula, Toddalia asiatica, Trifolium pratense, Trifolium repens, Xylocarpus granatum.
Constipation	37	*Acalypha indica, Alhagi maurorum, Aloe officinalis, Amaranthus spinosus, Artocarpus lakoocha, Basella alba, Berberis aristata, Berberis asiatica, Berberis lyceum, Bombax ceiba, Calotropis gigantea, Cassia fistula, Cassia italica, Centaurea cyanus, Clitoria ternatea, Coleus forskohlii, Ficus carica, Fraxinus ornus, Mimusops elngi, Momordica charantia, Morus rubra, Nicotiana tabacum, Nyctanthes arbor-tristis, Ocimum basilicum, Ocimum sanctum, Onosma echioides, Piper betle, Plantago ovata, Ricinus communis, Rosa centifolia, Sambucus nigra, Solanum stramoniifolium, Tamarindus indica, Tephrosia purpurea, Urtica dioica, Vitis vinifera, Ziziphus jujuba.*
Dewormer	33	*Aloe officinalis, Artemisia abrotanum, Artemisia absinthium, Artemisia maritima, Azadirachta indica, Breonia chinensis, Butea monosperma, Caesalpinia crista, Cajanus scarabaeoides, Cassia fistula, Combretum nanum, Corchorus capsularis, Crotalaria spectabilis, Curcuma longa, Cyperus rotundus, Eclipta prostrata, Enicostemma littorale, Euphorbia antiquorum, Euphorbia hirta, Glycosmis pentaphylla, Holarrhena antidysenterica, Imperata cylindrical, Justicia adhatoda, Nyctanthes arbor-tristis, Orthosiphon thymiflorus, Padus cornuta, Sapindus mukorossi, Seriphidium brevifolium, Solanum anguivi, Solanum erianthum, Tanacetum gracile, Tephrosia purpurea, Vitex negundo.*

Contd..

Appendix–IV–*Contd...*

Diseases	*No. of Species*	*Name of Plant*
Polio	31	*Asparagus adscenden, Asparagus racemosus, Bambusa arundinacea, Bombax ceiba, Cassia fistula, Colchicum luteum, Cotula anthemoides, Curculigo orchioides, Curcuma montana, Cuscuta reflexa, Dactylorhiza hatagirea, Embelia ribes, Gardenia gummifera, Gelsomium elegans, Helicteres isora, Hemionitis arifolia, Lannea coromandelica, Mesua ferrea, Mucuna pruriens, Myrica esculenta, Myristica fragrans, Onosma echioides, Parmelia tinctorium Pedalium murex, Pueraria tuberosa, Selaginella bryopteris, Tacca plantaginea, Tinospora cordifolia, Trachyspermum ammi, Tribulus terrestris, Withania somnifera.*
Chickenpox	30	*Abies spectabilis, Achyranthes aspera, Anemone vitifolia, Archidendron clypearia, Averrhoa carambola, Azadirachta indica, Bambusa tulda, Bombax ceiba, Coriandrum sativum, Costus speciosus, Crotalaria linifolia, Curcuma longa, Cyphostemma auriculatum, Dendrophthoe falcata, Descurainia sophia, Dichrostachys cinerea, Gloriosa superba, Lantana indica, Ocimum sanctum, Oroxylum indicum, Phlogacanthus thyrsiflorus, Phyllanthus amarus, Ricinus communis, Senna auriculata, Shorea robusta, Solanum torvum, Strychnos nux-vomica, Terminalia bellirica, Viola biflora, Ziziphus jujuba.*
Indigestion	30	*Acacia pennata, Acorus calamus, Aethusa cynapium, Aristolochia fontanesii, Aristolochia indica, Cadaba trifoliata, Cannabis sativa, Ceriscoides turgida, Citrus aurantifolia, Cocculus hirsutus, Cymbopogon jwarancusa, Cymbopogon jwarancusa subsp. olivieri, Cyperus rotundus, Desmodium triflorum, Desmodium velutinum, Emilia sonchifolia, Ferula narthex, Glycosmis pentaphylla, Hedyotis scandens, Heteropogon contortus, Kallstroemia pubescens, Luffa cylindrica, Malus pumila, Melastoma malabathricum, Phyla nodiflora, Piper betle, Prosopis farcta, Quercus incana, Smilax perfoliata, Smilax zeylanica.*

Contd..

Appendix–IV–*Contd...*

Diseases	*No. of Species*	*Name of Plant*
Bowel Complaints	28	*Anaphalis margaritacea, Aristolochia bracteolata, Aristolochia indica, Baliospermum solanifolium, Blumea balsamifera, Ceropegia candelabrum, Cissampelos pareira, Emilia sonchifolia, Euphorbia thymifolia, Foeniculum vulgare, Helicteres isora, Hybanthus enneaspermus, Melilotus indicus, Mikania micrantha, Ocimum basilicum, Paederia foetida, Peucedanum dhana, Phyllanthus fraternus, Plantago ovata, Pterospermum acerifolium, Quercus infectoria, Senna tora, Smilax perfoliata, Strychnos potatorum, Syagrus romanzoffiana, Tephrosia purpurea, Terminalia alata, Xanthium strumarium.*
Cold /Coryza	28	*Aconitum heterophyllum, Aconitum violaceum, Acorus calamus, Butea monosperma, Canna edulis, Cirsium wallichii, Coleus forskohlii, Dysphania ambrosioides, Euphorbia neriifolia, Glycosmis pentaphylla, Hyptis suaveolens, Justicia adhatoda, Leucas aspera, Mussaenda frondosa, Nicotiana rustica, Ocimum americanum, Ocimum sanctum, Orthosiphon thymiflorus, Oxalis corniculata, Plantago major, Plectranthus amboinicus, Plectranthus barbatus, Sambucus nigra, Solanum torvum, Terminalia bellirica, Toddalia asiatica, Vitex leucoxylon, Ziziphus abyssinica.*
Vomiting	28	*Aconitum heterophyllum, Aegle marmelos, Allium cepa, Angelica glauca, Anisomeles malabarica, Aquilaria malaccensis, Asparagus racemosus, Bridelia retusa, Cassia fistula, Citrus aurantifolia, Cryptocoryne spiralis, Cynodon dactylon, Cynoglossum wallichii, Cyperus rotundus, Dactyloctenium aegyptium, Euphorbia neriifolia, Indigofera tinctoria, Mentha arvensis, Mentha Canadensis, Meyna spinosa, Murraya koenigii, Mussaenda frondosa, Nelumbo nucifera, Ocimum gratissimum, Pistacia chinensis* Bunge ssp. *integerrima, Plectranthus barbatus, Rhus succedanea, Solanum anguivi.*
Rickets	27	*Acacia pennata, Acorus calamus, Allium sativum, Anisomeles indica, Asparagus racemosus, Asplenium ruta-muraria, Carica papaya, Cassytha*

Contd..

Appendix–IV–*Contd...*

Diseases	*No. of Species*	*Name of Plant*
		filiformis, Cicer arietinum, Clerodendrum indicum, Combretum indicum, Cryptolepis buchananii, Dioscorea hispida, Elytraria acaulis, Euphorbia hirta, Hedera helix, Helicteres isora, Holoptelea integrifolia, Leucas aspera, Madhuca indica, Mangifera indica, Momordica cochinchinesis, Osmunda regalis, Oxalis corniculata, Schefflera octophylla, Tadehagi triquetrum, Trigonella foenum-graceum.
Colic	25	*Achyranthes aspera, Allium stracheyi, Aloe officinalis, Asparagus racemosus, Balanites aegyptiaca, Coriandrum sativum, Erythrina variegata, Euphorbia hypericifolia, Foeniculum vulgare, Garcinia xanthochymus, Gymnosporia montana, Indigofera linifolia, Indigofera tinctoria, Justicia gendarussa, Lodoicea maldivica, Mentha piperita, Ocimum gratissimum, Ocimum sanctum, Pimpinella anisum, Piper longum, Plectranthus amboinicus, Prosopis cineraria, Prosopis farcta, Rhus semialata, Schrebera swieteniodes.*
Convulsions	25	*Allium sativum, Aneilema scapiflorum, Artemisia indica, Artemisia japonica, Caesalpinia crista, Calotropis gigantea, Cannabis sativa, Cerasus vulgaris, Cleome viscosa, Curcuma zedoaria, Desmodium triflorum, Euodia hortensis, Ferula asafoetida, Lygodium flexuosum, Matricaria chamomilla, Oxalis corniculata, Pinellia pedatisecta, Piper hongkongense, Pouzolzia zeylanica, Psidium guajava, Rorippa sarmentosa, Ruta graveolens, Scoparia dulcis, Senna occidentalis, Uncaria rhynchophylla.*
Sores	25	*Abutilon indicum, Acacia nilotica subsp. indica, Achyranthes aspera, Acorus calamus, Alangium salviifolium, Alocasia macrorrhizos, Azima tetracantha, Bauhinia purpurea, Bridelia stipularis, Calvatia cyathiformis, Cocos nucifera, Eucalyptus globules, Ficus religiosa, Hemidesmus indicus, Jatropha gossypifolia, Leonotis nepetifolia, Lepidagathis hamiltoniana, Murraya koenigii, Myrtus communis, Ophiorrhiza rugosa* Wall. var.

Contd..

Appendix–IV–*Contd...*

Diseases	*No. of Species*	*Name of Plant*
		argentea, Osbeckia nepalensis, Solanum surattense, Spondias pinnata, Syzygium malaccense, Uraria picta.
Bronchitis	22	*Abies spectabilis, Abrus precatorius, Ailanthus excelsa, Allium sativum, Alpinia galangal, Azadirachta indica, Croton tiglium, Curcuma longa, Euphorbia hirta, Ferula asafoetida, Ferula narthex, Ficus carica, Hedychium coronarium, Indigofera tinctoria, Ipomoea coptica, Jasminum arborescens, Melocalamus compactiflorus, Mucuna pruriens, Ocimum americanum, Opuntia dillenii, Pistacia chinensis* Bunge ssp. *integerrima, Ruta graveolens.*
Dentition Problem	21	*Abies spectabilis, Acorus calamus, Alocasia macrorrhizos, Anisomeles malabarica, Barleria prionitis, Berberis lycium, Capparis deciduas, Catunaregam spinosa, Echinops cornigerus, Evolvulus alsinoides, Gardenia resinifera, Matricaria chamomilla, Micromelum minutum, Myristica fragrans, Pistacia chinensis* Bunge ssp. *integerrima, Punica granatum, Rheum officinalis,. Syagrus romanzoffiana, Syzygium corynocarpum, Vitex negundo, Woodfordia fruticosa.*
Eye Complaints	20	*Argemone mexicana, Clematis gouriana, Cocculus hirsutus, Curculigo orchioides, Eclipta prostrata, Emilia sonchifolia, Entada phaseoloides, Erythrina variegata, Euphorbia hirta, Euphorbia thymifolia, Flemingia chappar, Heliotropium indicum, Indigofera linifolia, Jatropha multifida, Oryza rufipogon, Pouzolzia zeylanica, Senna occidentalis, Shorea robusta, Strychnos potatorum, Tephrosia purpurea.*
Enuresis	18	*Ageratum conyzoides, Bombax ceiba, Ceiba pentandra, Cocculus hirsutus, Cyperus rotundus, Ephedra gerardiana, Equisetum arvense, Ficus racemosa, Grewia rothii, Juglans regia, Rhus succedanea, Rubus moluccanus, Rubus niveus, Smilax perfoliata, Swertia chirayita, Thuja occidentalis, Trigonella foenum-graceum, Verbascum thapsus.*

Contd..

Appendix–IV–*Contd...*

Diseases	*No. of Species*	*Name of Plant*
Blood Dysentery	17	*Cassine glauca, Drosera burmanni, Embelia ribes, Euphorbia hirta, Euphorbia thymifolia, Glossocardia bidens, Glycosmis pentaphylla, Holarrhena antidysenterica, Indigofera linifolia, Indigofera tinctoria, Mikania micrantha, Nauclea orientalis, Peucedanum dhana, Tephrosia purpurea, Terminalia alata, Xanthium strumarium, Zanthoxylum armatum.*
Tetanus (Lockjaw)	16	*Achyranthes aspera, Ageratum conyzoides, Argemone mexicana, Atropa belladonna, Bambusa tulda, Cannabis sativa, Coscinium fenestratum, Eclipta prostrata, Jasminum grandiflorum, Lantana camara, Madhuca indica, Melanthera biflora, Moringa oleifera, Passiflora edulis, Pinellia pedatisecta, Solanum nigrum.*
Weakness	16	*Andrographis paniculata, Asparagus sarmentosus, Calotropis procera, Coix lacryma-jobi, Cuscuta reflexa, Cyperus rotundus, Dioscorea hispida, Elephantopus scaber, Fragaria nubicola, Glinus lotoides, Holarrhena antidysenterica, Hordeum vulgare, Leonotis nepetifolia, Sesamum indicum, Syzygium samarangense, Tournefortia montana.*
Pneumonia	14	*Acorus calamus, Aloe officinalis, Alstonia scholaris, Anisomeles indica, Balanites aegyptiaca, Blumea lacera, Cannabis sativa, Drymaria cordata, Ferula asafoetida, Ferula narthex, Gardenia gummifera, Plectranthus amboinicus, Sapindus mukorossi, Sesbania sesban.*
Catarrh	13	*Acalypha indica, Bacopa monnieri, Barleria prionitis, Basella alba, Crocus sativus, Eclipta prostrata, Foeniculum vulgare, Mussaenda frondosa, Ocimum americanum, Phyllanthus reticulatus, Piper betle, Ruta chalepensis, Ruta graveolens.*
Flatulence	12	*Acorus calamus, Andrographis paniculata, Anethum graveolens, Aristolochia indica, Cardiospermum helicacabum, Carum carvi, Cassia fistula, Embelia ribes, Foeniculum vulgare, Helicteres isora, Mentha arvensis, Prosopis farcta.*
Earache	11	*Bidens pilosa, Allium ascalonicum, Artemisia capillaris, Bidens pilosa, Caesalpinia crista,*

Contd..

Appendix–IV–*Contd...*

Diseases	*No. of Species*	*Name of Plant*
		Calamus travancoricus, Cardiospermum helicacabum, Costus speciosus, Ficus hispida, Gossypium hirsutum, Matricaria chamomilla.
Tonics	11	*Aconitum heterophyllum, Andrographis paniculata, Angelica glauca, Baccaurea ramiflora, Butea monosperma, Clerodendrum multiflorum, Cyamopsis tetragonoloba, Juglans regia, Mussaenda frondosa, Nepeta cataria, Senna alexandrina.*
Whooping Cough (Pertussis)	11	*Achyranthes aspera, Andrographis paniculata, Eschscholzia californica, Eucalyptus globules, Eucalyptus tereticornis, Limonia acidissima, Pergularia daemia, Pistacia chinensis* Bunge ssp. *integerrima, Pongamia pinnata, Pterolobium hexapetalum, Seriphidium thomsonianum.*
Asthma	9	*Aconitum violaceum, Bridelia stipularis, Dactyloctenium aegyptium, Euphorbia hirta, Indigofera tinctoria, Pergularia daemia, Phyllanthus reticulatus, Pistacia chinensis* Bunge ssp. *integerrima, Vernonia divergens.*
Pinworm Diseases (Threadworm)	9	*Abutilon indicum, Achyranthes aspera, Blumea lacera, Blumea membranacea, Centratherum anthelminticum, Erythrina stricta, Jasminum multiflorum, Rauvolfia serpentina, Wrightia tinctoria.*
Sleep Induction	9	*Desmodium gangeticum, Ficus hispida, Flueggea virosa, Lactuca virosa, Launaea sarmentosa, Phyllanthus urinaria, Primula veris, Zornia diphylla, Zornia gibbosa.*
Abdominal Troubles	6	*Aconitum heterophyllum, Acorus calamus, Citrullus colocynthis, Jatropha glandulifera, Seriphidium thomsonianum, Solanum nigrum.*
Allergic Reaction	6	*Allium sativum, Capsicum annum, Celastrus paniculatus, Glycine max, Mangifera indica, Phyllanthus fraternus.*
Boils (Furuncle)	6	*Blechnum orientale, Carex baccans, Celastrus paniculatus, Centella asiatica, Gloriosa superba, Tamilnadia uliginosa.*

Contd..

Appendix–IV–*Contd...*

Diseases	*No. of Species*	*Name of Plant*
Marasmus	6	*Benincasa hispida, Clerodendrum indicum, Commiphora mukul, Elytraria acaulis, Gynocardia odorata, Withania somnifera.*
Aphthae	5	*Coix lacryma-jobi, Cyperus rotundus, Ocimum gratissimum, Terminalia arjuna, Terminalia chebula.*
Candidiasis (Thrush)	5	*Coix lacryma-jobi, Cyperus rotundus, Psilotum nudum, Sauropus androgynus, Vitis vinifera.*
Cholera	5	*Elsholtzia blanda, Phyllanthus emblica, Psidium guajava, Spondias pinnata, Terminalia bellirica.*
Defective Limbs/ Weak Limbs	5	*Alangium salviifolium, Cassytha filiformis, Helicteres isora, Olea europaea, Salvadora persica.*
Emaciation	5	*Aegle marmelos, Asparagus sarmentosus, Holarrhena antidysenterica, Ipomoea mauritiana, Withania somnifera.*
Epileptic Fits	5	*Cymbopogon martinii, Cymbopogon schoenanthus, Dendrocnide harveyi, Hemidesmus indicus, Sapindus laurifolius.*
Loss of Appetite	5	*Andrographis paniculata, Aristolochia bracteolata, Boerhavia diffusa, Decaspermum fructicosum, Phyllanthus urinaria.*
Conjunctivitis	4	*Gossypium arboretum, Luffa acutangula, Origanum vulgare, Xanthium strumarium.*
Enlargement of Spleen	4	*Barringtonia acutangula, Embelia ribes, Nicotiana tabacum, Trichaurus ericoides.*
Headache	4	*Acorus calamus, Basilicum polystachyon, Lepionurus sylvestris, Santalum album.*
Hepatitis	4	*Andrographis paniculata, Cajanus cajan, Coptis teeta, Phyllanthus fraternus.*
Inflammation	4	*Jatropha curcas, Ocimum basilicum, Tarenna sambucina, Xanthium strumarium.*
Jaundice	4	*Andrographis paniculata, Boerhavia diffusa, Cissampelos pareira, Eclipta prostrata.*
Malaria	4	*Clerodendrum infortunatum, Justicia adhatoda, Scoparia dulcis, Tinospora cordifolia.*
Malnutrition	4	*Basella alba, Cicer arietinum, Helicteres isora, Mussaenda raiateensis.*

Contd..

Appendix–IV–*Contd...*

Diseases	*No. of Species*	*Name of Plant*
Memory Enhancer	4	*Gymnema sylvestre, Bacopa monnieri, Centella asiatica, Hydrocotyle sibthorpioides.*
Paralysis	4	*Arachis hypogaea, Curcuma caesia, Pinus roxburghii, Sida acuta.*
Prolapsus Ani	4	*Annona squamosa, Ichnocarpus frutescens, Psidium guajava, Uraria picta.*
Pulmonary Affections	4	*Clitoria ternatea, Ficus racemosa, Piper betle, Pisonia aculeata.*
Tympanitis	4	*Acorus calamus, Cucumis melo, Emilia sonchifolia, Piper betle.*
Anaemia	3	*Diospyros lanceifolia, Fagonia cretica, Solanum nigrum.*
Antiseptic	3	*Helicteres isora, Mangifera indica, Ocimum sanctum.*
Blood Purifier	3	*Azadirachta indica, Paeonia emodi, Tephrosia purpurea.*
Burn/Scalds	3	*Averrhoa carambola, Eupatorium cannabinum, Phlogacanthus thyrsiflorus.*
Cataract	3	*Clitoria ternatea, Emilia sonchifolia, Flemingia chappar*
Cramps	3	*Blumea lacera, Cassine glauca, Celosia argentea.*
Heat Eruptions	3	*Butea monosperma, Butea superba, Cissampelos pareira.*
Hyperthermia (Body High Temperature)	3	*Blumeopsis flava, Phaseolus lunatus, Sapindus mukorossi.*
Infected Navels	3	*Calvatia cyathiformis, Oxalis corniculata, Syagrus romanzoffiana.*
Nervous Disorders	3	*Gardenia gummifera, Gardenia resinifera, Premna tomentosa.*
Pimples	3	*Cissampelos pareira, Elephantopus scaber, Polianthes tuberosa.*
Refrigerant/Coolant	3	*Nelumbo nucifera, Pepromia pellucida, Stephania japonica.*
Restlessness and Excitability	3	*Lactuca serriola, Lactuca virosa, Primula veris.*

Contd..

Appendix–IV–*Contd...*

Diseases	*No. of Species*	*Name of Plant*
Rheumatism	3	*Acorus calamus, Alpinia galanga, Asystasia gangetica.*
Sinking of Fontanelle	3	*Breonia chinensis, Oxalis corniculata, Pometia pinnata.*
Throat Inflammation	3	*Marsilea minuta, Mimusops elngi, Stereospermum cholonoides.*
Wound/Cuts	3	*Mikania micrantha, Osbeckia stellata, Vernonia anthelmintica.*
Astringent	2	*Acacia catechu, Oxystelma esculentum.*
Biliousness	2	*Cassia fistula, Ocimum sanctum.*
Blindness	2	*Baliospermum solanifolium, Croton persimilis.*
Bodyache	2	*Helicteres isora, Mucuna pruriens.*
Diuretic	2	*Nelumbo nucifera, Thuja occidentalis.*
Dwarfism	2	*Amaranthus viridis, Boerhavia diffusa.*
Dyspnoea(Breathlessness)	2	*Aristolochia indica, Piper betle.*
Enlarged Scrotum	2	*Casearia tomentosa, Codiocarpus andamanicus.*
Excretory Troubles	2	*Carica papaya, Codiaeum variegatum.*
Fast Breathing	2	*Artocarpus heterophyllus, Madhuca indica.*
Fits	2	*Lantana indica, Ruta graveolens.*
Flue	2	*Nepeta cataria, Phlogacanthus thyrsiflorus.*
Gastropathy	2	*Madhuca indica, Pentapetes phoenicea.*
Hiccups	2	*Alysicarpus vaginalis, Diospyros malabarica.*
Improving Complexion	2	*Aerva lanata, Mussaenda raiateensis.*
Infantile Spasms/Sprains	2	*Pouzolzia zeylanica, Seriphidium thomsonianum.*
Influenza	2	*Euphorbia heyneana, Melocalamus compactiflorus.*
Nasal Bleeding	2	*Allium cepa, Urtica dioica.*
Nightblindness	2	*Amaranthus spinosus, Ricinus communis.*
Reduces Secretion of Saliva	2	*Monochoria hastata, Senna occidentalis.*
Stammering	2	*Acmella paniculata, Ficus religiosa.*
Umbilicus Infection	2	*Asparagus racemosus, Curcuma longa.*
Unconsciousness	2	*Curculigo orchioides, Physalis angulata.*
Viscid Phlegm	2	*Sansevieria roxburghiana, Tylophora asthmatica.*

Contd..

Appendix–IV–*Contd...*

Diseases	*No. of Species*	*Name of Plant*
Ascariasis	1	*Anaphalis margaritacea*
Bleeding Piles/ Haemorrhoids	1	*Shorea robusta*
Cholagogue	1	*Mentha piperita*
Continous Crying of Babies	1	*Xanthium strumarium*
Cystitis(Urinary Bladder Inflammation)	1	*Thuja occidentalis*
Dandruff	1	*Amaranthus spinosus*
Demulcent	1	*Althaea officinalis*
Diptheria	1	*Pongamia pinnata*
Dizziness	1	*Thunbergia fragrans*
Dog Bite	1	*Oryza sativa*
Emetic	1	*Acalypha indica, Clitoria ternatea, Terminalia catappa, Tylophora asthmatica*
Encephalopathy	1	*Leucas aspera*
Fungal Infection	1	*Rungia repens*
Gastroenteritis	1	*Solidago virgaurea*
Glossitis	1	*Osbeckia nepalensis*
Haemoptysis	1	*Madhuca indica*
Haemorrhage	1	*Acacia catechu*
Hepatopathy	1	*Madhuca indica*
Immunity Enhancer	1	*Ocimum americanum*
Infantile Leucorrhoea	1	*Cannabis sativa*
Inflamation of Skeletal Muscle Fibres	1	*Croton tiglium*
Insect Bites	1	*Barringtonia acutangula*
Internal Injury	1	*Tarenna sambucina*
Irritation	1	*Gardenia gummifera*
Kalazar/Black Fever	1	*Andrographis paniculata*
Kwashiorkor	1	*Acorus calamus*
Laryngitis	1	*Populus alba*
Leprosy (Hansen's Diseases)	1	*Terminalia catappa*

Contd..

Appendix–IV–*Contd...*

Diseases	*No. of Species*	*Name of Plant*
Leucoderma	1	*Justicia adhatoda*
Leukaemia	1	*Catharanthus roseus*
Lisping	1	*Argemone mexicana*
Misperistalsis	1	*Cassia fistula*
Mouth infection	1	*Oxalis corniculata*
Neuralgic Pains	1	*Matricaria chamomilla*
Nyctalopia	1	*Momordica cochinchinesis*
Otorrhoea	1	*Alocasia macrorrhizos*
Painful Urine	1	*Vernonia cinerea*
Pus Formation in the Ear	1	*Lepidium didymium*
Rhinitis	1	*Blepharispermum sub-sessile*
Scrofulous	1	*Calendula officinalis*
Snake Bites	1	*Radermachera xylocarpa*
Sneezing	1	*Andrographis paniculata*
Snuffles	1	*Apocynum cannabinum*
Sound of Snoring	1	*Terminalia chebula*
Sprue	1	*Salomonia cantoniensis*
Strangury/Painful Urination	1	*Madhuca indica*
Stomatitis	1	*Portulaca oleracea*
Stye	1	*Xanthium strumarium*
Swelling of Small Intestine	1	*Murraya paniculata*
Syphilis	1	*Pongamia pinnata*
Tuberculosis	1	*Lantana camara*
Typhoid (Enteric Fever)	1	*Ziziphus jujuba*
Umbilical Hernia	1	*Prosopis glandulosa*
Urinogenital Diseases	1	*Phyllanthus urinaria*
Verminosis	1	*Madhuca indica*
Vesicles	1	*Thunbergia fragrans* var. *laevis*
Visceral Obstruction	1	*Pavetta indica* L.var. *tomentosum*
Visceromegaly	1	*Citrullus colocynthis*
Warts	1	*Euphorbia hirta*
Xerophthalmia	1	*Momordica cochinchinesis*

Appendix V
Index to Botanical Names

Botanical Names	*Page No.*	*Botanical Names*	*Page No.*
Abies spectabilis	7	*Agrimonia eupatoria*	20
Abrus precatorius	7	*Ailanthus altissima*	21
Abutilon indicum	8	*Ailanthus excelsa*	21
Abutilon theophrasti	8	*Ailanthus triphysa*	21
Acacia caesia	9	*Ajuga integrifolia*	22
Acacia catechu	9	*Alangium salviifolium*	22
Acacia farnesiana	10	*Albizia amara*	23
Acacia nilotica subsp. *indica*	10	*Alcea rosea*	23
Acacia pennata	11	*Alhagi maurorum*	23
Acalypha indica	11	*Allium ascalonicum*	24
Achillea millefolium	12	*Allium cepa*	24
Achyranthes aspera	12	*Allium sativum*	25
Acmella paniculata	13	*Allium stracheyi*	25
Aconitum ferox	13	*Allium tuberosum*	26
Aconitum heterophyllum	14	*Allophylus cobbe*	26
Aconitum lethale	14	*Alocacia macrorrhizos*	26
Aconitum violaceum	15	*Aloe officinalis*	27
Acorus calamus	15	*Aloe vera*	27
Acranthera grandiflora	16	*Alpinia galanga*	28
Acronychia pedunculata	16	*Alpinia officinarum*	28
Adansonia digitata	17	*Alstonia scholaris*	28
Adiantum incisum	17	*Alternanthera sessilis*	29
Adiantum lunulatum	17	*Althaea officinalis*	29
Aeginetia indica	18	*Alysicarpus vaginalis*	30
Aegle marmelos	18	*Amaranthus spinosus*	30
Aerva lanata	19	*Amaranthus viridis*	31
Aeschynanthus parviflorus	19	*Amomum aromaticum*	31
Aethusa cynapium	20	*Amorphophallus sylvaticus*	31
Ageratum conyzoides	20	*Ampelocissus barbata*	32

Contd..

Appendix–V–*Contd...*

Botanical Names	*Page No.*	*Botanical Names*	*Page No.*
Amphineurion marginatum	32	*Aristolochia fontanesii*	44
Anacardium occidentale	32	*Aristolochia indica*	45
Anacolosa crassipes	33	*Aristolochia saccata*	45
Anacyclus pyrethrum	33	*Artemisia japonica*	47
Ananas comosus	33	*Artemisia abrotanum*	46
Anaphalis margaritacea	34	*Artemisia absinthium*	46
Andrographis echioides	34	*Artemisia capillaris*	46
Andrographis laxiflora	35	*Artemisia dracunculus*	47
Andrographis paniculata	35	*Artemisia indica*	47
Aneilema scapiflorum	36	*Artemisia maritima*	48
Anemone pulsatilla	36	*Artemisia persica*	48
Anemone vitifolia	36	*Artemisia sievarsiana*	48
Anethum graveolens	37	*Artocarpus heterophyllus*	49
Angelica glauca	37	*Artocarpus lakoocha*	49
Anisochilus carnosus	38	*Asclepias adscendens*	50
Anisomeles indica	38	*Asclepias curassavica*	50
Anisomeles malabarica	39	*Asclepias speciosa*	50
Annona squamosa	39	*Asparagus filicinus*	51
Anthemis odontostephana	39	*Asparagus gonoclados*	51
Antiaris toxicaria	40	*Asparagus racemosus*	51
Antidesma acidum	40	*Asparagus sarmentosus*	52
Apocynum cannabinum	40	*Asplenium ruta-muraria*	52
Aporosa cardiosperma	41	*Asystasia gangetica*	53
Aquilaria malaccensis	41	*Athyrium pectinatum*	53
Arachis hypogaea	41	*Atropa belladonna*	53
Archidendron clypearia	42	*Atylosia scarabaeoides*	54
Ardisia solanacea	42	*Averrhoa bilimbi*	54
Areca catechu	42	*Averrhoa carambola*	54
Argemone mexicana	43	*Avicennia officinalis*	55
Argyreia nervosa	43	*Azadirachta indica*	55
Aristolochia bracteolata	44	*Azima tetracantha*	56

Contd..

Appendix–V–*Contd...*

Botanical Names	*Page No.*	*Botanical Names*	*Page No.*
Baccaurea ramiflora	57	*Blumea membranacea*	69
Bacopa monnieri	57	*Blumeopsis flava*	69
Balanites aegyptiaca	58	*Boerhavia diffusa*	69
Baliospermum solanifolium	58	*Bombax ceiba*	70
Bambusa arundinacea	59	*Bombax insigne*	70
Bambusa tulda	59	*Breonia chinensis*	71
Baphicacanthus cusia	59	*Breynia vitis-idaea*	71
Baptisia tinctoria	60	*Bridelia retusa*	72
Barleria prionitis	60	*Bridelia stipularis*	72
Barringtonia acutangula subsp. *spicata*	61	*Bryophyllum pinnatum*	73
		Buchanania cochinchinensis	73
Basella alba	61	*Bulbophyllum fuscopurpureum*	74
Basilicum polystachyon	62	*Butea monosperma*	74
Bauhinia purpurea	62	*Butea superba*	74
Bauhinia racemosa	62	*Cadaba trifoliata*	75
Bauhinia variegata	63	*Caesalpinia crista*	75
Benincasa hispida	63	*Cajanus cajan*	76
Berberis aristata	64	*Cajanus crassus*	76
Berberis asiatica	64	*Cajanus scarabaeoides*	76
Berberis lycium	64	*Calamus rotang*	77
Berchemia lineata	65	*Calamus travancoricus*	77
Bergenia ciliata	65	*Calendula officinalis*	77
Bergenia pacumbis	65	*Callicarpa longifolia*	78
Bidens pilosa	66	*Callicarpa tomentosa*	78
Bischofia javanica	66	*Calotropis gigantea*	78
Bistorta macrophylla	66	*Calotropis procera*	79
Bixa orellana	67	*Calvatia cyathiformis*	79
Blechnum orientale	67	*Cananga odorata*	80
Blepharispermum sub-sessile	68	*Canna edulis*	80
Blumea balsamifera	68	*Cannabis sativa*	81
Blumea lacera	68	*Capparis deciduas*	81

Contd..

Appendix–V–*Contd...*

Botanical Names	*Page No.*	*Botanical Names*	*Page No.*
Capparis sepiaria	82	*Chrozophora rottleri*	95
Capsicum annum	82	*Cicer arietinum*	95
Caralluma aucheriana	82	*Cinchona calisaya*	96
Cardiospermum helicacabum	83	*Cinnamomum camphora*	96
Carex baccans	83	*Cinnamomum impressinervium*	97
Careya arborea	83	*Cinnamomum iners*	97
Carica papaya	84	*Cinnamomum tamala*	97
Carissa carandas	84	*Cinnamomum zeylanicum*	98
Carthamus tinctorius	85	*Cirsium wallichii*	98
Carum carvi	85	*Cissampelos pareira*	98
Cascabela thevetia	86	*Cissus quadrangularis*	99
Casearia tomentosa	86	*Citrullus colocynthis*	99
Cassia fistula	86	*Citrus aurantifolia*	100
Cassia italica	87	*Citrus aurantium* var. *bergamia*	100
Cassine glauca	88	*Cleistanthus collinus*	101
Cassytha filiformis	88	*Clematis gouriana*	101
Catharanthus roseus	89	*Clematis hexapetala*	101
Catunaregam spinosa	89	*Cleome gynandra*	102
Cedrus deodara	90	*Cleome viscosa*	102
Ceiba pentandra	90	*Clerodendrum indicum*	103
Celastrus paniculatus	90	*Clerodendrum infortunatum*	103
Celosia argentea	91	*Clerodendrum multiflorum*	104
Centaurea cyanus	91	*Clitoria ternatea*	104
Centella asiatica	92	*Coccinia grandis*	105
Centratherum anthelminticum	92	*Cocculus hirsutus*	105
Cerasus vulgaris	93	*Cocos nucifera*	106
Ceriscoides turgida	93	*Codiaeum variegatum*	106
Ceropegia candelabrum	93	*Codiocarpus andamanicus*	107
Chenopodium album	94	*Coffea benghalensis*	107
Christella dentata	94	*Coix lacryma-jobi*	107
Chrozophora plicata	95	*Colchicum luteum*	108

Contd..

Appendix–V–*Contd...*

Botanical Names	*Page No.*	*Botanical Names*	*Page No.*
Coleus forskohlii	108	*Curcuma zedoaria*	120
Combretum album	108	*Cuscuta reflexa*	121
Combretum indicum	109	*Cyamopsis tetragonoloba*	121
Combretum nanum	109	*Cyclea arnottii*	122
Commiphora mukul	109	*Cydonia oblonga*	122
Commiphora myrrha	110	*Cymbopogon citratus*	122
Coptis teeta	110	*Cymbopogon jwarancusa*	123
Corallocarpus epigaeus	111	*Cymbopogon jwarancusa* subsp. *olivieri*	123
Corchorus capsularis	111	*Cymbopogon martinii*	123
Coriandrum sativum	111	*Cymbopogon schoenanthus*	124
Coscinium fenestratum	112	*Cynodon dactylon*	124
Costus speciosus	112	*Cynoglossum wallichii*	125
Cotula anthemoides	113	*Cyperus longus*	125
Crocus sativus	113	*Cyperus rotundus*	125
Crossandra infundibuliformis	114	*Cyphostemma auriculatum*	126
Crotalaria juncea	114	*Dactylicapnos macrocapnos*	126
Crotalaria linifolia	114	*Dactyloctenium aegyptium*	127
Crotalaria prostrata	115	*Dactylorhiza hatagirea*	127
Crotalaria spectabilis	115	*Dalbergia lanceolaria* subsp. *paniculata*	128
Croton persimilis	115	*Dalbergia volubilis*	128
Croton tiglium	116	*Datura innoxia*	128
Cryptocoryne spiralis	116	*Datura metel*	129
Cryptolepis buchananii	117	*Decaspermum fructicosum*	129
Cucumis melo	117	*Dendrocnide harveyi*	129
Cucurbita pepo	117	*Dendrophthoe falcata*	130
Curculigo orchioides	118	*Descurainia sophia*	130
Curcuma angustifolia	118	*Desmodium gangeticum*	130
Curcuma aromatica	119	*Desmodium laxiflorum*	131
Curcuma caesia	119	*Desmodium triflorum*	131
Curcuma longa	119		
Curcuma montana	120		

Contd..

Appendix–V–*Contd...*

Botanical Names	*Page No.*	*Botanical Names*	*Page No.*
Desmodium velutinum	132	*Entada phaseoloides*	144
Dichrostachys cinerea	132	*Enydra fluctuans*	145
Dicliptera bupleuroides	132	*Ephedra gerardiana*	145
Dillenia indica	133	*Equisetum arvense*	145
Dioscorea hispida	133	*Equisetum giganteum*	146
Diospyros lanceifolia	134	*Eranthemum roseum*	146
Diospyros malabarica	134	*Erythrina stricta*	146
Diplazium esculentum	134	*Erythrina suberosa*	147
Diplocyclos palmatus	135	*Erythrina variegata*	147
Diplopenta odorata	135	*Eschscholzia californica*	148
Dipteracanthus prostratus	135	*Eucalyptus globules*	148
Dischidia benghalensis	136	*Eucalyptus tereticornis*	148
Drosera burmanni	136	*Euodia hortensis*	149
Drymaria cordata	136	*Eupatorium cannabinum*	149
Dysphania ambrosioides	137	*Euphorbia antiquorum*	150
Echinacea angustifolia	137	*Euphorbia heyneana*	150
Echinops cornigerus	137	*Euphorbia hirta*	150
Eclipta prostrata	138	*Euphorbia hypericifolia*	151
Elaeocarpus sphaericus	138	*Euphorbia neriifolia*	151
Elatostema platyphyllum	139	*Euphorbia nivulia*	152
Elephantopus scaber	139	*Euphorbia plumerioides*	152
Eleusine coracana	140	*Euphorbia serpens*	153
Elsholtzia blanda	140	*Euphorbia thymifolia*	153
Elsholtzia ciliata	141	*Evolvulus alsinoides*	154
Elsholtzia communis	141	*Fagonia cretica*	154
Elytraria acaulis	141	*Ferula asafoetida*	154
Embelia ribes	142	*Ferula narthex*	155
Embelia subcoriacea	142	*Ficus benghalensis*	155
Embelia tsjeriam-cottam	143	*Ficus benjamina*	156
Emilia sonchifolia	143	*Ficus carica*	156
Enicostemma littorale	144	*Ficus hispida*	157

Contd..

Appendix–V–*Contd...*

Botanical Names	*Page No.*	*Botanical Names*	*Page No.*
Ficus racemosa	157	*Grewia microcos*	170
Ficus religiosa	158	*Grewia rothii*	170
Flacourita indica	158	*Guaiacum officinale*	170
Flacourtia jangomas	159	*Gymnema sylvestre*	171
Flagellaria indica	159	*Gymnosporia montana*	171
Flemingia chappar	159	*Gynocardia odorata*	172
Flemingia strobilifera	160	*Haematoxylum campechianum*	172
Flueggea virosa	160	*Haldina cordifolia*	172
Foeniculum vulgare	160	*Hedera helix*	173
Forsythia suspensa	161	*Hedera nepalensis*	173
Fragaria nubicola	161	*Hedychium coronarium*	173
Fraxinus ornus	162	*Hedychium spicatum*	174
Galium rotundifolium	162	*Hedyotis scandens*	174
Garcinia mangostana	162	*Helicteres isora*	175
Garcinia xanthochymus	163	*Heliotropium indicum*	175
Gardenia gummifera	163	*Hemidesmus indicus*	176
Gardenia resinifera	167	*Hemionitis arifolia*	176
Gardenia taitensis	167	*Heteropogon contortus*	177
Gelsomium elegans	165	*Hibiscus rosa-sinensis*	177
Geranium walliachinum	165	*Holarrhena antidysenterica*	177
Givotia moluccana	165	*Holoptelea integrifolia*	178
Glinus lotoides	166	*Hordeum distichon*	178
Glochidion ellipticum	166	*Hordeum vulgare*	179
Gloriosa superba	166	*Houttuynia cordata*	179
Glossocardia bidens	167	*Huperzia phlegmaria*	179
Glycine max	167	*Hybanthus enneaspermus*	180
Glycosmis pentaphylla	168	*Hydrastis canadensis*	180
Gmelina arborea	168	*Hydrocotyle javanica*	180
Gnaphalium uliginosum	169	*Hydrocotyle sibthorpioides*	181
Gossypium arboreum	169	*Hygrophila auriculata*	181
Gossypium hirsutum	169	*Hyptis suaveolens*	182

Contd..

Appendix–V–*Contd...*

Botanical Names	*Page No.*	*Botanical Names*	*Page No.*
Ichnocarpus frutescens	182	*Lactuca serriola*	194
Ilex paraguariensis	183	*Lactuca virosa*	195
Imperata cylindrica	183	*Lagenaria siceraria*	195
Incarvillea emodi	183	*Lannea coromandelica*	195
Indigofera cassioides	184	*Lantana camara*	196
Indigofera cordifolia	184	*Lantana indica*	196
Indigofera linifolia	184	*Lasianthus japonicas* subsp. *longicaudus*	197
Indigofera tinctoria	185	*Launaea nudicaulis*	197
Inula helenium	185	*Launaea procumbens*	197
Ipomoea coptica	185	*Launaea sarmentosa*	198
Ipomoea mauritiana	186	*Lawsonia inermis*	198
Ipomoea turbinata	186	*Lens culinaris*	198
Iris domestica	187	*Leonotis nepetifolia*	199
Iris germanica	187	*Lepidagathis cristata*	199
Ixora coccinea	187	*Lepidagathis hamiltoniana*	200
Jasminum arborescens	188	*Lepidium didymum*	200
Jasminum grandiflorum	188	*Lepionurus sylvestris*	200
Jasminum multiflorum	189	*Leucas aspera*	201
Jasminum rottlerianum	189	*Leucas helianthemifolia*	201
Jatropha curcas	189	*Limonia acidissima*	202
Jatropha glandulifera	190	*Litchi chinensis*	202
Jatropha gossypifolia	190	*Lithospermum officinale*	203
Jatropha multifida	191	*Lobelia angulata*	203
Juglans regia	191	*Lodoicea maldivica*	203
Justicia adhatoda	191	*Lonicera japonica*	204
Justicia gendarussa	192	*Ludwigia octovalvis*	204
Kaempferia galanga	192	*Luffa acutangula*	204
Kaempferia rotunda	193	*Luffa cylindrica*	205
Kalanchoe lanciniata	193	*Lycopodium clavatum*	205
Kallstroemia pubescens	194	*Lygodium flexuosum*	206
Kickxia ramosissima	194		

Contd..

Appendix–V–*Contd...*

Botanical Names	*Page No.*	*Botanical Names*	*Page No.*
Macrotyloma uniflorum	206	*Morinda pubescens*	219
Madhuca indica	206	*Moringa oleifera*	219
Mallotus philippensis	207	*Morus rubra*	220
Malus pumila	207	*Mucuna pruriens*	220
Mangifera indica	208	*Murraya koenigii*	220
Manilkara kauki	208	*Murraya paniculata*	221
Marsilea minuta	209	*Musa balbisiana*	221
Martynia annua	209	*Mussaenda frondosa*	222
Matricaria chamomilla	210	*Mussaenda raiateensis*	222
Melanthera biflora	210	*Myrica esculenta*	223
Melastoma malabathricum	210	*Myristica fragrans*	223
Melilotus indicus	211	*Myrtus communis*	223
Melocalamus compactiflorus	211	*Myxopyrum serratulum*	224
Melochia corchorifolia	212	*Naravelia zeylanica*	224
Mentha arvensis	212	*Nauclea orientalis*	224
Mentha canadensis	212	*Nelsonia canescens*	225
Mentha piperita	213	*Nelumbo nucifera*	225
Mentha spicata	213	*Nepeta cataria*	226
Mesua ferrea	213	*Nicotiana rustica*	226
Meyna laxiflora	214	*Nicotiana tabacum*	227
Meyna spinosa	214	*Nyctanthes arbor-tristis*	227
Microglossa pyrifolia	215	*Nymphaea alba*	228
Micromelum minutum	215	*Ocimum americanum*	228
Mikania micrantha	215	*Ocimum basilicum*	229
Mimosa pudica	216	*Ocimum gratissimum*	229
Mimusops elngi	216	*Ocimum sanctum*	230
Momordica charantia	217	*Olea europaea*	230
Momordica cochinchinesis	217	*Onosma echioides*	231
Monochoria hastata	218	*Ophiorrhiza rugosa* Wall. var. *argentea*	231
Monochoria vaginalis	218	*Opuntia dillenii*	231
Morinda citrifolia	218		

Contd..

Appendix–V–*Contd...*

Botanical Names	*Page No.*	*Botanical Names*	*Page No.*
Orchis mascula	232	*Phaseolus lunatus*	243
Origanum vulgare	232	*Phlogacanthus thyrsiflorus*	244
Oroxylum indicum	232	*Phyla nodiflora*	244
Orthosiphon rubicundus	233	*Phyllanthus acidus*	245
Orthosiphon thymiflorus	233	*Phyllanthus airy-shawii*	245
Oryza rufipogon	234	*Phyllanthus amarus*	245
Oryza sativa	234	*Phyllanthus emblica*	246
Osbeckia nepalensis	234	*Phyllanthus fraternus*	246
Osbeckia stellata	235	*Phyllanthus reticulatus*	247
Osmunda regalis	235	*Phyllanthus simplex*	247
Oxalis corniculata	235	*Phyllanthus urinaria*	248
Oxystelma esculentum	236	*Physalis angulata*	248
Padus cornuta	236	*Phytolacca acinosa*	248
Paederia foetida	237	*Phytolacca americana*	249
Paeonia emodi	237	*Pimpinella anisum*	249
Pandanus odoratissimus	238	*Pinellia pedatisecta*	249
Panicum antidotale	238	*Pinus roxburghii*	250
Papaver somniferum	238	*Piper betle*	250
Parmelia tinctorium	239	*Piper hongkongense*	251
Parochetus communis	239	*Piper longum*	251
Passiflora edulis	239	*Pisonia aculeata*	251
Pavetta crassicaulis	240	*Pistacia integerrima*	252
Pavetta indica L.var. *tomentosum*	240	*Plantago major*	252
Pedalium murex	241	*Plantago ovata*	253
Pegolettia senegalensis	241	*Platycodon grandiflorum*	253
Pentanema indicum	241	*Plectranthus amboinicus*	253
Pentapetes phoenicea	242	*Plectranthus barbatus*	254
Prepromia pellucida	242	*Pleurospermum angelicoides*	254
Pergularia daemia	242	*Pogostemon benghalense*	255
Persicaria bistorta	243	*Polianthes tuberosa*	255
Peucedanum dhana	243	*Polycarpon prostratum*	255

Contd..

Appendix–V–*Contd...*

Botanical Names	*Page No.*	*Botanical Names*	*Page No.*
Polygonum aviculare	256	*Remusatia vivipara*	268
Polygonum chinensis	256	*Rhazya stricta*	268
Polyscias fruticosa	256	*Rheum australe*	268
Pometia pinnata	257	*Rheum officinalis*	269
Pongamia pinnata	257	*Rheum spiciforme*	269
Populus alba	258	*Rhododendron arboreum*	269
Portulaca oleracea	258	*Rhus coriaria*	270
Pothos scandens	258	*Rhus semialata*	270
Pouzolzia zeylanica	259	*Rhus succedanea*	270
Premna tomentosa	259	*Ricinus communis*	271
Primula veris	260	*Rorippa sarmentosa*	271
Prinsepia utilis	260	*Rosa centifolia*	272
Prosopis cineraria	260	*Rosa damascena*	272
Prosopis farcta	261	*Roylea cinerea*	272
Prosopis glandulosa	261	*Rubia cordifolia*	273
Prunus persica	261	*Rubus idaeus*	273
Prunus serotina	262	*Rubus moluccanus*	273
Psidium guajava	262	*Rubus niveus*	274
Psilotum nudum	263	*Rungia pectinata*	274
Pteris multifida	263	*Rungia repens*	275
Pterolobium hexapetalum	263	*Ruta chalepensis*	275
Pterospermum acerifolium	264	*Ruta graveolens*	275
Pterospermum canescens	264	*Saccharum bengalense*	276
Pueraria tuberosa	264	*Salomonia cantoniensis*	276
Punica granatum	265	*Salvadora persica*	277
Quercus incana	266	*Salvia officinalis*	277
Quercus infectoria	266	*Sambucus nigra*	277
Quercus robur	266	*Sansevieria roxburghiana*	278
Radermachera xylocarpa	267	*Santalum album*	278
Ranunculus muricatus	267	*Sapindus laurifolius*	279
Rauvolfia serpentina	267	*Sapindus mukorossi*	279

Contd..

Appendix–V–*Contd...*

Botanical Names	*Page No.*	*Botanical Names*	*Page No.*
Sauropus androgynus	280	*Solanum nigrum*	292
Sauropus quadrangularis	280	*Solanum stramoniifolium*	292
Schefflera octophylla	280	*Solanum surattense*	293
Schleichera oleosa	280	*Solanum torvum*	293
Schrebera swieteniodes	281	*Solidago virgaurea*	294
Scleria lithosperma	281	*Sphaeranthus indicus*	294
Scoparia dulcis	281	*Spondias pinnata*	294
Scrophularia ningpoensis	282	*Stachytarpheta jamaicensis*	295
Selaginella bryopteris	282	*Stephania japonica*	295
Semecarpus anacardium	282	*Stereospermum cholonoides*	296
Senna alexandrina	283	*Streblus asper*	296
Senna auriculata	283	*Strychnos nux-vomica*	297
Senna occidentalis	284	*Strychnos potatorum*	297
Senna tora	284	*Stylosanthes fruticosa*	297
Seriphidium brevifolium	285	*Swertia chirayita*	298
Seriphidium thomsonianum	285	*Syagrus romanzoffiana*	298
Sesamum indicum	286	*Syzygium aromaticum*	299
Sesbania grandiflora	286	*Syzygium corynocarpum*	299
Sesbania sesban	287	*Syzygium cumini*	299
Setaria italica	287	*Syzygium malaccense*	300
Shorea robusta	287	*Syzygium samarangense*	300
Sida acuta	288	*Tabernaemontana alternifolia*	301
Sida rhombifolia	288	*Tacca plantaginea*	301
Skimmia laureola	289	*Tadehagi triquetrum*	301
Smilax perfoliata	289	*Tagetes minuta*	302
Smilax zeylanica	289	*Tamarindus indica*	302
Solanum americanum	290	*Tamilnadia uliginosa*	303
Solanum anguivi	290	*Tanacetum gracile*	303
Solanum dulcamara	291	*Taraxacum mongolicum*	303
Solanum erianthum	291	*Tarenna sambucina*	304
Solanum giganteum	291	*Tectaria coadunata*	304
		Tephrosia purpurea	304

Contd..

Appendix–V–*Contd...*

Botanical Names	*Page No.*	*Botanical Names*	*Page No.*
Terminalia alata	305	*Valeriana jatamansi*	317
Terminalia arjuna	305	*Ventilago madraspatana*	318
Terminalia bellirica	306	*Veratrum viride*	318
Terminalia catappa	306	*Verbascum Thapsus*	318
Terminalia chebula	307	*Vernonia anthelmintica*	319
Thuja occidentalis	307	*Vernonia cinerea*	319
Thunbergia fragrans	308	*Vernonia divergens*	319
Thunbergia fragrans var. *laevis*	308	*Vicia sativa* subsp. *nigra*	320
Tinospora cordifolia	308	*Vigna marina*	320
Toddalia asiatica	309	*Vigna umbellata*	320
Toona ciliate	309	*Viola biflora*	321
Tournefortia Montana	310	*Viola odorata*	321
Trachyspermum ammi	310	*Viola pilosa*	321
Trachyspermum roxburghianum	310	*Vitex leucoxylon*	322
Tragia involucrate	311	*Vitex negundo*	322
Tribulus terrestris	311	*Vitex trifolia*	323
Trichaurus ericoides	311	*Vitis vinifera*	323
Trichodesma indicum	312	*Volutarella divaricata*	323
Trichosanthes tricuspidata	312	*Withania somnifera*	324
Trifolium pratense	313	*Woodfordia fruticosa*	324
Trifolium repens	313	*Wrightia arborea*	325
Trigonella foenum-graceum	313	*Wrightia tinctoria*	325
Tylophora asthmatica	314	*Xanthium strumarium*	325
Tylophora flexuosa	314	*Ximenia americana*	326
Typha australis	314	*Xylocarpus granatum*	326
Typha elephantine	315	*Xylosma longifolia*	327
Typhonium trilobatum	315	*Zanthoxylum armatum*	327
Uncaria rhynchophylla	315	*Zingiber officinale*	327
Uncaria sessilifructus	316	*Ziziphus abyssinica*	328
Uraria picta	316	*Ziziphus jujuba*	328
Urtica dioica	316	*Zornia diphylla*	329
Vaccinium myrtillus	317	*Zornia gibbosa*	329

Appendix VI
Index to English Names

English Names	*Page No.*	*English Names*	*Page No.*
Aaron's Rod	295, 318	American Black Nightshade	290
Able	258	American Corn Mint	212
Abrus	7	American Cotton	169
Absinthi	46	American Hellebore	318
Acid Lime	100	American Indigo	60
Acrid Lettuce	195	American Mulberry	220
Adamant Creeper	99	American Night Shade	249
Aerica Nut	42	American Rope	215
African Baobab Tree	17	American Saffron	85
African Marigold	77	American Tobacco	227
African Millet	140	American Worm Wood	137
African Saffron	85	Amy Root	40
Agallochum	41	Ananas	33
Agar Wood	41	Anet	37
Agrimony	20	Angelica	37
Ailanto	21	Anise	249
Air Plant	73	Aniseed	249
Air Plant	136	Annual Lion's Ear	199
Ajenjo	46	Annual Yellow Sweet Clover	211
Alamo Vine	185	Antiaris	40
Alexandria Senna	87	Appa Grass	20
Algarobo	260	Appel-leaf	321
Alkanet	231	Apple	207
Allseed	256	Apple Mangrove	326
Almondette Tree	73	Apple of Peru	129
Aloe Wood	41	Arabian Costus	112
Alpinia Chinensis of Chemist	28	Arabian Manna Plant	23
Alyce Clover	30	Aralia Ming	256
American Arborvitae	307	Arbor-Vitae	307

Contd..

Appendix–VI–*Contd...*

English Names	*Page No.*	*English Names*	*Page No.*
Areca Palm	42	Bal Kino	74
Argans	232	Ballon Flower	253
Arjun Tree	305	Ballon Vine	83
Arnatto	67	Balsam Apple	217
Arrow Leaf Pondweed	218	Bamboo	59
Arrow-Leaf Monochoria	218	Ban Oak	266
Asafoetida	154	Banyan Tree	155
Asafoetida	155	Baobab	17
Ash Coloured Fleabane	319	Barbados Aloe	27
Ash Pumpkin	63	Bark cloth tree	40
Ashgourd	63	Barmuda Grass	124
Ashy Roylea	272	Barren Ivy	173
Asian Scalystem	141	Bastard Agrimony	20
Asiatic Bitter Yam	133	Bastard Brandy Bush	170
Asparagus	50	Bastard Nervain	295
Assam Indigo	59	Bastard Oleander	86
Assam Rum	59	Bastard Saffron	85
Asteracantha	181	Bastard Teak	74
Asthma Weed	150	Bastard Teak	259
Atis Root	14	Bauhinia	62
Ave Grace	275	Bauhinia	63
Aztec Tobacco	226	Beach Bean	320
Babul	10	Beach Launaea	198
Babul Acacia	10	Beach Sunflower	210
Babuna	113	Beach Wiregrass	127
Bacon Weed	94	Bead Tree	138
Bael Tree	18	Bearded Iris	187
Baghnokhi	209	Beddanuts	306
Bahama Grass	124	Beebread	313
Bairwort	321	Belladona	53
Bakul Tree	216	Bellary Grass	177

Contd..

Appendix–VI–*Contd...*

English Names	*Page No.*	*English Names*	*Page No.*
Belleric Myrobalan	306	Bitter Bottle Gourd	195
Bellyache Bush	190	Bitter Calbash Gourd	195
Bengal Arum	315	Bitter Cress	200
Bengal Cardamom	31	Bitter Cucumber	99
Bengal Coffee	107	Bitter Gourd	217
Bengal Gram	95	Bitter Oleander	177
Bengal Kino	74	Bitter Sweet	291
Bengal Pogostemone	255	Black Babul	10
Bengal Quince	18	Black Berry	273
Bengal Rose	177	Black Catechu	9
Bergamotte Orange	100	Black Currant	40
Bergenia	65	Black Datura	129
Berry	81	Black Indian Hemp	40
Betal Nut	42	Black Jack	66
Betel Leaf Pepper	250	Black Murdah	305
Betel Leaf Vine	250	Black Musala	36
Betel Nut Palm	42	Black Musle	118
Betel Pepper	250	Black Musli	118
Bian Xu	256	Black Nightshade	292
Bidi-Leaf Tree	62	Black Pepper	182
Big Egg Plant	291	Black Plum	299
Bilberry	317	Black Thorn	192
Bile Tree	294	Black Zedoary	119
Bindwood	173	Blackberry Lily	187
Bird Cherry	236	Blady Grass	183
Birthwort	45	Blister Creeper	83
Bishop Wood	66	Blond Psyllium	253
Bishop's Weed	310	Blood Flower	50
Bistort	243	Blood Wort	256
Bitter Albizia	23	Blue Eranthemum	146
Bitter Apple	99	Blue Ginger	28

Contd..

Appendix–VI–*Contd...*

English Names	*Page No.*	*English Names*	*Page No.*
Blue Globe Thistle	137	Broom Jute Sida	288
Blue Gum	148	Broomstick	66
Blue Oxalis	239	Brown Hemp	114
Blue Panic Grass	238	Brownbread Rice	234
Blue Pussyleaf	225	Buckhorn	235
Bluebottle	91	Buddhist	63
Blue-Gum Eucalypt	148	Bulb Onion	24
Blumea	68	Bullet Wood	216
Bo Tree	158	Bullock's Lungwort	318
Bohe	212	Bullseyes	29
Bold-Leaf Launaeae	197	Burma Bean	243
Bolivian Bark	96	Burman's Sundew	136
Bombay Senna	283	Burmese Grape	57
Bonduc Nut	75	Burr Tree	224
Bone Setter	99	Burra Gokhru	241
Bottle-Gourd	195	Burweed	325
Bow String Hemp	78	Bush Cane	159
Box Berry	223	Bush Tea	182
Box Myrtle	223	Butter Tree	206
Boy's Love	46	Butterfly Ginger	173
Bracteated Birthwort	44	Butterfly Pea	104
Bracted Bungleweed	22	Butterfly Tree	62
Branched Cancerwort Myanmar	194	Butterprint	8
Branched Fluellen	194	Buttonweed	8
Brandy Mint	213	Cabbage Rose	272
Brazil Wood	170	Calamus	15
Brazilian Tea	295	Calcutta Cane	59
Bread and Cheese	29	Calenduls	77
Bread Tree	7	Californian Poppy	148
Broad-leaf bramble	273	Calumba Wood	112
Broom Creeper	105	Camapu	248

Contd..

Appendix–VI–*Contd...*

English Names	*Page No.*	*English Names*	*Page No.*
Camel Grass	124	Catalonion Jasmine	188
Camel Thorn	23	Catch Thorn	328
Camp Siege	263	Catechu	9
Campeachy Wood	172	Catechu Nigrum	9
Camphor Cinnamomum	96	Catmint	226
Camphor Tree	96	Catnip	226
Canadian Mint	212	Caupatiya	209
Canary Cheesewood	224	Cedar	66
Candahar Tree	168	Cedar Mangrove	326
Candiflower	104	Ceylon Calumba Root	112
Cane	77	Ceylon Cinnamon	98
Cannonball Mangrove	326	Ceylon Iron-wood	213
Cape York Leichardt	224	Ceylon Tea	88
Caper	81	Ceyolon Oak	280
Carambola Apple	54	Chameleon Plant	179
Caranda	84	Chebulic Myrobalan	108
Caraway	85	Chebulic Myrobalan	307
Caribbean Caltrop	194	Cheesewood	225
Carilla Fruit	217	Cheranjee	73
Carry-Me-Seed	246	Chicaquil	191
Carum	310	Chick Pea	95
Cashew	32	Chicken Weed	153
Cashew Apple	32	Chicken-Pea	95
Cashew Nut	32	China Jute	8
Cashmere Tree	168	China Rose	177
Cassia Cinnamon	97	Chinese Almond	41
Cassia Indigo	184	Chinese Bell Flower	253
Cassie Flowers	10	Chinese Chaste Tree	322
Castor	271	Chinese Creeper	215
Castor Oil Plant	271	Chinese Date	328
Castor Seed	271	Chinese Dye Tree	67

Contd..

Appendix–VI–*Contd...*

English Names	*Page No.*	*English Names*	*Page No.*
Chinese Figwort	282	Climbing Dalbergia	128
Chinese Gelsemium	165	Climbing Entada	144
Chinese Hibiscus	177	Climbing Lily	166
Chinese Honey Suckle	109	Climbing Mountain Bean	320
Chinese Knotweed	256	Climbing Pear Bamboo	211
Chinese Laurel	40	Climbing Staff Tree	90
Chinese Leeks	26	Clitoria	104
Chinese Moon-creeper	237	Cloud Apple	300
Chinese Rhubarb	269	Clove Tree	299
Chinese Senna	284	Cluster Bean	121
Chinese Sumac	270	Cluster Fig	157
Chinese Sumach	21	Coarse Tassel Fern	179
Chinese Violet	53	Coast Button Grass	127
Chinese Water Lily	225	Cobbler's Tack	167
Chinese Worm Wood	46	Cochin Turmeric	119
Chir Pine	250	Cochin Turmeric	120
Chiretta	35	Cock's Comb	91
Chiretta Plant	298	Cocklebur	20
Chocolate Weed	212	Cocklebur	325
Choir-bottomcane	77	Cockspur	251
Christ's Tears	107	Cocoa Nut	106
Christ's Thorn	84	Cocoanut Palm	106
Christmas Bush	129	Coco-Grass	125
Christmas Candlestick	199	Coconut Palm	106
Churl's Treacle	25	Coconut Tree	106
Cinchona Bark	96	Coffee Senna	284
Cinnamon	98	Cogongrass	183
Citron Lime	100	Colocynth	99
Clawflowered Laurel	16	Columbo Wood	112
Clearing Nut	297	Comb Fringe Grass	127
Clearing-nut Tree	297	Comb Rungia	274

Contd..

Appendix–VI–*Contd...*

English Names	*Page No.*	*English Names*	*Page No.*
Common Bamboo of Bengal	59	Common Walnut	191
Common Barley	179	Common Water Lily	228
Common Basil	228	Conch Grass	124
Common Basil	229	Conessi Bark	177
Common Cari	85	Congest-Flowered Dinochloa	211
Common Club Moss	205	Congo Tobacco	81
Common Cowitch	220	Congopea	76
Common Curculigo	118	Coomb Teak	168
Common Elder	277	Cooslip	260
Common Emetic Nut	89	Coptis	110
Common Fig	156	Coral Jasmine	227
Common Ginger Lily	173	Coral Plant	191
Common Grape Vine	323	Coral Tree	146
Common Gromwell	203	Coriander	111
Common Guava	262	Cork-Leaved Bayur	264
Common Indigo	185	Corn Mint	212
Common Jasmine	188	Cornflower	91
Common Jasmine	189	Coromandel Gooseberry	54
Common Jujube	328	Cotton	169
Common Marigold	77	Country Borage	253
Common Milk Hedge	151	Country Fig	157
Common Mullein	318	Country Ipecacuanha	314
Common Myrrh Tree	110	Country Mallow	8
Common Myrtle	223	Country Mallow	288
Common Oak	266	Country Sarasaparilla	176
Common Purslane	258	Country Senna	87
Common Quince	122	Cow Clover	313
Common Rattan	77	Cow Grass	313
Common Rue	275	Cow's Lungwort	318
Common Sesban	287	Cowhage	220
Common Tassel Fern	179	Cowslip	260

Contd..

Appendix–VI–*Contd...*

English Names	*Page No.*	*English Names*	*Page No.*
Crab's Claw	252	Deodar	90
Crab's eye	7	Desert Date	58
Cream of Tartar Tree	17	Devil's Claw	251
Creeping Butea	74	Devil's Dung	155
Creeping Fig	157	Devil's Fig	293
Creeping Foxglove	53	Devil's Flower	91
Creeping Launaea	197	Devil's Needle	66
Creeping Pratia	203	Devil's Trumpet	129
Creeping Rungia	275	Devils Tree	28
Creeping Tickfoil	131	Dhub Grass	124
Crested Late-Summer Mint	141	Dillenia	133
Crested Lepidagathis	199	Dilly	37
Crimson Seeded Sedge	83	Ditch Bur	325
Crossandra	114	Ditta Bark Tree	28
Croton	106	Diverging Vernonia	319
Croton Oil Plant	116	Diverse Trifoliate Sumach	26
Crowfoot Grass	127	Dog Mustard	102
Cuckoo's Joy	208	Dog's Mercury	101
Cucumber Tree	54	Dog's Parsley	20
Cuddapah Almond	73	Dog's Tongue	152
Curacao Aloe	27	Dogbane	40
Curry Leaf Tree	220	Dogbane	268
Curssavian Shallow-Wort	50	Doorweed	256
Custard Apple	39	Double Bean	243
Cutch	9	Double Coconut Palm	203
Cutleaf Groundcherry	248	Downy Jasmine	189
Cypress	125	Downy Wood Fern	94
Cypress Root	125	Dragon Mugwort	47
Cypress Shrub	198	Drooping Sea Wormweed	285
Damask Rose	272	Drumstick Tree	219
Deadly Nightshade	53	Drymaria	136

Contd..

Appendix–VI–*Contd...*

English Names	*Page No.*	*English Names*	*Page No.*
Duch White Clover	313	Elephant Ear Taro	26
Duck's eye	42	Elephant Grass	314
Dulcamara	291	Elephant Grass	315
Dwarf Bush Willow	109	Ellliptical-Leaf Ardisia	42
Dwarf Cherry	262	Embelia	142
Dwarf Cherry	93	Embelia	143
Dwarf Lid Brinjal	293	Embelia Fruit	142
Dwarf Mesquite	261	Emblic Myrobalan	246
Dwarf Morning Glory	154	Emetic Nut Tree	89
Dyer's Bugloss	231	Emetic Swallow-wort	314
Dyer's Saffron	85	English Ivy	173
Eagle Wood	41	English or European Oak	266
East Himalayan Fir	7	Englishman's Foot	252
East Himalayan Silver Fir	7	Ephedra	145
East India Galingale	192	Erica Tamarisk	311
East Indian Arrowroot	118	Eschallot	24
East Indian Globe-thistle	294	Estragon	47
East Indian Screw Tree	175	Eucalyptus	148
East Indies Coral Tree	147	Eucalyptus	148
Easter Tree	177	Euphorbia	151
Echinacea	137	European Elder	277
Edible-Stemmed Vine	99	European Goldenrod	294
Egg Tree	163	European Grape	323
Egyptian Beans	225	European Gromwell	203
Egyptian Morning Glory	185	European Pasque Flower	36
Egyptian Priven	198	European Raspberry	273
Egyptian Rattle Pod	287	European Walnut	191
Elecampane	185	European White Water-lily	228
Elephant apple	133	Exile Oleander	86
Elephant Apple	202	Eye Balm	180
Elephant Creeper	144	Eye Root	180

Contd..

Appendix–VI–*Contd...*

English Names	*Page No.*	*English Names*	*Page No.*
False Brandy Bush	170	Flame of the forest	74
False Columba	112	Flame of the Woods	187
False Ipecae	50	Fleabane	47
False Iroko	40	Flixweed	130
False Mvule Sacking Tree	40	Florida Keys Nutrush	281
False Pareira	98	Floss Flower	20
False Primrose	204	Fluxweed	130
False Rattan	159	Foetid Cassia	284
False Sandle Wood	326	Fool's Parsley	20
False Waterwillow	34	Forest Ghost Flower	18
Farkle Berry	317	Forest-Pepper	309
Feathery Coconut Palm	298	Fox Grape	323
Fennel	160	Fox-Grape	126
Fenugreek	313	Fragrant Onion	26
Fern Asparagus	51	French Mulberry of Western Ghats	78
Fern Brake	235		
Fever Nut	75	French Physic Nut	191
Fever Nut	78	Fringed Rue	275
Field Mint	212	Galanga	125
Field-Horsetail	145	Galingale	125
Fig Tree	156	Gambir Vine	316
Fijian longan	257	Ganges Primerose	53
Filiformis Cassytha	88	Garber Bean	144
Finger Millet	140	Garden Croton	106
Finkel	160	Garden Lettuce	194
Firecracker Flower	114	Garden Marigold	77
Fire-flame Bush	324	Garden Mint	213
Fish Fern	67	Garden Onion	24
Fishwort	179	Garden Purslane	258
Five Leaved Chaste Tree	322	Garden Rue	275
Flag Root	15	Garden Sage	277

Contd..

Appendix–VI–*Contd...*

English Names	*Page No.*	*English Names*	*Page No.*
Garderobe	46	Governor's Plum	158
Garland Flower	173	Graceful Pouzolzsbush	259
Garlic	25	Graceful spurge	151
Garlic Chives	26	Gram	95
Gaubpersimmon	134	Grape Leaf Anemone	36
Geranium Grass	124	Grasshopper Tree	42
Geranium Tree	62	Grassllike Zornia	329
German Chamomile	210	Great Plantain	252
German Iris	187	Greater Cat's Tail	315
Giant Dodder	121	Greater Galangal	28
Giant Horsetail	146	Greater Grasshopper Tree	42
Giant Potato	186	Greek Hayes	313
Giant Spine Gourd	217	Greek Milkweed	50
Giant Taro	26	Greek Myrtle	223
Gigantic Swallow-Wort	79	Green Ginger	46
Gingelly	286	Green Hellebore	318
Ginger	327	Green or Wild Amaranth	31
Giradol	95	Green Pigweed	31
Giriba Palm	298	Grey Emetic Nut	303
Glabrous Goglet Flower	93	Grey Oak	266
Globe Thistle	137	Gromwell	203
Glory Tree	103	Ground Ivy	173
Goat Weed	20	Groundnut	41
Gold Thread	110	Groundpine	205
Golden Apple	18	Guava	262
Golden Collyrium	108	Guimauve	29
Golden Eye Grass	118	Guinea Pepper	82
Golden Grain	318	Gulancha Tinospora	308
Golden Seal	180	Gular Fig	157
Golden Shower	86	Gum Guaiacum (resin)	170
Gourian Clematis	101	Gum Tree	10

Contd..

Appendix–VI–*Contd...*

English Names	*Page No.*	*English Names*	*Page No.*
Gummy Cape Jasmine	163	Himalayan Paeony	237
Hairy Onosma	231	Himalayan Prinsepia	260
Hairy-Fruited Pea-Eggplant	292	Himalayan Rhubarb	268
Hall's Honeysuckle	204	Himalayan Strawberry	161
Happy Tree	309	Himalayan White Oak	266
Haroboi Myrrh	110	Hlapi-E-Nyenyane or Tuingras	113
Hartshorn Bush	235	Hoary Basil	228
Hastate-Leaf Pondweed	218	Hogplum	294
Healing-Leaf Tree	291	Hogweed	256
Heart Fern	176	Hollyhock	23
Heart Leaf	179	Holy Basil	230
Heart Leaf Indigo	184	Holy Fruit Tree	18
Heart Orchid	127	Holy Milk Hedge	152
Heart Pea	83	Honey Mesquite	261
Heart Shaped Mikania	215	Horehound	182
Heartleaf Drymary	136	Hornbean Leaved Sida	288
Heart-Leaved False Pickerel Weed	218	Horse Radish Tree	219
Hemp Agrimony	149	Horsegram	206
Hemp Dogbane	40	Hound's Berry	292
Hemp Eupatorium	149	Humble Plant	216
Hen-fat	94	Hundred Husbands	52
Henna	198	Hundred-leaved Rose	272
Herb Christopher	235	Hurtlesickle	91
Herb of Grace	275	Hyptis	182
Herba Schcenanth of Pharmacists	123	Illipe	206
High Taper	318	India Bean	147
Himalayan Cedar	90	India Coral-bean	147
Himalayan Hounds-Tongue	125	India Quince	18
Himalayan Incarvillea	183	India Water Lily	245
Himalayan Ivy	173	Indian Abutilon	8
Himalayan Long-leaved	250	Indian Acalypha	11

Contd..

Appendix–VI–*Contd...*

English Names	*Page No.*	*English Names*	*Page No.*
Indian Aconite	13	Indian Indigo	185
Indian Aconite	15	Indian Ipecacuanha	116
Indian Allophylus	26	Indian Ipecacunha	314
Indian Almond	306	Indian Jujube	328
Indian Aloe	27	Indian Kudzu	264
Indian Asparagus	51	Indian Laburnum	86
Indian Barberry	64	Indian Lilac	55
Indian Bdellium Tree	109	Indian Liquorice Root	7
Indian Bead	80	Indian Long Pepper	251
Indian Bedellium	109	Indian Lotus	225
Indian Beech	257	Indian Madder	273
Indian Birthwort	45	Indian Mahogany	309
Indian Bombax	70	Indian Mallow	8
Indian Borage	253	Indian Mangrove	55
Indian Borage	312	Indian Medlar Tree	216
Indian Bowstring Hemp	278	Indian Mourner	227
Indian Bryonia	111	Indian Mulberry	218
Indian Cadaba	75	Indian Nightshade	290
Indian Caper	82	Indian Oak	61
Indian Cassia Lignea	97	Indian Pellet Shrub	240
Indian Catmint	38	Indian Pellet Shrub	240
Indian Coleus	254	Indian Pennywort	92
Indian Coral-Tree	147	Indian Poke	248
Indian Dammer	287	Indian Privet	322
Indian Dill	37	Indian Redwing	263
Indian Elm	178	Indian Reed	80
Indian gentian	144	Indian Reed-Mace	315
Indian Goseberry	246	Indian Rhododerdron	210
Indian Gum Arabic Tree	10	Indian Rhubarb	268
Indian Hemp	81	Indian Rhubarb	269
Indian Hemp	114	Indian Root	14

Contd..

Appendix–VI–*Contd...*

English Names	*Page No.*	*English Names*	*Page No.*
Indian Saffron	119	Italian Senna	87
Indian Sarsaparilla	117	Ivory Tree	177
Indian Sarsaprilla	176	Ivy	173
Indian Senna	283	Ivy Gourd	105
Indian Shot	80	Iwarancusa Grass	123
Indian Silver Fir	7	Jaca Tree	49
Indian Smilax	289	Jack Fruit	49
Indian Snowberry	71	Jack Wood	49
Indian Sorrel	235	Jaffarabad Aloe	27
Indian Spinach	61	Jalpaiguri Cardamom	31
Indian Spring	23	Jamaica False Vervain	295
Indian Stinging-Nettle	311	Jamaica Loostrife	204
Indian Summer Raspberry	273	Jaman	299
Indian Sweet Fennel	160	Jambolan	299
Indian Traveller's Joy	101	Japan Varnish Tree	21
Indian Trumpet Flower	232	Japanese Bell Flower	253
Indian Turmeric	180	Japanese Honeysuckle	204
Indian Turn Sole	175	Japanese Mint	212
Indian Valerian	317	Japanese Wormwood	47
Indian Wild Pepper	323	Jasmine Flowered Carrisa	84
Indian Worm Wood	47	Java Apple	300
Ink Berry	105	Java Galangal	28
Intellect Tree	90	Java Pennywort	180
Intoxicating Yam	133	Javanese Wood Plant	19
Iron-grass	256	Jhingam	195
Iron-wood of Assam	213	Joan Silverpin	238
Iron-wood Tree	213	Job's Tears	107
Island lychee	257	Joseph's coat	106
Island Musk	149	Jubbalpore Hemp	114
Ispaghula	253	Jujube	328
Italian Millet	287	Jujube	328

Contd..

Appendix–VI–*Contd...*

English Names	*Page No.*	*English Names*	*Page No.*
Jungle Flame Ixora	187	Leafy Milk Hedge	152
Jute	111	Leichhardt Pine	225
Kaffir Potato (Wild variety)	108	Leichhardt Tree	225
Kamala Tree	207	Lemon Basil	229
Kanju	178	Lemon grass	124
Karanj	257	Lentil	198
Karaunda	84	Leopard Flower	187
Kenanga	80	Leopard Lily	187
Khorasan Thorn	154	Lesser Swinecress	200
Khruea Sai Tan	32	Lichen	239
King of Bitter	35	Life Giving Herb	282
King of Desert	260	Lignum Vitae (Wood)	170
King's Fern	235	Lima Bean	243
King's Tonic	237	Lime Berry	215
King's-fruit	162	Lindley's Aporosa	41
Knot Wort	256	Lion's Ear	199
Knotgrass	256	Lipstick Tree	67
Kool Tee	206	Litchi	202
Kumaon Oak	266	Lizard Tail	179
Kurchi Bark	177	Logwood Tree	172
Lac Tree	280	Loked Leaf Typhonium	315
Laciniate Kalanchoe	193	Lollipop Climber	135
Lain Qiao	161	Lomba	140
Laka Wood	16	Long Leaved Beauty Berry	78
Lamb Mint	213	Long Turmeric	119
Land-caltrops	311	Long Wort	37
Lantana Weed	196	Long-leaved Barleria	181
Large Basil	229	Long-Rooted Birthwort	44
Large Leaved knotweed	66	Loose Flowered Desmodium	131
Lavender Moonvine	186	Lopez Tree	309
Lawn Pennywort	181	Lotus Lily	228

Contd..

Appendix–VI–*Contd...*

English Names	*Page No.*	*English Names*	*Page No.*
Lotus Senna	283	Man's Plaything	316
Lotus Sweetjuice	166	Mango	208
Louse-bur	325	Mango Tree	208
Lovage	310	Mangosteen	162
Love Nut	203	Manna Ash	162
Lover's Plant	46	Mantie	256
Lovestone	173	Many Branched Horsetail	146
Luck Plant	160	Mapacho	226
Lucky Nut Tree	86	Maple-Leaved Bayur Tree	264
Lukrabo Seeds	172	March Violet	321
Lychee	202	Margosa Tree	55
Madagascar Plum	158	Marigold	77
Madar Milk Weed	79	Marigold	302
Maddar Wood	46	Marijuma	81
Madderwort	46	Marking Nut Tree	282
Madras Carpet	113	Marsh Cudweed	169
Madras Gram	206	Marsh Herb	145
Madras Hemp	114	Marsh Mallow	29
Mahabaleshwar Raspberry	274	Marsh Orchid	127
Mahot	288	Marshy Betelvine	218
Mahua	206	Mate	183
Maiden Hair Creeper	206	Matted Sandmat	153
Malabar Almond	306	Maxican Tea	137
Malabar Catmint	38	Maypop	239
Malabar Glory Lily	166	Medicinal Senna	283
Malabar Mint	39	Melon	117
Malabar Night Shade	61	Melon Tree	84
Malabar Nut	191	Mesua	213
Malay Almond	306	Mexican Cotton	169
Malay Apple	300	Mexican Poppy	43
Malay Blumea	68	Mexican Prickly Poppy	43

Contd..

Appendix–VI–*Contd...*

English Names	*Page No.*	*English Names*	*Page No.*
Microcos	170	Mugwort	47
Mile-a-Minute	215	Mukul Myrrh Tree	109
Milfoil	12	Munj Sweetcane	276
Milk Weed	50	Musk Basil	62
Mint Weed	182	Musk Jasmine	189
Mistletoe	130	Musk Melon	117
Mistletoe Berrythorn	56	Mustard Tree	277
Mochi Wood Tree	147	Muyna	214
Mongolian Dandelion	303	Myrrh	110
Monkey Bread Tree	17	Myrtle Root	15
Monkey Face Tree	207	Mysore Raspberry	274
Monkey Jack	49	Nacighty	316
Monkey nut	41	Narrow Leaf Purple Coneflower	137
Monkey Tree	67	Narrow Leaved Vetch	320
Monkey-Pod	42	Narrow-Leaf Echinacea	137
Monkshood	13	Narrow-Leaf Indigo	184
Monochoria	218	Narrow-leaved Turmeric	118
Morelle	292	Narrwo Leaf Arelia	256
Morinda Tree	219	Native Raspberry	273
Motherwort	47	Nectarine	261
Mountain Apple	300	Neem Tree	55
Mountain Apple	300	Negro Coffee	284
Mountain Ebony	62	Nepal Ivy	173
Mountain Ebony	63	Nepal Osbeckia	234
Mountain Grass	281	Ner	289
Mountain Knot Grass	19	Nettle	316
Mountain Plum	326	Ngai Camphor	68
Mountain Tournefortia	310	Nickar Bean	78
Mountain Turmeric	120	Nicker Bean	144
Mouth Root	110	Night Jasmine	227
Mowra	206	Nilofar	228

Contd..

Appendix–VI–*Contd...*

English Names	*Page No.*	*English Names*	*Page No.*
Ning-Po-Figwort	282	Paddy	234
Noched Cow-Pea	320	Padri Tree	267
Noon Plant	242	Paeony Rose	237
Nut-Grass	125	Painted- Leaved Uraria	316
Nutmeg Tree	223	Pala Indigo Plant	325
Nux-vomica	297	Pale Rose	272
Oak Fern	94	Palma Christi	271
Oblong-Leaved Croton	115	Palmarosa	123
Octophyllous Aralia	280	Palmata-leaf Pinellia	249
Official Camphor	96	Panakenake	203
Oil Cake Tree	23	Panicled Spot Flower	13
Okambodi or Oumbodi	241	Papaw	84
Old Maid	89	Papaya Tree	84
Old Man	46	Paraguay Tea	183
Old Man's Flannel	318	Pareira Brava Root	98
Oldman Wormwood	46	Pasque Flower	36
Olive	230	Passion Flower	239
Olive Tree	230	Patana Oak	83
Onion	24	Pataroot	122
Opium Poppy	238	Paw-Paw	84
Orange Crossandra	114	Peach Tree	261
Orange Jasmine	221	Peacock's Tail	258
Orange Wood	49	Peanut	41
Orchid Tree	62	Peanut Grass	76
Orchid Tree	63	Pearly Everlasting	34
Orgament	232	Pedatisect Pinellia	249
Oriental Cashew	282	Peepal Tree	158
Orris	187	Pegu Catechu	9
Osbeckia	235	Pellitory	33
Otto of Roses	272	Pepper Elder	242
Oval-Leaved Pondweed	218	Pepper Mint	213

Contd..

Appendix–VI–*Contd...*

English Names	*Page No.*	*English Names*	*Page No.*
Perfumed Cherry	78	Prickly Amaranth	30
Pergularia	242	Prickly Chaff Flower	12
Periploca of the Woods	171	Prickly Hound's-Tongue	125
Persian Fennel-gaint	155	Prickly Lettuce	194
Persian Manna Plant	23	Prickly Nightshade	293
Persian Walnut	191	Prickly Pear	231
Peruvian Bark	96	Prickly Poppy	43
Petty Morel	292	Prickly-Leaved Elephant's Foot	139
Phyllanthus	246	Prostrate Rattlepod	115
Physic Nut	75	Prostrate Wild Petunia	135
Physic Nut	189	Pullback	251
Pie-marker	8	Pumpkin	117
Pigeon Pea	76	Puncture-vine	311
Pig-weed	256	Puneala Palm	159
Pinang Palm	42	Pungent Shrub	149
Pineapple	33	Purging Fistula	86
Pink Ladies Slipper	180	Purging Nut	189
Pink Periwinkle	89	Purple Fleabane	92
Poison Berry	290	Purple Fleebane	319
Poison Nut	189	Purple Jatropha	190
Poke Root	249	Purple Lippia	244
Polynesian Cress	271	Purple mangosteen	162
Pomegranate Tree	265	Purple Moonflower	186
Pompkin	117	Purple Orchid	232
Pongam Oil Tree	257	Purple Spored Puffball	79
Poor Man's Treacle	25	Purple Tephrosia	304
Pot Marigold	77	Purslane	258
Pot Marjoram	232	Puzzle Nut	326
Pot Purslane	258	Pythagorean Bean	225
Potato Plant	247	Quail Grass	91
Potato Tree	291	Queen of the shores	64

Contd..

Appendix–VI–*Contd...*

English Names	*Page No.*	*English Names*	*Page No.*
Queen Palm	298	Red Tassle Flower	143
Quince	122	Red Weed	212
Rag Jag	94	Redball Snakegourd	312
Ragi	140	Red-Fruited Nightshade	292
Raisin Berberry	64	Reinwardtia	225
Rangoon Bean	243	Resurrecting Plant	282
Rangoon Creeper	109	Resurrection Lily	193
Raspberry	274	Rhizomatous Iris	187
Rat Ear	242	Ribbed Gourd	204
Rattan	77	Riber Ebony	134
Rattan Chair	77	Ribgrass	252
Rattle Pod	114	Rice	234
Rauvolfia Root	267	Ridged Gourd	204
Red Bitter-Apple	291	Roast Potato Plant	247
Red Campion	91	Rockfoil	65
Red Cedar	309	Roman Wormweed	285
Red Cherry	93	Rosary Pea	7
Red Chilli	82	Rose Apple	300
Red Clover	313	Rose Tree	269
Red Cotton Tree	70	Rosha Grass	123
Red Creeper	318	Rosy Eranthemum	146
Red Grass	76	Rosy Milkweed Vine	236
Red Java Tea	233	Rosy-flowered Indian Periwinkle	89
Red knotweed	66	Rottler's Chrozophora	95
Red Kwao Kruva	74	Roucou	67
Red Legs	256	Rough Chaff Flower	12
Red Mulberry	220	Rough Fruited Buttercup	267
Red Periwinkle	89	Round- Leaved Bedstraw	162
Red Physic Nut	58	Round Rooted Galangal	193
Red Rice	234	Round Zedoary	120
Red Sage	277		

Contd..

Appendix–VI–*Contd...*

English Names	*Page No.*	*English Names*	*Page No.*
Roundleaf Spurge	153	Sea Daisy	210
Roxburgh's Foldwing	132	Sea Sedge	15
Royal Fern	235	Sea Worm Seed	48
Rue Plant	275	Seeded Under Leaf	247
Rusa Grass	123	Selaginella	282
Russ Grass	123	Senegal Senna	87
Rusty Mimosa	11	Senna	283
Sacred Basil	230	Sensitive Plant	216
Sacred Fig Tree	158	Serpentine	267
Sacred Ixora	187	Serpentine Root	267
Sacred Lotus	225	Sesame	286
Sacred Tobacco	226	Sesban	286
Safflower	85	Sessile Alternanthera	29
Saffron	113	Seychelles Nut	203
Saffron Crocus	113	Shallot	24
Saffron Teak	172	Shalot	24
Sage	277	Shamrock Pea	239
Sage Brush	48	Sharp Leaved Gambir	315
Sage- Leaved Alangium	22	Shatterstone	248
Sal Tree	287	Sheathed Alysicarpus	30
Salep	127	Shield Fern	67
Salep	232	Shield Leaf Ariopsis	268
Salt Bush	277	Shiny Bush	242
Sandal Neem	309	Shiranjitia	324
Sandal Tree	278	Shocking Blue	165
Sappan Wood Tree	172	Shoe Button Ardesia	42
Scented Babul	10	Shoe Flower	177
Scholar's Tree	28	Short-Leaved Worm Seed	285
Scotch Marigold	77	Showy Milkweed	50
Screw Pine	238	Showy Pigeon	76
Sea Coconut Palm	203	Showy Rottlepad	115

Contd..

Appendix–VI–*Contd...*

English Names	*Page No.*	*English Names*	*Page No.*
Showy Silk Cotton Tree	70	Soft Leichhardt	225
Shrubby Basil	229	Sonkadi	241
Shrubby Pencil Flower	297	Sorrowful Tree	227
Siamease Ginger	28	Sour Cherry	93
Siamese Rough-bush	296	Sour Lime	100
Sickle Bush	132	South African Basil	62
Sickle Senna	284	South European Flowering Ash	162
Sicklepod	284	South India Soapnut	279
Sida	288	Southernwood	46
Sikkim Blushwort	19	Sow Thistle Flower	143
Silk Cotton Tree	70	Sowa	37
Silk Plant	23	Soyabean	167
Silk Weed	50	Spade Flower	180
Silkcotton Tree	70	Spanish Garlic	24
Silverbush	242	Spanish Jasmine	188
Sinhaporekapok	90	Spanish Needle	66
Sita's Yarn	88	Spanish Pellitory	33
Sitherwood	46	Spanish Saffron	113
Six-Petalled Clematis	101	Sparkle Berry	317
Skeleton Fork Fern	263	Sparrow	256
Slender Rattan	77	Spear Grass	177
Slipper Thorn	231	Spearmint	213
Small Flower Nightshade	290	Spider Brake	263
Small Flowered Melilot	211	Spider Fern	263
Small Indian Ipecacuanha	171	Spider Flower	102
Small Indian Oak	61	Spiked Ginger Lily	174
Smooth-Leaf White Voilet	321	Spiked Rhubarb	269
Snake's Head	209	Spinous Kino Tree	72
Snakeroot	243	Spiny Bamboo	59
Snake-wood	297	Spiny Fruit Buttercup	267
Soap Nut Tree of North India	279	Spiny Fruited Crowcup	267

Contd..

Appendix–VI–*Contd...*

English Names	*Page No.*	*English Names*	*Page No.*
Spiny Pigweed	30	Sweet Broom	281
Spiral Ginger	112	Sweet Cinnamon	15
Spiralis	116	Sweet Clock Vine	308
Spogel Seeds	253	Sweet Flag	15
Sponge Gourd	205	Sweet Galingale	125
Sponge Tree	10	Sweet Melon	117
Spreading Hogweed	69	Sweet Pepper	82
Spurge Cactus	150	Sweet Violet	321
St. Thomas Bean	144	Sylvaticus Voodoo Lily	31
Staghorn Clubmoss	205	Sylvaticus Voodoo Lily	32
Star Gooseberry	280	Syrian Mesquite	261
Stickle Wart	20	Tacca	301
Sticky Cleome	102	Tahitian Gardenia	164
Stinking Cedar	21	Tailed Maiden Hair	17
Stinking Weed	284	Tallow Gourd	63
Stinking-roger	302	Tallow Wood	326
Stone Breaker	248	Tamarind Tree	302
Stone Flower	239	Tanner's Cassia	283
Strawberry Fern	176	Tape Vine	295
Strinking Opalberry	237	Tarragon	47
Strychnine Tree	297	Tea Bush	229
Sugar Apple	39	Teaweed	288
Sun Hemp	114	Tellicherry Bark	177
Supper Lily	166	Tentwort	52
Supple -Jack	65	Thatch Grass	183
Swamp Basil	62	The Bergamot	100
Swamp Pea	286	The Chaulmugra Oil	172
Swedish Myrtle	223	The China Box Tree	221
Sweet Apple	39	The Country Gooseberry	245
Sweet Basil	229	The Creat	35
Sweet Belladona	248	The Elephant Creeper	43

Contd..

Appendix–VI–*Contd...*

English Names	*Page No.*	*English Names*	*Page No.*
The Fig	156	Tiger Palm	249
The Gall or Dyers Oak	266	Tinospora	308
The Juncus Odoratus	123	Tobacco	227
The Lemon Grass	122	Tobacco Tree	291
The Lesser Galangal	28	Togari Wood of Madaras	219
The Lesser Galangal of Ainslie	174	Tongue Fern	176
The Light Galangal	29	Toon	309
The Marsh Mint	212	Tooth Brush Tree	277
The Obtuse Leaved Mimusops	208	Toothache Tree	327
The Pellitory of Spain	33	Toothbrush	281
The Purging Croton	116	Tooth-Brush Plant	168
The Salep Orchid	232	Toothed Leaf Chilla	86
The Spiral Water Trumpet	116	Tortora	191
The Stinging Nettle	316	Touch-me-not	216
The Sumach Tree of Europe	270	Trailing Eclipta	138
The Togari Wood of Madras	218	Tree Cotton	169
Thick- leaved Lavender	38	Tree Jasmine	188
Thick-leaved Pennywort	92	Tree Of Heaven	21
Thorn Apple	128	Tree of Life	73
Thorny Amaranth	30	Tree of the Gods	21
Thorny Bamboo	59	Tree Rhododendron	269
Thorny Bug-Tree	291	Tree Turmeric	64
Thorny Staff Tree	171	Tree Turmeric	112
Three-Leaf Soapberry	279	Trefle Gros	301
Three-leaved Pine	250	Triangular Spurge	150
Thumbe	201	Trijugate-Leaved Sop-nut	279
Thyme Leaved Gratiola	57	Tropical Chickweed	136
Thyme-leaved Spurge	153	Tropical Clover	76
Tiare	164	True Cedar	90
Tick Tree	130	True Kopok Tree	90
Tiger Lily	166	True Myrtle	223

Contd..

Appendix–VI–*Contd...*

English Names	*Page No.*	*English Names*	*Page No.*
True Passion Flower	239	Vine-leaved Anemone	36
Trumpet-Flower	296	Virgate Leaf-Flower	247
Tsi	179	Virgin Mary's Candle	318
Tua-Tua	190	Virginian Tobacco	227
Tuber-Flower	103	Walking Maiden Hair Fern	17
Tuberose	255	Wall Rue	52
Turkey Berry	293	Wallich Cranesbill	165
Turmeric	119	Wallich's Thistle	98
Turnsole	95	Warer Willow	192
Twin Flowered Violet	321	Wart Cress	200
Twisted Acacia	9	Wash Sponge	205
Two Coloured Grewia	170	Water Apple	300
Two-Rowed Barley	178	Water Cress	145
Umbrella Tree	238	Water Fern	235
Upas Tree	40	Water Hemp	149
Upland Cotton	169	Water Hyssop	57
Utrasum Bead Tree	138	Water Rattan	77
Variable Combretum	108	Water Rose	228
Varnish Tree	282	Wax Apple	300
Vegatable Marrow	117	Wax Flower	301
Vegetable Fern	134	Wax Jambu	300
Vegetable Gold	110	Wax Tree	270
Vegetable Sponge	205	Wax-Leaved Climber	117
Velvet Bean	220	Waybroad	252
Velvet Dock	318	Weaver's Beam Tree	281
Velvetleaf	8	Wedelia	210
Velvet-Leaf	98	Weeping Fig	156
Velvet-Leaf Desmodium	132	Weeping Forsythia	161
Villous leaf Desmodium	132	West Indian Chickweed	136
Vine	171	West Indian Ipecacuanha	50
Vinegar Wood	66	Western Pearly Everlasting	34

Contd..

Appendix–VI–*Contd...*

English Names	*Page No.*	*English Names*	*Page No.*
Whip Vine	159	White-Wooded Wallrothia	322
Whisk Fern	263	Wild Apricot	239
White Bean	21	Wild Asparagus	51
White Berry- Bush	160	Wild Black Cherry	262
White Catamaran Tree	165	Wild Castor	58
White Cedar	307	Wild Celery	310
White Clover	313	Wild Chamomile	210
White Cotton Tree	90	Wild Cherry	262
White Emetic Nut	164	Wild Cinchona	71
White Ginger	173	Wild Cinnamom	97
White Gourd	63	Wild Clove	204
White Jute	111	Wild Clover	313
White Lady	222	Wild Cotton	169
White Maiden Hair	52	Wild Croton	58
White Mallow Wymote	29	Wild Ginger	112
White Mangrove	55	Wild Guava	83
White Murdah	305	Wild Hops	160
White Nightshade	290	Wild Indigo	60
White Onion	24	Wild Indigo	304
White Palle	21	Wild Jasmine	188
White Pavetta	240	Wild Killer	104
White Pavetta	240	Wild Kuthi	54
White Poplar	258	Wild Lettuce	195
White Poppy	238	Wild Lucerne	297
White Pumpkin	195	Wild Mango	294
White Sandal Tree	278	Wild Mangosteen	134
White Siris	21	Wild Marjoram	232
White Teak	168	Wild Mustard	102
White Thunbergia	308	Wild Mustard	102
White Weed	20	Wild Orange-Tree	309
White-Wood Chaste Tree	322	Wild Passion Vine	239

Contd..

Appendix–VI–*Contd...*

English Names	*Page No.*	*English Names*	*Page No.*
Wild Peach	261	Woolly Morning Glory	43
Wild Plum	326	Wooly Dyeing Rosebay	325
Wild Saffron	85	Worm Killer	44
Wild Sage	196	Worm Seed	48
Wild Sage	196	Worm Wood	46
Wild Spikenard	182	Wormseed	285
Wild Spinach	94	Woundwort	294
Wild Tobacco	226	Yarrow	12
Wild Tomato	248	Ye gu	18
Wild Turmeric	119	Yellow Bark	96
Wild Varnish Tree	270	Yellow Bleeding Heart	126
Willow Primrose	204	Yellow Broom	60
Wilol Cherry	93	Yellow Cheesewood	225
Wind Flower	36	Yellow Indigo	60
Wine-grape	323	Yellow Oleander	86
Winter cherry	83	Yellow Oxalis	235
Winter Cherry	248	Yellow Teak	172
Winter Cherry	324	Yellow Thistle	43
Wire Bush	212	Yellow Zedoary	119
Wodier	195	Yellow-berried Nightshade	293
Wolf's Foot Clubmoss	205	Yellow-nail-dye plant	60
Wolfsbane	13	Yerba Mate	183
Wonder Tree	260	Ylang-Ylang	80
Wood Apple	202	Zachum Oil Plant	58
Wood-bind	173	Zedoary	120
Woody Nightshade	291		

Appendix VII
Index to Hindi Names

Hindi Names	Page No.	Hindi Names	Page No.	Hindi Names	Page No.
Aak	79	Ajowanj	310	Amarbel	88
Aakada	79	Ajwain	310	Amarbeli	88
Aam	208	Ak	218	Amb	208
Aankuri Bankuri	18	Ak	78	Ambodha	295
Abi	122	Akada	78	Ambuj	225
Aboli	114	Akanadi	98	Ameta	256
Aburok	40	Akarkara	33	Amli	302
Acc	219	Akasgaddah	111	Ampli	302
Ach	218	Akashabela	121	Amra	295
Ach	219	Akashbel	121	Amrita	308
Achar	73	Akhaul	22	Amroda	253
Ada	327	Akhor	191	Amrool	300
Adhasisi	326	Akhrot	191	Amrud	262
Adrak	327	Akol	22	Amrul	40
Adusa	192	Akola	22	Amrul	235
Afim	238	Akorkara	33	Amrut	262
Afsantin	46	Akra	291	Ananas	33
Afyun	238	Akra	320	Anannas	33
Ag	79	Akta	320	Anantmul	176
Agali	36	Al	218	Anar	265
Agar	41	Al	219	Anboti	235
Agasti	286	Alkaddu	195	Anda	84
Agni Khair	200	Alubalu	93	Andhahuli	312
Aila	9	Am	208	Angur	323
Aila	11	Am	262	Angurshefa	53
Ajavayana	48	Amala	246	Anjir	156
Ajavayana	285	Amalbel	126	Ankora	22
Ajmod	310	Amalbel	323	Ankra	320
Ajmot	310	Amaltas	87	Anola	246
Ajmud	310	Amarabela	121	Antamul	314

Contd..

Appendix–VII–*Contd...*

Hindi Names	*Page No.*	*Hindi Names*	*Page No.*	*Hindi Names*	*Page No.*
Anvula	246	Babuna	113	Ban jayapal	190
Anwal	283	Babuna	210	Ban Long	204
Apang	12	Babur	10	Ban Tamakhu	291
Aparajita	104	Babutulsi	228	Ban Tulsi	229
Apmarga	12	Bacchang	13	Ban Tulsi	229
Aprajit	104	Bach	15	Banafshah	322
Arand	271	Badahara	49	Banaphasha	321
Arar	89	Badami	306	Banasaa	192
Areetha	279	Badi Indrayan	99	Banda	130
Arhun	305	Badri	161	Bandhani	167
Arlu	232	Bagun	70	Bandian	249
Arni	104	Bahabirang	143	Banfsa	321
Aru	261	Bahari	228	Ban-haridra	119
Arusha	192	Bahera	306	Banillachi	31
Asgandh	324	Bahupatra	135	Banj	266
Asluna	305	Baibirang	143	Banjan	42
Asmania	145	Baichi	158	Banjira	92
Assarsauna	232	Baikal	171	Bankau	198
At	39	Bailewa	297	Bankhara	128
Atasitaphal	39	Bairang	142	Banmethi	211
Atis	14	Bairi	86	Bannalia	256
Ativisha	14	Bak	286	Ban-natia	256
Atvika	14	Bakam	172	Ban-Nimbu	168
Aungra	246	Bakarcha	160	Bano	173
Aura	246	Bakra	88	Banokra	326
Aval	283	Baksi	319	Banpalas	281
Baberang	142	Bakul	216	Bans	59
Baberang	143	Ballari	226	Bar	155
Babla	10	Ban	266	Bara Gokhru	241
Babui Tulsi	229	Ban Ber	328	Bara Guma	199
Babul	10	Ban champak	80	Baralunia	258

Contd..

Appendix-VII-*Contd...*

Hindi Names	*Page No.*	*Hindi Names*	*Page No.*	*Hindi Names*	*Page No.*
Baranda	230	Belatak	282	Bheri	86
Baranimbu	100	Beli	202	Bhilara	282
Bare (Bayr)	262	Belkangu	101	Bhillar	66
Bargad	155	Belumbu	54	Bhitura	289
Barhal	49	Bemgsag	92	Bhiunli	288
Barhanta	290	Bemt	77	Bhoja	15
Barhanta	311	Bendarli	205	Bhui aonla	247
Bari	293	Beri	86	Bhuichampa	193
Barial	63	Beri	160	Bhuikhakhasa	283
Barial	62	Bet	77	Bhuiokra	244
Bariara	288	Bethuasag	94	Bhungi	143
Barielachi	31	Bhadradanti	191	Bhuyikohada	186
Barikasaumdi	284	Bhaira	306	Bibsar	221
Barikateri	290	Bhalumas	291	Bichhloo	241
Bariktel	286	Bhander	328	Bichhu Booti	317
Barisounf	161	Bhang	81	Bichu	209
Barre	85	Bhangra	138	Bidang	142
Barsanga	220	Bhans	59	Bidar Kand	186
Baru	107	Bhant	103	Bihi	122
Bash	15	Bharangi	103	Bihi	262
Basna	286	Bharani	303	Bijaura	100
Bastra	78	Bharband	43	Bilaikand	186
Bathua	94	Bhat	103	Bilangru	158
Bauna	210	Bhat	167	Bili	18
Bebina	222	Bhatbhamt	43	Bimb	105
Bedina	222	Bhatia	128	Bina	55
Bekara	184	Bhekal	260	Binari	227
Bekkra	260	Bheki	226	Biranjasif	12
Bel	18	Bhela	282	Bish	13
Bela	188	Bhendu	175	Bisi	42
Bela Palas	74	Bherenda	190	Biskhafra	69

Contd..

Appendix–VII–*Contd...*

Hindi Names	*Page No.*	*Hindi Names*	*Page No.*	*Hindi Names*	*Page No.*
Biswul	11	Chakki	49	Chhokara	260
Bithua	175	Chakramarda	284	Chhota Gokhru	311
Bodha	74	Chakunda	284	Chhotadudhilata	171
Bojhidan	51	Chakunda	284	Chhotakulpha	312
Bol	110	Chalmeri	245	Chhotalunia	258
Bolsari	216	Chalmugra	172	Chhotataroda	87
Bor	155	Chalta	133	Chhuari	48
Bora	315	Chameli	188	Chhuari	285
Borraal	218	Chameli	188	Chhuchhuni	115
Brahmaduni	43	Chameli	189	Chian	144
Brahmamanduki	92	Champa	193	Chibbige	297
Brahmi	57	Chana	95	Chihumt	74
Brahmi	92	Chanaka	95	Chilara	86
Buinowla	247	Chandal	278	Chilbil	178
Bujjerbhang	227	Chandarjot	189	Chilkan	156
Bukhar Jadi	199	Chandni	301	Chilla	86
Bundahia	212	Chandrabhaga	267	Chimine	308
Bundava	212	Chandramula	192	Chir	250
Bungandana	26	Chandui	301	Chirayata	35
Bungsee	98	Charas	81	Chirayita	298
Bur	123	Chauli	30	Chirchita	12
Burans	269	Chaupatia	209	Chirchitta	66
But	95	Chavulmungri	172	Chireta	105
Cakki	49	Chaya	19	Chirghas	175
Canda	37	Cheel	94	Chiriya-Kanda	51
Caturang	196	Chena	287	Chirmi	7
Caval	234	Cherara	260	Chirmiti	7
Chaaval	234	Cherootpathi	296	Chironji	73
Chachibet	77	Chhalmugra	172	Chita Dhan	234
Chagulbati	224	Chhatim	28	Chitra	64
Chakavat	284	Chhikura	260	Chitra Kushmal	64

Contd..

Appendix–VII–*Contd...*

Hindi Names	*Page No.*	*Hindi Names*	*Page No.*	*Hindi Names*	*Page No.*
Cholai	30	Dagurin	157	Dehan	309
Cholayi	31	Daheya	296	Dekamari	164
Choli	30	Dahu	49	Deobans	59
Chora	37	Daira	325	Deodar	90
Chorota	102	Dakh	323	Deshi-badam	306
Chorota	102	Dalap	147	Desi Lanatan	196
Choru	37	Dalchini	98	Devdarru	90
Chota Chirayita	144	Dalme	160	Dhaak	74
Chota Kirayita	144	Damahan	154	Dhai	324
Chota-chand	267	Damar	287	Dhak	74
Chota-dhatura	326	Dampel	163	Dhalim	265
Chotadudhi	153	Dandal	327	Dhan	234
Chota-gokhru	326	Dandotpala	319	Dhania	111
Chota-italkusa	201	Danti	58	Dhanya	111
Chotalalkura	201	Danti	116	Dhao	49
Choti-dudhi	153	Darchini	97	Dharauli	325
Chotra	64	Darhald	64	Dharimb	265
Chouri	88	Dariya-i-nariyal	203	Dhatela	260
Chucka	115	Darmar	327	Dhattura	43
Chuka tripati	235	Daru Haldi	64	Dhatura	129
Chulta	133	Daruchini	98	Dhaula	324
Chunna	95	Darya-ka-nariyal	203	Dhaura	324
Churanhar	101	Dasmari	141	Dhava	49
Cid	250	Dasmuli	146	Dhebri	88
Cil	250	Dauna	48	Dhera	22
Codhara	39	Dauna	47	Dhobein	128
Coffee	107	Davana	47	Dhol Dak	146
Dabh	183	Davana	48	Dhou	49
Dabhir	107	Dawi	324	Dhudhi	182
Dabra	316	Debkoti	91	Dhuna	287
Daduri	157	Dedwar	90	Dier	105

Contd..

Appendix–VII–*Contd...*

Hindi Names	*Page No.*	*Hindi Names*	*Page No.*	*Hindi Names*	*Page No.*
Dikamali	163	Gadepurna	69	Ghatyari	123
Dikamali	164	Gagandhul	238	Ghemar	238
Dikemali	164	Gahat	206	Ghikawar	27
Dikmali	163	Gamathi Pudina	213	Ghikumar	27
Diwana Babul	10	Gambhari	168	Ghikumari	27
Dolan Champa	174	Gand Kaphal	161	Ghiya Tori	205
Dolu	268	Gandan	44	Ghoda Tulsi	281
Dona	47	Gandana	12	Ghont	328
Dopahariya	242	Gandana	24	Ghorbach	15
Dub	124	Gandati	44	Ghorumba	99
Dubra	124	Gandbabul	10	Ghugachi	7
Dudhi	117	Gandha Trina	122	Ghunchi	7
Dudhi	151	Gandhabel	123	Ghunghanchi	7
Dudhi	153	Gandhali	237	Ghurga	93
Dudhi	236	Gandhapalashi	174	Ghurgia	93
Dudhi	325	Gandhela	220	Ghurush	320
Dudhi	325	Gandibudi	166	Gidar Tamaku	318
Dudhia-Bis	15	Ganga Sedi	227	Gidardrak	126
Dudhialata	236	Gangerua	217	Gila	144
Dudhikalave	151	Ganindri	302	Giloy	308
Dudlutta	236	Ganja	81	Girgiti	168
Dugdhica	236	Garari	101	Girimala	87
Dumpa	259	Garbi	144	Gobara	38
Ekdania	72	Garrar	101	Gobhi	139
Ek-Kanda-Lasun	24	Gathivana	47	Gobla	157
Elachi	31	Gauli	72	Gokhru	184
Erandi	271	Gazpipal	251	Gokhru	311
Eucalyptus	148	Gelsimium	165	Gokhru	326
Faridbuti	241	Gewani	329	Gokshri	311
Fasli Gulab	272	Ghaneri	196	Gokshura	181
Gab	134	Ghato	103	Gol Kaddu	63

Contd..

Appendix–VII–*Contd...*

Hindi Names	*Page No.*	*Hindi Names*	*Page No.*	*Hindi Names*	*Page No.*
Golkaddu	195	Gurial	62	Hatiya	286
Goncha	220	Gurkkamai	292	Hattajuri	175
Gondai	72	Gurlu	107	Hattian	90
Gopali	38	Gurmal	108	Hazarmani	248
Gorakh Amli	17	Gurmar	171	Hazarmuli	50
Gorakh Mundi	294	Gwiar	62	Hejurchei	199
Gorbach	15	Hadjod	99	Henna	198
Gorkhabundi	19	Hadjora	99	Hetamundiya	312
Gouli	185	Hakna	184	Hier	105
Govar	121	Hakshardana	151	Hijgal	61
Gramirch	82	Halad	119	Hindirevandchini	268
Gudrisag	29	Haldi	119	Hindisana	283
Gugal	109	Haldu	172	Hing	154
Gugal	110	Hamsagar	73	Hing	155
Guggul	109	Hamsagar	193	Hingan	58
Guggul Dhup	21	Hankha	136	Hingol	58
Guhbabul	10	Hansapadi	17	Hingot	58
Gulab	272	Hansapagi	17	Hingra	155
Gulab	272	Har Singar	227	Hingu	58
Gulabzurdi	272	Hara	108	Hinguna	154
Gulal Tulsi	229	Harad	108	Hirabol	110
Gulanca	308	Harad	307	Hirankhuri	143
Gular	157	Harara	108	Hirantutiya	108
Gulcheri	255	Harara	307	Hiun	82
Gul-e-Asharfi	77	Harfarauri	245	Hulpul	102
Gulhar	177	Harra	307	Hunraj	256
Gulkakra	217	Haruch	145	Hurhur	102
Gulkhaira	29	Hathajori	209	Hurhureh	102
Gulshabbo	255	Hathi Kathian	17	Hussuk	311
Gunara	238	Haththathoira	231	Ijal	61
Gundali	237	Haththazori	67	Imli	302

Contd..

Appendix–VII–*Contd...*

Hindi Names	*Page No.*	*Hindi Names*	*Page No.*	*Hindi Names*	*Page No.*
Indrajau	178	Jamur	238	Jawa	179
Indrayan	99	Jand	260	Jawakha	179
Indrayan	312	Jangali Amla	245	Jawasa	23
Ingar	61	Jangi Gobi	197	Jayapatri	223
Irum	66	Jangli Amla	246	Jhak	277
Isabgal	253	Jangli Amli	246	Jhampi	8
Isabgul	253	Jangli Angur	32	Jhar-i-huldi	112
Isarmul	44	Jangli Aushbah	289	Jhatela	260
Isbgol	252	Jangli Hala	200	Jhau	311
Isharmul	45	Jangli Haldi	119	Jhinti	60
Ishermul	45	Jangli Kalimirch	309	Jhunjhunia	115
Issufgul	253	Jangli Lakuch	40	Jinga	204
Isvaralingi	135	Jangli Muli	68	Jingan	195
Isvarmul	44	Jangli Tulsi	140	Jonkaphal	175
Jagru	132	Jangli Tur	54	Jotri	223
Jaimangal	267	Jangli-badam	306	Juhi	188
Jainti	287	Jangli-Darchini	97	Juilata	255
Jaiphal	223	Janglie-Randi	190	Jujar	61
Jait	287	Jangli-gainda	302	Juliakanta	181
Jaitun	230	Jangliharrar	102	Jungli Aam	295
Jalapapili	244	Jangli-pikvam	314	Jungli Baler	76
Jalayamani	105	Janumarak	30	Jungli Errand	189
Jalnim	57	Jar-Amla	246	Junwasa	23
Jamalgota	116	Jasum	177	Juti	221
Jaman	299	Jasun	177	Jutuk	242
Jambhal	299	Jasund	40	Juwasa	23
Jamona	236	Jati	188	Kaashmira	185
Jamrassi	88	Jau	179	Kabaiya	292
Jamrul	300	Javansa	23	Kabbar	124
Jamti-Ki-Bel	105	Javasa	23	Kachnar	62
Jamun	299	Javitri	223	Kachnar	63

Contd..

Appendix-VII-*Contd...*

Hindi Names	*Page No.*	*Hindi Names*	*Page No.*	*Hindi Names*	*Page No.*
Kachura	120	Kakra-singi	271	Kamini	221
Kacora	192	Kakri	117	Kamrup	156
Kadam	71	Kakun	287	Kamtakaliya	181
Kadamba	71	Kala Bhangra	38	Kanak	129
Kadami	172	Kala Hinsalu	274	Kanakchampa	264
Kadimah	118	Kala Tulsi	228	Kananir	86
Kado	140	Kala Zira	85	Kanchari	29
Kadvagokhru	241	Kaladhatura	129	Kandan	63
Kadvilauki	195	Kalangbans	59	Kandara	137
Kaephal	223	Kalatendu	134	Kandhara	327
Kaghzi Nimbu	100	Kala-Tulsi	230	Kandhi	8
Kagsha	157	Kale Hinure	274	Kandiari	43
Kahiree	150	Kali Haldhi	119	Kandiari	328
Kahu	305	Kali Musli	118	Kandika	328
Kahu	194	Kali Tori	204	Kanduri	105
Kahua	305	Kali Tulsi	229	Kangach	305
Kaikayi	109	Kalidudhi	117	Kanghani	8
Kaimil	195	Kalidudhi	182	Kanghi	8
Kaiphal	223	Kalihaladi	120	Kanghi	137
Kaj	72	Kalijhamp	17	Kangi abel	72
Kajra	297	Kalijhant	17	Kaniar	62
Kaju	32	Kalijira	92	Kaniar	264
Kajubadam	32	Kalijira	319	Kanj	309
Kakatundi	50	Kalizer	104	Kanji	257
Kakein	158	Kalmegh	35	Kanju	178
Kakni	117	Kamal	225	Kankra	240
Kakni	287	Kamarakh	55	Kankra	240
Kakor	328	Kamaramga	55	Kanni	287
Kakoranda	68	Kamba	83	Kanphuta	160
Kakra	252	Kambar	168	Kanphuti	83
Kakrashingi	252	Kamila	207	Kanphytia	102

Contd..

Appendix–VII–*Contd...*

Hindi Names	*Page No.*	*Hindi Names*	*Page No.*	*Hindi Names*	*Page No.*
Kantabans	59	Karka	72	Kateli	293
Kantabhaji	30	Karmal	55	Kateli	293
Kantela	137	Karnikara	240	Katgularia	157
Kantenatia	30	Karnikara	240	Katha	9
Kanthal	49	Karola	217	Kathachampa	240
Kanthari	82	Karosli	227	Kathachampa	264
Kantisenbal	70	Karpas	169	Kathal	49
Kanturi	105	Karpasi	169	Kathar	49
Kanval	225	Karpur	96	Kathbel	202
Kapalphoti	83	Karra	178	Kathnim	220
Kapas	169	Karrah	85	Kat-karanja	75
Kapasi	175	Karu	81	Katmauli	62
Kapur	96	Karund	31	Katpatra	327
Kapurakachari	174	Karyartharani	295	Katsareya	60
Kapurijadi	19	Kasai	72	Katsol	273
Karalia	102	Kasai	257	Kattang	59
Karamba	93	Kasail	43	Katul	303
Karanga	260	Kashkash	238	Katvinayi	144
Karanji	257	Kashmal	64	Kauha	305
Karanju	75	Kasondi	284	Kaunch	220
Karanta	117	Kassi	72	Kauri dodi	50
Karaunda	84	Kassunt	160	Kauributi	22
Karchanua	296	Kasturchara	289	Kausum	280
Karchi	178	Kasumba	85	Kavitha	202
Karel	81	Katagurkamai	56	Kedarpatta	289
Karela	217	Katar	158	Kedgi	238
Kareli	217	Katari	327	Keli	80
Karer	81	Katathuhar	152	Kemuka	112
Kareta	288	Katbel	202	Kena	112
Karihari	166	Kateli	43	Keore-Ke-Mul	187
Karil	81	Kateli	290	Ker	81

Contd..

Appendix–VII–*Contd...*

Hindi Names	*Page No.*	*Hindi Names*	*Page No.*	*Hindi Names*	*Page No.*
Kesar	113	Khoi	123	Kowa	104
Kesari	67	Khoja	72	Krishna Fal	239
Keu	112	Kholi	11	Krishnasiris	23
Kevda	238	Khor	191	Krungora	260
Khair	9	Khubkallana	130	Kuar	178
Khaira	29	Khulakhudi	92	Kuba	83
Khaira-Kajhor	29	Khurbuj	117	Kuchla	297
Khairbabul	9	Khurfa	258	Kuenine	96
Khaj	72	Khuriari	93	Kukaraundha	68
Khaji	72	Khurrur	93	Kukronda	68
Khanda	145	Khursa	258	Kukronda	68
Khans	183	Kidamari	44	Kulanjan	28
Kharbuja	84	Kikar	10	Kulinjan	28
Kharbuja	117	Kikoa	97	Kulthi	54
Kharenta	288	Kimul	195	Kulthi	206
Kharif	114	Kingani	171	Kulujnath	35
Kharjal	277	Kinnab	81	Kumarika	289
Kharkath	232	Kiplaya	161	Kumbhi	83
Kharmar	275	Kirala	87	Kumeria	177
Kharmor	275	Kiramar	44	Kumra	118
Kharsing	267	Kirata	35	Kumuda	228
Khas	194	Kiravat	35	Kunch	7
Khatua	137	Kirayat	35	Kundaliya	131
Khavi	123	Kirchi	86	Kunderi	105
Khejri	260	Kirmala	285	Kundo	189
Kher	177	Kiryato	35	Kundphul	189
Kheri	132	Kiwach	220	Kuppi	11
Khetua	40	Kokan	91	Kural	173
Khirni	208	Konrar	118	Kurchi	178
Khirni	325	Koobhi	20	Kureya	178
Khitmigajhar	29	Kotsemba	66	Kurfa	258

Contd..

Appendix–VII–*Contd...*

Hindi Names	*Page No.*	*Hindi Names*	*Page No.*	*Hindi Names*	*Page No.*
Kuri	227	Lancheri	29	Magrabu	176
Kurilo	51	Languli	166	Maha Nim	21
Kurry Patta	220	Lasan	25	Mahadebjata	206
Kusht	112	Latjira	12	Mahagotukola	180
Kushtha-Bheda	185	Latkan	67	Mahakal	312
Kussum	85	Lau	195	Mahanim	309
Kust	112	Laue	195	Maharukha	21
Kusum	280	Lauka	195	Mahi	270
Kut	112	Laung	299	Mahoor	13
Kutchumku	280	Launkia	218	Mahua	206
Kuthari	166	Lavanavalli	53	Mainphal	89
Kwel	22	Lebu	100	Maja	266
Kyirin	236	Lelka	157	Majtari	47
Lablab	173	Leteku	57	Majuphul	266
Ladiva	172	Lichi	202	Makandi	254
Lagavang	132	Lil	185	Makhi	247
Lahsan	25	Lim	309	Makkal	99
Lajalu	216	Limbado	21	Makoi	292
Lajari	231	Limbu	100	Makra	127
Lajjavati	216	Limu	100	Makurkendi	134
Lajwanti	216	Limun	100	Malatulsi	229
Lakuch	49	Limun	100	Malay jamun	300
Lal Dudhi	150	Lippa-Pank	132	Mali	163
Lal Guma	199	Luckmuna	53	Malkangani	90
Lal Mirch	82	Ludra	244	Malkunki	90
Lalbaclu	61	Luhuriya	252	Mamira	110
Lal-Bhuin Anvalah	248	Lunia	258	Mamiran	110
Laljhari	165	Machoti	256	Mamri	228
Lallei	23	Madan	89	Mandara	146
Lalmurghka	91	Madar	78	Mandara	147
Lamjak	123	Madar	79	Mandara	147

Contd..

Appendix–VII–*Contd...*

Hindi Names	*Page No.*	*Hindi Names*	*Page No.*	*Hindi Names*	*Page No.*
Mandukaparni	181	Mirzanjosh	232	Mushkbala	317
Mandva	140	Mitha	325	Muslikand	118
Mangala	124	Mitha- Bis	14	Mustaka	125
Mangustan	162	Mitha Neem	220	Mutchli	270
Manjit	273	Mitha Telia	15	Mutha	125
Manjith	273	Mitha Zahar	13	Muthi	313
Maraphali	175	Mithi Patti	281	Muyna	214
Marchula	221	Mithiari	173	Muyuna	214
Marorphali	175	Mochkand	138	Nagadanti	175
Marosi	175	Mohin	195	Nagakesar	213
Marshikha	17	Mohwa	206	Nagdana	47
Marua	229	Moina	214	Naghar	213
Marul	278	Moka	281	Nagphana	231
Maruva	140	Morang-elachi	31	Nagphani	231
Masser	198	Morpankhi	307	Nanda	124
Mastaru	47	Motha	125	Narcha	111
Masuphal	266	Mothitrina	315	Narel	106
Masur	198	Moyen	195	Nari	145
Masuri	198	Muchkand	264	Narial	106
Matazor	248	Mudana	214	Nariyal	106
Matta-ara	40	Muddhe Dhup	21	Nariyalkaper	106
Matti	21	Mukhajali	136	Narkachura	119
Maulser	216	Mukul	109	Navamallika	188
Maulsiri	216	Mundi	294	Nebu	100
Mauwa	206	Mungna	219	Neel Kanthi	95
Mehndi	198	Mung-phali	41	Neelolpal	218
Merasimgi	171	Munja	276	Neimal	297
Methi	313	Mura	146	Nelmal	297
Mewri	322	Murad	223	Nenua	205
Mhaner	93	Murva	278	Neora	61
Michai	186	Musel	124	Nepali Dhaniya	327

Contd..

Appendix–VII–*Contd...*

Hindi Names	*Page No.*	*Hindi Names*	*Page No.*	*Hindi Names*	*Page No.*
Nibari	227	Palaas	74	Paral	296
Nibu	100	Palakhari	124	Paresiyavasan	17
Nibu	100	Palandu	24	Pari	98
Nikuncham	9	Palas	74	Parja	227
Nil	184	Palas Lata	74	Paroa	157
Nil	185	Pamad	284	Parur	296
Nil Phool	20	Pamkara	147	Parvel	122
Nili Tulasi	182	Pan	250	Parwa	177
Nilinargandhi	192	Panar	303	Passi	128
Nilkatte	175	Panevar	284	Pat	111
Nim	55	Pangra	147	Patala	160
Nimb	55	Pani ki Sambhalu	323	Patalatumbi	93
Nimbu	100	Pani Kumhra	63	Patalkand	265
Nirmali	297	Paniah	303	Patang	172
Nisinda	322	Paniala	66	Pater	315
Nora	289	Paniala	159	Patera	314
Nugduna	47	Panikisambhala	322	Path	122
Nuli	302	Panjirikapat	38	Patha	98
Odhul	177	Panjoli	247	Patha	122
Ondhaphuli	312	Panjuli	247	Patharchatta	141
Ota	163	Pankain	66	Pathi	122
Padal	296	Panni	123	Pathorchur	253
Padaria	296	Panniari	61	Pathri	198
Padiala	296	Papar	257	Pati	47
Padrian	62	Paparaminta	213	Patis	14
Pagun	70	Papari	240	Patkarru	272
Pahari Errand	189	Papita	84	Pattaajavain	253
Pahari Pudina	213	Papri	240	Patton-Ki-Send	151
Paiman	299	Papurijadi	19	Penaura	303
Pakhanabheda	65	Par	295	Pendua	303
Pakhanbed	65	Paral	267	Perumaram	21

Contd..

Appendix-VII-*Contd...*

Hindi Names	*Page No.*	*Hindi Names*	*Page No.*	*Hindi Names*	*Page No.*
Petha	63	Piyal	73	Ramchana	126
Petha	118	Piyar	73	Ramchana	323
Phalinda	299	Podina	212	Ramdatun	295
Phangla	255	Poi	61	Ramsar	276
Pharad	147	Pongilyam	21	Randhuni	310
Pharamgi	43	Post	238	Rangan	187
Pharhad	147	Potali	168	Rangoon	109
Pharna	25	Potari	8	Rangoon-Ki-Bel	109
Phenil	279	Prisniparni	316	Rasabharii	273
Phok	145	Priyadarsha	114	Rasant	64
Phutki	210	Ptee	47	Rasuat	64
Pila	86	Pudina	213	Rasun	25
Pilacameli	224	Pudinah	212	Ratanjot	189
Pilu	277	Punernava	69	Ratanjot	231
Pindora	303	Punir	324	Ratanpurush	180
Pinja Phool	294	Punk	108	Rati	7
Pinwal	156	Pur	296	Ratmundiya	312
Pipal	158	Purula	205	Ratti	7
Pipal	251	Purula	296	Rawasan	287
Pipar	158	Puthkanda	12	Reetha	279
Piparisari	259	Qualamidarchini	98	Reetha	279
Piplamul	251	Ragatsemal	70	Remgani	293
Pipli	158	Ragatsenbal	70	Revatchini	268
Pipli	251	Rajana	187	Risht	279
Pipulka	13	Rajmari	12	Rishtak	279
Pirun	104	Rakas-Gaddah	111	Ritha	279
Pismaram	275	Rakta	63	Rithoul	160
Pismarum	275	Rakti	7	Rohisha	124
Pitni-ber	328	Ral	287	Rohni	207
Pitti	318	Ram Kurthi	167	Romghas	124
Piyaj	24	Ram Tulsi	229	Rookmini	187

Contd..

Appendix–VII–*Contd...*

Hindi Names	*Page No.*	*Hindi Names*	*Page No.*	*Hindi Names*	*Page No.*
Rousaghas	124	Sahasimuli	51	Sanamoti	210
Rudraki	139	Sahinjna	219	Sanay	283
Rudraks	139	Sairil	211	Sani	69
Rudraksha	139	Saittan Ka Jat	28	Sanjivini	282
Rula	207	Sakena	184	Sannai	114
Rupi	11	Sal	287	Santal	278
Rusa	296	Salab-Misri	232	Santha	324
Rusa Ghas	123	Salad	194	Saphed Simal	90
Rusaghas	124	Salam	127	Saptala	188
Sabbajaya	80	Salampanja	127	Saptarna	28
Sadab	275	Salbiasetakuss	277	Sarala	250
Sadab	275	Salep	127	Sariva	176
Sadabahar	89	Salep Misri	232	Sarivan	131
Sadadhatura	129	Salpan	131	Sarkanda	276
Sadaphul	89	Salpan	159	Sarphonka	304
Sadasuhagen	89	Salpani	131	Sarpmool	243
Sadberg	77	Salsa	176	Sataavar	51
Sadowani	242	Salun	131	Sataparna	28
Safed Arand	189	Salwan	131	Satari	275
Safed Chandan	278	Samandar Ka Pat	43	Satari	275
Safed Datura	128	Samandarphaind	43	Satavar	50
Safed Murgha	91	Samandarsof	43	Satavare	51
Safed Safari	262	Samandarsokh	43	Satavari	52
Safeda	148	Sambhalu	322	Satawar	50
Safedak	79	Sami	260	Satawar	51
Safedmusli	50	Samkru	107	Sathra	232
Sag-angur	53	Samraj	319	Satmuli	51
Sagona	306	Samudulan	139	Satpura	128
Sagovani	242	Samundarphal	61	Satwin	28
Sahadabi	288	San	123	Satyanasi	43
Sahadevi	319	Sanai	283	Satyanasi	293

Contd..

Appendix–VII–*Contd...*

Hindi Names	*Page No.*	*Hindi Names*	*Page No.*	*Hindi Names*	*Page No.*
Sauna	232	Shariphal	39	Sitakipanjiri	38
Saundhiya	124	Sharmpate	216	Sitaphal	39
Saunf	161	Shatmula	124	Siyah Musli	118
Saunf	249	Shembal	70	Siyahmusli	36
Sauri	30	Sheng	41	Soa	37
Sawolf	249	Sheng-Dana	41	Sodori	319
Seb	207	Shewat	176	Soi	37
Seh	235	Shia-Jira	85	Somaraji	237
Sehund	151	Shiral	170	Somraj	92
Sein	305	Shirkhist	162	Somraji	92
Sem	243	Shitta-rattai	28	Somrapani	329
Semal	70	Shivet	138	Sonali	294
Semul	70	Shiwan	168	Sonamukhi	283
Semur	70	Shumuk	270	Sonkadi	241
Setakala Arak	102	Shyamakranta	154	Sonp	161
Sev	207	Shyona	232	Sont	161
Sewan	168	Sialu	327	Sonth (when dry)	327
Seyla	219	Siamalata	182	Sowa	37
Shadevi	95	Siddhi	81	Soya	37
Shahdevi	95	Sij	151	Sudharg	272
Shaitan	28	Sij	152	Sufaidmurga	91
Shajna	219	Silarumba	80	Sugandhabala	135
Shakakul	52	Silkanti	97	Sugandhavacha	192
Shakakul	51	Silvalingi	135	Sugandhbach	28
Shakakul	51	Sinar	87	Sugandhbala	317
Shakaloo	51	Singli	328	Sundwar	268
Shambalu	322	Siora	296	Sunn	114
Shanavalli	95	Sirhu	183	Sunsuniya	209
Shankpushpi	154	Siriphal	18	Supari	43
Shaqaqul	52	Sirphal	18	Supyari	43
Sharifah	39	Sita bani	46	Sura	177

Contd..

Appendix–VII–*Contd...*

Hindi Names	*Page No.*	*Hindi Names*	*Page No.*	*Hindi Names*	*Page No.*
Sural	265	Thanella	93	Tuni	309
Surasarabi	280	Thawi	324	Tur	76
Surasaruni	71	Thohar	151	Tuvar	76
Sureta	255	Thor	152	Tuver	76
Survari	91	Thuja	307	Udisambhalu	192
Sutopsha	37	Thunera	103	Udsalap	237
Sutri	320	Thunjhunia	115	Uloo	183
Swarnapatri	283	Tidhara-Sehnr	150	Umar	157
Swetbarela	288	Tikhar	71	Undarbibi	190
Tagar	317	Tikhur	118	Untkataru	181
Tajkalam	97	Til	286	Urni	104
Talamkhana	181	Tilkatel	286	Ustarkhar	154
Talishpatra	7	Tilla	15	Usturgar	154
Talispatri	159	Tindhara Sehund	150	Utranajutuka	242
Tallispatra	7	Tirra	265	Vacha	15
Tamaku	227	Titaiya Baur	215	Vajradanti	60
Tamal	163	Titalauki	195	Vakamba	83
Tambakhu	226	Titapat	111	Vamini	124
Tambaku	227	Titpatti	272	Van-Gobi	197
Tambuli	250	Tondarsaijhad	171	Vantulsi	228
Tamul	163	Tongol Lati	149	Vapchi	92
Tarwar	283	Toon	309	Vartuli	132
Tatrak	270	Torbi	184	Vasaka	244
Tatri	270	Tori	204	Vatanasini	224
Tavakhira	118	Torki	184	Vatkana	67
Tejpatt	97	Tripatra	313	Vatsanabh	14
Tej-phal	327	Tsulesi	235	Veerataru	132
Telia Kachang	15	Tulsi	230	Verenda	190
Tendu	134	Tumba	99	Vidang	142
Tepari	8	Tumri	195	Vidari Kaud	265
Tesu	74	Tumru	327	Vilatti	202

Contd..

Appendix–VII–*Contd...*

Hindi Names	*Page No.*	*Hindi Names*	*Page No.*	*Hindi Names*	*Page No.*
Vilayati Babul	10	Visnukranti	104	Yukeliptas	148
Vilayatiafranthin	46	Wakollu	54	Zafran	113
Vilayatimehndi	223	Wangru	82	Zakhm-e-Hayat	73
Vilayatri Bakes	192	Wawrung	142	Zergul	77
Vilayeti Mung	41	Wilayati Tulasi	182	Zir	156
Vindangah	143	Yavasa	23	Zira	85
Vingar	171				

Appendix VIII
Index to Sanskrit Names

Sanskrit Names	*Page No.*	*Sanskrit Names*	*Page No.*	*Sanskrit Names*	*Page No.*
Aaluka	261	Agnidamanaka	48	Ajagandhi	62
Abda	125	Agnijarah	203	Ajaji	157
Abdhinarikelah	203	Agnijvala	244	Ajaka	228
Abdhiphal	61	Agnijwala	324	Ajaka	229
Abhaya	307	Agnikrita	32	Ajaka	230
Abheerupatri	51	Agnimantha	104	Ajamoda	262
Abhipriya	296	Agnimanthah	104	Ajamoda	310
Abhiru	51	Agnimukhi	166	Ajaoda	310
Acchaka	218	Agnishekhara	113	Ajapada	38
Achari	145	Agnishikha	85	Ajashringi	171
Achchhuka	219	Agnisikha	166	Ajasringi	195
Achchuka	218	Agragrahi	33	Ajata	248
Adhaki	76	Agrimakhya	39	Ajuta	248
Adhanani	231	Agudagandha	155	Akara	33
Adhapushpi	312	Agurah	41	Akara	246
Adhomukha	312	Aharbandhava	78	Akarakarabhah	33
Adhyand	248	Aheru	51	Akashavalli	88
Adiphala	246	Ahichhatra	161	Akashvalli	121
Aditya	78	Ahifen	239	Akasvalli	88
Adityabhakta	102	Ahiganda	44	Akhoda	191
Agaralohita	94	Ahigandha	45	Akhuparnika	189
Agaru	41	Ahikhanda	288	Aklika	185
Agaru	109	Ahula	283	Aksah	306
Agasti	286	Ailni	166	Akschoda	191
Agastyah	286	Airawata	49	Akshadhara	296
Agati	286	Ajabaksha	10	Akshata	179
Aghata	12	Ajaballi	171	Akshiba	219
Aghori	158	Ajabhakshya	154	Aksotah	191
Agnibija	92	Ajagandha	102	Alabu	195

Contd..

Appendix-VIII-*Contd...*

Sanskrit Names	*Page No.*	*Sanskrit Names*	*Page No.*	*Sanskrit Names*	*Page No.*
Alankrita Kesara	114	Amri	246	Aphasantin	46
Alarka	79	Amrinala	123	Apina	309
Alaskapha	172	Amrita	14	Apiyadruma	160
Alpamarisha	30	Amritavha	117	Apmarga	12
Ama	33	Amritphala	122	Aprithaktvacha	262
Amala	248	Amrutaphala	122	Aragvadha	87
Amar	46	Amrutaphala	323	Arala	233
Amara	99	Amrutdaru	170	Aralu	21
Amaravallari	121	Anannasa	33	Araluka	233
Amaravela	121	Ananta	24	Aranya	223
Amari	124	Ananta	60	Aranyaharidra	119
Ambhoja	225	Anantamulah	176	Aranyajayapala	190
Ambudhiphala	61	Ananto	154	Aranyajiraka	319
Ambuja	61	Aneshta	119	Aranyavasini	323
Ambu-prasada	297	Angaka	216	Ardrakam	327
Amburuha	180	Angaravalli	257	Ari	11
Ambuvallika	217	Angarvriksha	58	Arimaedah	10
Amlajambir	100	Angulidala	58	Arimarda	284
Amlaka	49	Anilaghnaka	306	Arimeda	10
Amlanisha	174	Anilsara	134	Arimedaka	10
Amlaparni	269	Anjalikaraka	216	Arishta	279
Amlavetasa	269	Anjalikarika	216	Arishta	326
Amlee	54	Anjanakeshi	231	Arishtak	279
Amlika	235	Anjira	156	Arishtha	25
Amlika	246	Ankola	22	Arista	279
Amlika	302	Ankusha	97	Aritamanjari	11
Amlotaja	235	Anshumati	131	Arjakan	229
Amogha	142	Apangaka-Kharaman-Jari	12	Arjunah	305
Amova	296	Aparajita	104	Ark	79
Amra	208	Aphala	248	Arka	78
Amratakan	295			Arkamula	44

Contd..

Appendix–VIII–*Contd...*

Sanskrit Names	*Page No.*	*Sanskrit Names*	*Page No.*	*Sanskrit Names*	*Page No.*
Arkamula	45	Asphota	104	Ayudham	284
Arkapriya	177	Asra	113	Ayugmaparna	29
Arkaragah	203	Asthibhaksha	73	Babbula	10
Arkavallabhah	242	Asthisrnkhala	99	Babbulaka	10
Arnoda	125	Astmabayday	19	Babuna	210
Aru	261	Asumodhagam	313	Babunaj	210
Aruha	248	Asurasa	229	Badari	220
Aruna	7	Asvakarna	287	Badhaphala	257
Aruna	14	Asvatthah	158	Bahala	37
Aruna	28	Asyapatra	225	Bahalchakshu	171
Aruna	113	Ataicha	14	Bahira	306
Aruna	177	Atamagupta	220	Bahugandha	188
Aruna	273	Atarusha	21	Bahukantaka	311
Aruni	71	Atibala	8	Bahupada	155
Aruni	280	Atichhatra	181	Bahupallava	59
Arusharah	283	Atikesara	272	Bahuparni	313
Asana	72	Atimanjula	272	Bahupatra	245
Asara	41	Atimukha	189	Bahupatra	248
Ashaya	49	Atisaraghni	14	Bahupatra	264
Ashmaghna	65	Atisarahita	22	Bahupatrica	140
Ashmantaka	62	Ativisha	14	Bahupraja	276
Ashokari	71	Atmagupta	51	Bahupushpa	147
Asholam	18	Atmamuli	154	Bahushalya	9
Ashvagola	252	Atmaraksha	100	Bahushikha	244
Ashvamaraka	86	Atripya	39	Bahusutaa	51
Ashvashakota	168	Atyamlaporni	323	Bahuvaraka	132
Ashveta	168	Atyuha	227	Bahuvirya	124
Ashwagandha	324	Avakapshpi	161	Bahuvirya	30
Ashwa-Puchha	145	Avartani	175	Bakuchee	92
Ashyuka	218	Avighnah	84	Bakuchi	121
Asmaghni	281	Avika	163	Bakulah	216

Contd..

Appendix–VIII–*Contd...*

Sanskrit Names	*Page No.*	*Sanskrit Names*	*Page No.*	*Sanskrit Names*	*Page No.*
Bala	135	Bastantri	43	Bhasmaka	142
Bala	288	Bastishodhana	89	Bhavana	163
Balapatra	9	Bebula	10	Bhavya	133
Balapatra	24	Behera	306	Bhavya	163
Balaputra	9	Bha	226	Bhedini	85
Balatanaya	9	Bhadra	19	Bhekaparni	92
Balaya	75	Bhadra	15	Bhimayojini	65
Balhika	155	Bhadradanti	191	Bhindi	60
Balhika	155	Bhadradaru	250	Bhisangmata	192
Balika	8	Bhadraparni	168	Bhissata	166
Ballikhadira	11	Bhadrasri	278	Bhringara	138
Balobhojya	95	Bhadrendrani	83	Bhringaraja	138
Balya	8	Bhakshyapatra	250	Bhringi	14
Balya	237	Bhallatakah	283	Bhringi	155
Ban Tamaku	318	Bhandi	226	Bhringi	44
Bana	60	Bhandi	273	Bhringiphalla	295
Bana	276	Bhandira	30	Bhu Lavangah	204
Banah	304	Bhandira	103	Bhucanakah	41
Banamethika	211	Bhanga	81	Bhuchampaca	193
Banapunkhah	304	Bhangura	14	Bhudhatri	246
Bandhujivaka	187	Bhantaka	103	Bhumi Kusmanda	265
Bandhuka	187	Bhanti	103	Bhumiari	283
Bandhukah	242	Bhantriya	225	Bhumichampa	193
Banfsa	321	Bhantuka	233	Bhumikushmanda	186
Barbura	10	Bhanu	78	Bhumyahuli	87
Barha	47	Bharadwaji	50	Bhumyamalaki	246
Barhibarha	103	Bharangi	93	Bhunimba	35
Barhichuda	103	Bhargi	103	Bhunimba	298
Barhichuda	91	Bharija	133	Bhunimbah	35
Barhipushpa	47	Bharngi	93	Bhupadma	283
Basl	24	Bhaskareshta	102	Bhuribala	8

Contd..

Appendix–VIII–*Contd...*

Sanskrit Names	*Page No.*	*Sanskrit Names*	*Page No.*	*Sanskrit Names*	*Page No.*
Bhuripushpa	161	Bramhagarbha	102	Chanadruma	311
Bhurundi	175	Bramhatirtha	112	Chanaka	95
Bhustana	182	Bramhi	145	Chanda	220
Bhustrina	122	Brhati	290	Chanda	326
Bhutabhna	25	Bridhamitra	170	Chandra	96
Bhutalapota	87	Brihadagala	288	Chandramulika	192
Bhutali	118	Brihaddala	55	Chandraparni	237
Bhutapala	88	Brihatee	290	Chandrika	313
Bhutari	155	Brihatphala	100	Chara	150
Bhuti	124	Brihatpilu	277	Chara	73
Bhutika	124	Brihattrina	59	Charati	180
Bhutumbi	94	Brihavaruni	100	Charati	248
Bibhitaka	306	Brinda	230	Charavella	102
Bidanga	142	Burbura	10	Chataki	7
Bidhanta	144	Canaka	95	Chauhara	48
Bijapura	100	Candanah	278	Chebra	317
Bijapura	265	Caturangi	196	Chetaki	188
Bijasneha	74	Caturdharalata	224	Chhardini	105
Bilva	18	Catuspatri	209	Chhatr	37
Bimbi	105	Chaityadru	158	Chhatra	122
Bimbika	105	Chakramarda	284	Chhatra	273
Bimbora	79	Chakranai	308	Chilhaka	86
Biranjasipha	12	Chakrangi	145	Chilihintah	105
Bodha	22	Chakrashreni	171	Chilika	94
Bodhidruma	158	Chakshushya	297	Chinaka	287
Bohah	110	Chalusha	219	Chinaka	286
Brahmadani	17	Chamarapushpa	238	Chipita	297
Brahmadani	43	Chamari	63	Chira	155
Brahmavriksha	157	Chamarika	63	Chirabilva	257
Brahmi	92	Champeya	213	Chirapushpa	216
Brahmuni	103	Chamrika	62	Chirayu	90

Contd..

Appendix-VIII-*Contd...*

Sanskrit Names	*Page No.*	*Sanskrit Names*	*Page No.*	*Sanskrit Names*	*Page No.*
Chirbita	84	Dandi	48	Devi	100
Chitra	142	Danta	48	Devi	131
Chitradandakah	169	Dantadhavana	9	Dhamani	231
Chitraka	264	Dantaghata	100	Dhamargavah	204
Chitrala	100	Dantaharshana	100	Dhanaka	111
Chitrapada	17	Dantakarshana	100	Dhananjaya	305
Chitrapatri	244	Danti	58	Dhanavalli	224
Chitraphala	100	Darbha	183	Dhanvi	154
Chitravalli	100	Daruharidra	64	Dhanvi	216
Chosa	239	Daruharidra	112	Dhanya	234
Chudala	7	Daruharidra	64	Dhanyaka	111
Chukra	302	Daruharidra	64	Dharakadambah	172
Chundrika	267	Darupita	64	Dharttarashtrapadi	17
Chuta	208	Darusita	98	Dhastura	129
Cinca	302	Darvi	64	Dhataki	324
Cirabilvah	178	Darvi	112	Dhatri	246
Dadamari	284	Darvika	139	Dhatriphala	61
Dadima	265	Darvika	312	Dhattura	128
Dahana	309	Dasa	71	Dhatushya	59
Dahu	49	Dasi	60	Dhavan	146
Dakshinatriya	106	Desabadama	306	Dhavanika	293
Dala	97	Deva	71	Dhira	113
Dalaphala	206	Devadaru	90	Dhirghapatolika	205
Damana	48	Devajagdha	124	Dhiva	327
Damana	189	Devakili	80	Dhruva	155
Damanah	47	Devakusumam	299	Dhumala	28
Damanakah	47	Devasasha	319	Dhumapatra	44
Damani	48	Devatamalla	303	Dhumrapatrika	227
Damara	46	Devatma	158	Dhupavrikshaka	250
Damnak	232	Devavriksha	29	Dhurta	129
Danda	317	Devi	94	Dhusara	160

Contd..

Appendix–VIII–*Contd...*

Sanskrit Names	*Page No.*	*Sanskrit Names*	*Page No.*	*Sanskrit Names*	*Page No.*
Dhustoora	129	Duralambha	24	Gandhamalati	223
Dhvankshamachi	292	Durga	109	Gandhapatri	324
Dhyama	124	Durlabha	24	Gandhapushpika	312
Dhyamakah	123	Durlabha	311	Gandhari	24
Dinesavalli	318	Durlabha	120	Gandhari	62
Dipaka	113	Durva	124	Gandhasshati	174
Dipana	317	Eiendri	57	Gandhini	268
Dipani	313	Ekadivi	72	Gandhira	291
Dipta	155	Ekangi	221	Gandhotkata	48
Dirgha	58	Ekavira	72	Gandira	108
Dirghamula	132	Ela	31	Gandira	152
Dirghapatra	24	Eraka	314	Gandirah	126
Dirghapatra	297	Eraka	315	Gandiri	273
Dirghapatraka	25	Erand	271	Gangati	303
Dirghavalli	100	Eranda	84	Ganjka	81
Divabhishta	250	Eranda	271	Ganrupaka	79
Divakar	78	Erandhaphala	58	Garbhada	292
Divya	92	Forskohlii	254	Garbhaghatini	166
Divya	179	Gadadhara	285	Gargumaru	135
Divyatumbi	94	Gahara	142	Garikera	106
Draksha	323	Gajapatra	158	Garikulay	167
Dravanti	189	Gajeshta	265	Gartalabu	94
Dridhabija	10	Galani	15	Garudapu	206
Dridhabjja	121	Gambhira	228	Garudi	105
Drishopadma	225	Gandamalika	216	Gaura	257
Drona Pushpi	201	Gandana	12	Gauri	119
Dronika	185	Gandha	48	Gauriphal	273
Dugdhi	236	Gandha	285	Gauriphal	274
Dugdhika	151	Gandhadhika	161	Gaurtvaka	58
Dugdhika	153	Gandhali	237	Gavakshi	296
Duralabha	175	Gandhalu	327	Gavayodbhava	118

Contd..

Appendix-VIII-*Contd...*

Sanskrit Names	*Page No.*	*Sanskrit Names*	*Page No.*	*Sanskrit Names*	*Page No.*
Gayatri	9	Gouhi	105	Guggla	110
Gerdruma	297	Grahdruma	306	Guggulu	109
Ghalanta	22	Grahini	154	Guha	316
Ghambhari	168	Grahini	311	Gula	231
Ghana	245	Gramya	185	Gulmamula	327
Ghanaghana	292	Gramyakunkuma	85	Gundra	125
Ghanta	8	Granat	265	Gunja	7
Ghantarava	115	Granthi	199	Gunjiku	7
Ghartkumari	27	Granthika	47	Gurubija	198
Ghonta	72	Granthika	199	Gurunchi	308
Ghonta	328	Granthila	30	Guvaka	43
Ghosha	37	Granthila	81	Gvedhukah	107
Ghosha	142	Granthiparna	199	Haima	298
Ghosha	252	Granthiparni	199	Halina	238
Giajjhra	29	Gridhrani	44	Hamasagara	193
Gidar Tamaku	318	Gridhrapatra	44	Hamsagar	73
Giribhita	65	Grihadruma	171	Hamsapadi	131
Girija	62	Grihini	155	Hansaghri	17
Girikadambah	172	Grinjana	25	Hansapadi	17
Girikarnika	83	Gubak	43	Hapusa	294
Girimallika	178	Gucchapatra	275	Harichandana	113
Gojihva	139	Guchha	140	Haridra	119
Gokshura	241	Guchhala	122	Harijan-manjari	11
Gokshura	311	Guchhapatra	275	Harini	273
Golomi	15	Guchnaphala	297	Haripriya	71
Gondhangri	17	Guda	152	Harita	103
Gopakanta	158	Guda	169	Haritaki	307
Goraksaphalini	121	Guda	231	Haritaksha	157
Gorakshi	17	Gudatvacha	98	Harita-manjari	11
Gorani	121	Gudhapatra	81	Harivallabha	177
Gosthabartaku	293	Guggilam	287	Harra	307

Contd..

Appendix–VIII–*Contd...*

Sanskrit Names	*Page No.*	*Sanskrit Names*	*Page No.*	*Sanskrit Names*	*Page No.*
Hastipadi	139	Indiralaya	225	Jambir	100
Hastiparna	205	Indivarah	218	Jambu	25
Hastisunda	175	Indivari	51	Jambuh	299
Hastyaluka	133	Indrani	322	Jambuka	238
Hayamaraka	325	Indrasana	81	Jamhira	100
Hayapriya	179	Indravalli	83	Janesha	188
Helanchi	145	Indravriksha	90	Jantuka	164
Hema	213	Indrayava	178	Jantumari	100
Hema Pushpi	118	Indrayava	101	Jantuphala	157
Hemakanti	64	Ingudee	58	Japa	177
Hemamalati	224	Ingudi	306	Jarana	85
Hemapatri	283	Ishadgola	253	Jarana	284
Hemaragi	119	Ishana	260	Jargadi	107
Hemashikha	43	Ishta	260	Jata	45
Hijjala	61	Ishvara	45	Jataphala	106
Hilamochi	145	Ishvari Mul	45	Jatha	104
Hinghka	155	Itora	326	Jati	188
Hingu	155	Jadyari	100	Jatila	15
Hinguputra	58	Jaguda	113	Jatila	48
Hiranyatutha	108	Jaiphala	223	Jatiphala	246
Homa	9	Jaitun	230	Jatuka	155
Homadhanya	286	Jajiphalam	223	Javitri	223
Hoya	324	Jala	71	Jaya	8
Hridya	85	Jalaja	15	Jayanti	287
Hridya	98	Jalajantu	58	Jayantika	287
Hulkul	102	Jalaproya	219	Jayapala	116
Ijjala	61	Jambavam	299	Jayarini	104
Ikshugandha	181	Jambha	100	Jayavaha	191
Ikshugandha	311	Jambhaka	100	Jeevdaru	170
Iksurah	181	Jambhala	78	Jhingini	195
Indhana	46	Jambhala	100	Jhinti	60

Contd..

Appendix–VIII–*Contd...*

Sanskrit Names	*Page No.*	*Sanskrit Names*	*Page No.*	*Sanskrit Names*	*Page No.*
Jingi	273	Kalasaka	111	Kantalu	10
Jingini	195	Kalaskandha	10	Kantangur	56
Jirnaparna	71	Kalhora	22	Kanthari	82
Jivana	95	Kalikari	166	Kapali	142
Jivanthika	308	Kalinga	178	Kapikacchu	220
Jvarantaka	298	Kaliyugalaya	306	Kapitana	158
Jvarantika	119	Kamboja	105	Kapitana	295
Jvarapaha	321	Kamlottama	85	Kapittha	202
Jyotir	83	Kampillaka	86	Kapitthah	202
Jyotishmati	90	Kampillaka	207	Kapura	96
Kaashmira	185	Kamu	246	Karahi	229
Kachhara	236	Kana	251	Karaka	74
Kadamba	71	Kanakaphala	116	Karaka	81
Kajutaka	32	Kananaerand	189	Karamaranga	54
Kakachedi	240	Kananari	260	Karamardah	84
Kakachedi	240	Kanchana	213	Karanjah	257
Kakadani	82	Kanchano	309	Karaskara	297
Kakamachi	292	Kandarohaka	231	Karava	33
Kakasphurja	297	Kandarpa	233	Karavella	217
Kakilakshya	181	Kandera	30	Karavi	161
Kakmachi	291	Kangeri	235	Karavi	37
Kaknasika	209	Kangu	287	Karaviraka	305
Kakodumbar	156	Kanguni	287	Karchura	120
Kakodumbara	157	Kankati	8	Karira	81
Kaktundi	50	Kankchchi	7	Kariyartharani	295
Kakubhah	305	Kanta	309	Karkata	252
Kala Bhangra	38	Kantaki	9	Karkata Sringee	271
Kalameshi	159	Kantaki	24	Karkata sringi	252
Kala-meshika	273	Kantaki	147	Karkataka	217
Kalana	77	Kantakikaranjah	75	Karkati	9
Kalanja	227	Kantalaka	156	Karkati	117

Contd..

Appendix–VIII–*Contd...*

Sanskrit Names	*Page No.*	*Sanskrit Names*	*Page No.*	*Sanskrit Names*	*Page No.*
Karmaraka	54	Katrina	124	Kilima	90
Karmarangah	55	Kattukanni	16	Kini	156
Karnavati	64	Katuka	105	Kinkirata	60
Karnika	171	Katukanda	25	Kirata	35
Karnikara	264	Katurasa	100	Kiratatikah	35
Karpasamu	169	Katutumbi	195	Kirata-tikta	298
Karpasi	169	Kaudigam	101	Kirramata	17
Karpasika	169	Kautakari	293	Kitali	59
Karpura	96	Kautukasanjaka	33	Kitamari	17
Karpurakachali	174	Kaya	100	Kitamari	44
Karshaphala	306	Kekishikha	91	Kitari	44
Karshya	49	Kemuka	112	Kokila	296
Karuka	216	Kesara	155	Kooshmanda	63
Karukah	55	Kesaraja	138	Koraja	28
Karukha	54	Keshara	113	Koshataki	204
Karunta	60	Keshya	138	Koushikaha	109
Karvellak	217	Ketaki	238	Koushi-kaha	287
Kashaya	10	Kevala	142	Kovidara	63
Kashaya	154	Khadira	9	Krama	54
Kashayi	49	Khadirapatrika	11	Kramuka	43
Kashmeera	112	Khadiraprani	159	Krimighni	44
Kashmira	14	Kharjuri	118	Krimighni	227
Kashmira	112	Kharopatraka	227	Krishanirgunda	192
Kashtaki	10	Kharva	272	Krishna	85
Kasiravidari	186	Kharvuja	117	Krishna Haridra	119
Kasmarda	284	Khasa	239	Krishnabija	39
Kasmari	168	Khatmi	29	Krishnajaji	85
Kastidanti	115	Khatmi	37	Krishna-Kamboji	247
Kata	64	Khavalli	121	Krishna-nemba	220
Katakatphalam	297	Khirni	208	Krishnasariva	117
Katphala	223	Khulkhuri	181	Krishnasirisha	23

Contd..

Appendix–VIII–*Contd...*

Sanskrit Names	*Page No.*	*Sanskrit Names*	*Page No.*	*Sanskrit Names*	*Page No.*
Krishnatvaka	305	Kushmanda	63	Langoli	106
Kriti	304	Kushmanda	118	Lanji	80
Krsnaguruh	41	Kushtha	112	Lanka	82
Krtamalah	87	Kushtha-Bheda	185	Lashuna	25
Ksarasresthah	281	Kushthaghni	157	Lata-karanj	75
Kshapa	119	Kusthapa	172	Lataksiri	314
Kshetravastuka	137	Kustumburi	111	Latapalas	74
Kshiravi	236	Kusumbha	85	Latapalasha	74
Kshiri	236	Kutaja	178	Latavamsa	77
Kshiri	301	Kutaja	325	Lavalee	245
Kshirini	43	Kutajah	325	Lavana-valli	53
Kshudra	107	Kutajastri	325	Lavangam	299
Ksudrasimbi	121	Kutasalmali	70	Laya	123
Kuberakshi	75	Kutasalmali	90	Lekhanatmaka	297
Kukkura	103	Kuthera	228	Likucha	49
Kukkuradry	68	Laghu	123	Lingini	135
Kukundara	68	Laghududhika	153	Lodhrapushpa	206
Kukundarh	68	Laghukashmarya	223	Lohita	85
Kukura	47	Laghupatha	98	Lohitanga	207
Kukuradru	68	Laghuvishnukranta	154	Lona	258
Kulanjan	28	Lajja	216	Loni	258
Kulaththa	206	Lajjalu	216	Lonika	258
Kulisha	99	Lajjika	216	Lunia	258
Kumari	27	Laksavrksah	281	Madala	21
Kumuda	228	Lakshapushpa	272	Madana	22
Kundah	189	Lakshmi	119	Madana	89
Kundali	56	Lakshmi	260	Madaniyahetu	324
Kunga	15	Lalbarila	288	Madayantika	198
Kuni	156	Lamphala	163	Madhava	206
Kunta	107	Langali	244	Madhavi	37
Kurkaru	118	Langli	166	Madhu	206

Contd..

Appendix–VIII–*Contd...*

Sanskrit Names	*Page No.*	*Sanskrit Names*	*Page No.*	*Sanskrit Names*	*Page No.*
Madhuchhada	91	Mamajjakah	144	Medhini	198
Madhuindrayava	325	Mamaskari	216	Medhya	9
Madhuka	206	Mana	155	Medhya	260
Madhupanjara	216	Mandapi	41	Methica	161
Madhuparni	308	Mandar	147	Methika	313
Madhurica	161	Mandara	79	Mishmitikta	110
Maha vriksha	277	Mandari	283	Misi	37
Mahabhita	216	Mandirriksha	156	Moca	70
Mahadronah	39	Mandukparni	92	Moch	90
Mahahala	105	Mangalya	18	Mocha	70
Mahakanda	24	Mangalya	158	Mochi	145
Mahakanda	25	Manjari	195	Mogha	142
Mahakapitha	132	Manjari	230	Monohara	189
Mahakosataki	205	Manjistha	273	Monomoda	189
Mahamoda	189	Manjula	156	Morata	101
Mahamundi	294	Manodna	188	Mrdvika	323
Mahanimba	21	Manojna	250	Mrigashinga	175
Mahanimbu	22	Marakatincles	297	Muchukunda	264
Mahaphala	106	Marich	82	Mukhya	120
Mahapindi	93	Markandi	283	Muktapushpa	189
Maharakha	21	Markara	138	Mukula	216
Mahavira	72	Markundi	283	Mukulakah	281
Mahija	327	Marura	278	Mundi	294
Majuphul	266	Masaparni	54	Mundirika	294
Makaranda	189	Mastadaru	90	Munja	276
Mala	248	Masurah	198	Munjariki	229
Malati	188	Matsyaksi	29	Munjatakah	127
Malayaja	278	Maurvi	171	Mura	146
Malkangni	90	Maurvi	278	Murva	278
Mallika	188	Mausala	262	Mushukunda	264
Mallikapushpa	178	Medhasingi	171	Muskakah	281

Contd..

Appendix–VIII–*Contd...*

Sanskrit Names	*Page No.*	*Sanskrit Names*	*Page No.*	*Sanskrit Names*	*Page No.*
Mustaka	125	Nar-Kachura	119	Nilotpalam	218
Nadayi	287	Narttakah	140	Nimbah	55
Nadi-Hingu	163	Nasanu	223	Nimbataru	147
Nadivasa	207	Nata	233	Nimbu-ka	100
Nagabandhu	158	Nati	231	Nipa	71
Nagadamani	47	Navamallica	188	Nira Brahmi	57
Nagadru	231	Nayagrodhi	189	Nirgundi	322
Nagajihva	144	Neelanirgundi	323	Nirjara	308
Nagakesarah	213	Nichula	61	Nishula	61
Nagapushpah	213	Nichula	77	Nisomali	256
Nagarjuni	150	Nidigdhika	293	Nissaraphalaka	305
Nagarjuni	153	Nidralu	229	Nityakalyani	89
Nagatumbi	94	Nikucha	49	Nivara	234
Nagavalli	250	Nikumbha	58	Nogaya	213
Nahi	144	Nikunjika	9	Nrtyakundalah	140
Nahika	111	Nila	85	Nyagrdhah	155
Naikadala	84	Niladurva	124	Odhrapushpa	177
Naipala	298	Nilaghiria	104	Okharadi	166
Nakharanjani	198	Nilanirgundi	192	Oshthaphala	39
Nakharotila	86	Nilaniryasa	148	Oshthi	105
Nakuli	45	Nilapushpa	321	Padama	225
Nalada	123	Nilapushpa	322	Padangi	17
Nalakunkuma	227	Nilapushpa	322	Padapotpala	264
Nali	231	Nilapushpi	154	Padmacarini	180
Nalika	231	Nilataru	106	Padmaka	262
Nalini	231	Nilavriksha	304	Padma-Pushkara	187
Nanamadhusnahi	289	Nilini	185	Palaasha	74
Nandi	101	Nilishila	160	Palagaruda	29
Nandi	301	Nilkantha Shikha	17	Palandu	24
Nandiviriksha	309	Nilkanthi	22	Palasha	74
Nandivruksha	171	Nilkumba	190	Palindi	182

Contd..

Appendix–VIII–*Contd...*

Sanskrit Names	*Page No.*	*Sanskrit Names*	*Page No.*	*Sanskrit Names*	*Page No.*
Panasa	49	Pashanabheda	281	Phanasa	49
Panasah	49	Pashanabhedi	253	Phenila	279
Pansa	49	Pasugandha	102	Phenila	279
Papata	240	Patala	122	Phulla	318
Papata	240	Patalagarudi	105	Pichhila	181
Paphanah	219	Patali	296	Pikaksha	207
Paranti	187	Patang	172	Pilu	277
Pararu	138	Patanga	138	Piluh	277
Parasi	48	Patha	98	Pindalu	303
Paravalli	182	Patha	122	Pindapushpa	265
Paravatapadi	322	Patha	295	Pindava	163
Paravatay Eranda	189	Pathi	291	Pindi	274
Paravati	33	Pathya	307	Pindi	214
Parevata	262	Patoli	296	Pinditaka	214
Paribhadra	146	Patra	97	Pindituka	214
Paribhadrah	146	Patraka	97	Pindu	214
Parijata	147	Patraka	97	Pinjara	213
Parijata	147	Patrasnuhi	152	Pippali	251
Parijatah	227	Patri	48	Pisasha	21
Parisha	84	Patrollekhataruh	296	Pishtika	118
Parjani	64	Pattaranjaka	172	Pita	22
Parnabija	73	Patturah	29	Pita	85
Parnapeejam	73	Paura	124	Pita	110
Parnayavanu	253	Pauskaram	185	Pitachandana	278
Parpata	275	Pavitra	286	Pitadaru	64
Parpatakah	275	Pavitraka	48	Pitadru	64
Parpatha	275	Pavitraka	158	Pitadru	112
Parsniparni	131	Payodhara	106	Pitaka	10
Parvaphala	63	Phalashaka	63	Pitakalika	283
Parvati	32	Phalguni	157	Pitamulika	268
Parvati	195	Phalum	43	Pitapushpi	283

Contd..

Appendix–VIII–*Contd...*

Sanskrit Names	*Page No.*	*Sanskrit Names*	*Page No.*	*Sanskrit Names*	*Page No.*
Pitataila	90	Punyagandha	188	Rakta	85
Pitberela	288	Pura	60	Raktabija	67
Piyala	73	Pushkara	112	Raktabindu	153
Poag	43	Pushkara	225	Raktaeramada	190
Pooga	43	Pushpi	177	Raktamrah	281
Poogi	43	Pusitoa	150	Raktapadi	17
Pootiha	213	Putaki	61	Raktapadi	216
Potaki	61	Putiha	212	Raktaphala	202
Potika	161	Putiha	213	Raktappittaghni	192
Prabala	237	Putikaranhah	178	Raktata	187
Prabhadrah	55	Putikaranja	75	Raktavalli	318
Pramodini	195	Putikashtha	250	Ramachhardanaka	89
Pranada	244	Putimayusika	102	Raman	220
Praputrata	284	Putivata	18	Ramasenaka	298
Prasarani	237	Putrapada	293	Rangaka	138
Prasniparni	316	Putrashringi	171	Rangapatri	185
Pratanika	233	Raajaadani	150	Rangoon-Ki-Bel	109
Pratika	177	Ragadoli	198	Ranjaka	198
Prativishaa	15	Rahutrishta	25	Rasadhya	295
Prishniparni	316	Raja	287	Rasagandha	110
Priya	71	Rajakoshatoki	205	Rasala	208
Priyaka	71	Rajani	119	Rasaphala	106
Priyakari	324	Rajanikasa	227	Rasayani	292
Priyala	73	Rajapatha	295	Rasna	28
Priyala	206	Rajapriya	24	Rasna	89
Priyalah	73	Rajatarih	222	Rasona	25
Priyangu	78	Rajbala	288	Ratha	77
Prshniparni	132	Rajika	157	Rava	134
Pryamkari	291	Rajnigandha	255	Ravana	134
Pudina	213	Rajphala	299	Ravi	78
Punarnava	69	Rakta	7	Rechanaka	207

Contd..

Appendix–VIII–*Contd...*

Sanskrit Names	*Page No.*	*Sanskrit Names*	*Page No.*	*Sanskrit Names*	*Page No.*
Rechani	58	Sahasravirya	52	Satamulika	139
Revatchini	268	Sainjna	219	Satapatri	272
Rewatavakra-Sodhi	100	Sakachara	198	Satapushpi	37
Rithakaranja	279	Sakralata	83	Satavari	51
Rocani	213	Sala Parni	131	Satavari	51
Rochani	213	Salabmisri	232	Satavari	52
Rochi	145	Salah	287	Sati	120
Rochni	246	Salampamisri	127	Sati	174
Rodanika	24	Sali	234	Saumyagandha	272
Rodini	154	Salmali	70	Savya	305
Roga	177	Samanga	273	Sehunda	151
Rohina	155	Samudranta	24	Sephalika	227
Rohini	223	Samudrapalaka	43	Shadanatmaka	297
Rohini	273	San	114	Shahastaraptri	12
Rohisah	123	Sana	114	Shailabija	283
Rohisha	124	Sandananika	11	Shakhotaka	296
Roma	97	Sanjeevani	282	Shalaparni	131
Romalu	220	Sanjeevani Bhoothi	282	Shalashura	49
Ruchishya	297	Sapitaka	205	Shalatu	18
Ruchya	85	Saptaparna	29	Shaleya	161
Rudrajata	44	Saptashira	250	Shalini	37
Rudraksha	139	Saptola	188	Shaliparni	316
Rujakara	55	Saralan	250	Shalya	18
Rukma	213	Sarasvati	57	Shankpushpi	154
Sachi	305	Sarba	176	Shanta	246
Sadampuspa	89	Saria	182	Sharadi	244
Sadaphala	18	Sariva	176	Sharadika	216
Safed Musli	50	Sarpagandha	267	Sharani	237
Sagarodhurga	55	Sarvajaya	80	Shari	315
Sahadeva	288	Sasasruti	143	Sharpunkha	304
Sahadevi	319	Sasasrutih	143	Shatapadapa	90

Contd..

Appendix–VIII–*Contd...*

Sanskrit Names	*Page No.*	*Sanskrit Names*	*Page No.*	*Sanskrit Names*	*Page No.*
Shatapatri	272	Shwetakusuma	326	Sriparnah	222
Shatawari	50	Shyaamkandaa	15	Sriphalah	18
Shati	174	Sidhapushpa	86	Sripriya	208
Shaundi	251	Silarumba	80	Sthirraga	64
Shetapushapa	249	Silikhadira	11	Sthulagoksurah	241
Shigra	58	Simbitika	207	Sthulpushpa	302
Shigru	219	Simhi	290	Stuya	231
Shikha	91	Simhunda	150	Subha	39
Shikhini	91	Sinduri	67	Suchi	53
Shiknari	252	Sinhala	98	Suda	39
Shimbi	315	Sinhi	293	Sudala	264
Shiraphala	106	Siralam	54	Sudha	231
Shirnaroma	103	Sirisika	128	Sudhavriksha	171
Shita	8	Sitaphalam	39	Sugandha	85
Shita	77	Sivadrumah	18	Sugandha	95
Shitangi	17	Snehaphala	286	Sugandhavacha	28
Shitapushpa	8	Snigdhabijah	253	Sugandhavacha	192
Shiva	246	Snigdhajirakah	253	Sugandhika	102
Shiva	248	Snuha	231	Sukanasa	111
Shivpriya	81	Snuhi	150	Sukemari	188
Shobhanjana	219	Snuhi	151	Sukshmapatra	321
Sholi	119	Snuhi	231	Sukta	302
Shophahari	229	Snuka	231	Sulabha	44
Shosha	61	Soma	145	Sunanda	44
Shuka	47	Somalata	275	Sunanda	45
Shuklika	235	Somavalli	105	Sunisannak	209
Shulanashana	155	Somraji	92	Sunn	114
Shunthi	327	Soubira	328	Suphala	168
Shveta	65	Srigalakanta	43	Supita	110
Shvetapushpi	312	Srihastini	175	Supushpa	264
Shwet Brihati	293	Srikhandam	278	Sura	198

Contd..

Appendix–VIII–*Contd...*

Sanskrit Names	*Page No.*	*Sanskrit Names*	*Page No.*	*Sanskrit Names*	*Page No.*
Surabhi	62	Swargistha	283	Tantri	308
Surabhi	71	Swarnpatri	283	Tanukshiri	295
Surabhi	216	Swetachilli	137	Tanvi	131
Surabhi	260	Syamalata	182	Tapana	104
Surabhidaruka	250	Syandana	134	Tapasapatri	48
Surabhininiba	220	Taga	284	Tapinjha	163
Surabhininiba	221	Tagara	317	Tapona	283
Surasa	97	Tailanga	216	Tarapushpa	189
Surasa	229	Tailaparnah	148	Tejpatra	97
Sursa	230	Tajana	59	Tikaphala	297
Surupa	188	Talakshira	118	Tiksnaksah	309
Suryavarta	95	Talavrinta	208	Tiktagoksurah	241
Suryavarta	95	Talemnlika	118	Tiktalabu	195
Suryavarta	102	Talisha	159	Tila	286
Susara	134	Talishapatra	159	Tilah	286
Sushaka	30	Talishpatra	7	Tilaka	97
Sushavi	85	Tallispatra	7	Tilashigru	219
Sushavi	217	Tamabija	206	Tilparni	102
Suvaha	322	Tamakhuh	227	Tinduki	134
Suvaka	17	Tamala	163	Tindula	134
Suvarnaka	87	Tamalapatra	97	Tinisah	210
Suvarnavarna	64	Tamalapatram	97	Tintrini	302
Svadikhallari	11	Tamalapatram	97	Tiriakphala	240
Svadukantaka	158	Tambula	250	Tishya	246
Svadukosataki	204	Tamracuda	68	Triakphala	240
Svaduparni	236	Tamrika	273	Tribija	41
Svarnaksiri	43	Tandidik	270	Tridharaka	150
Svetabija	206	Tandikodebaha	154	Trikantajata	56
Svetakutaja	325	Tandula	30	Trinashita	244
Svetamula	237	Tandula	234	Tripadi	131
Svetasalmali	90	Tanduliya	31	Triparna	265

Contd..

Appendix–VIII–*Contd...*

Sanskrit Names	*Page No.*	*Sanskrit Names*	*Page No.*	*Sanskrit Names*	*Page No.*
Tripatra	18	Unmatta	129	Vanaharidra	119
Triphala	246	Upakalya	251	Vanahasa	189
Triputah	26	Upakunchika	31	Vanajayapala	190
Trisandhya	177	Upas	40	Vana-Kulathha	54
Trivapushpi	67	Upodika	61	Vanamalini	326
Tulaphala	78	Utkata	98	Vanamalliga	189
Tulsi	230	Utpalam	228	Vanamethika	211
Tumburu	327	Utpalam	228	Vana-Methika	313
Tunah	309	Uttaravaruni	242	Vananimbuka	168
Tundi	105	Vaboula	10	Vanaprastha	206
Tungi	62	Vadadru	168	Vanaraj	62
Tungi	229	Vaikunthah	39	Vanatiktika	295
Tunna	156	Vajana	37	Vanda	130
Turangi-gandha	324	Vajra	151	Vanira	77
Tuvara	55	Vajra	181	Vanodbhava	286
Tuvaraka	55	Vajra	231	Vansa	59
Tuvari	75	Vajrabhringi	227	Vansh	59
Tuvarka	172	Vajrakantaka	150	Vanya kahu	194
Tuwari	76	Vajravalli	99	Vapra	273
Tvaka	98	Vajri	150	Vara	113
Udadhiphala	61	Vajri	231	Vara	142
Uddala	11	Vakra	252	Varahakareni	324
Udumbarah	157	Vakula	216	Varangi	119
Udumbaraparni	58	Vakushi	92	Varaphala	106
Ugra	15	Valakah	135	Varisona	225
Ugragandha	102	Valkala	40	Varshakali	85
Ujaya	81	Valliparna	206	Varvara	10
Ulapalabhenda	38	Vamaka	22	Varvara	229
Ulpalasariba	182	Vanabarbarika	228	Vasaka	192
Uma	119	Vanacchedi	196	Vasanavhaya	97
Umrita	246	Vanachampaka	80	Vasantaduta	208

Contd..

Appendix–VIII–*Contd...*

Sanskrit Names	*Page No.*	*Sanskrit Names*	*Page No.*	*Sanskrit Names*	*Page No.*
Vasantitikta	105	Vijaya	104	Vitika	250
Vashira	244	Vijaya	185	Vitunna	91
Vastukah	94	Vikacha	103	Vrddhadarukah	43
Vasuka	79	Vikankata	8	Vriddhutulsi	229
Vasuka	94	Vikankata	171	Vrihi	234
Vataghni	192	Vikasa	273	Vrikshabhaksha	130
Vatah	155	Vikinaroma	103	Vrikshadani	258
Vatari	25	Viktaka	298	Vriksharaja	158
Vatari	271	Vimba	105	Vriksharuha	130
Vatika	171	Vindangah	143	Vrikshikali	171
Vatori	283	Vira	30	Vrischikali	175
Vatsanava	13	Vira	60	Vrishyagandha	8
Vatu	233	Viradru	132	Vrittaphala	265
Vatya	288	Viravriksha	132	Vrittapornia	257
Vavvola	10	Virupa	154	Vrittapushpa	71
Vayasi	292	Virupa	311	Vrscikali	317
Vella	142	Visakantaki	217	Vujri	151
Venul	59	Visamustih	20	Vysalnkaha	209
Verta	77	Visha	13	Wadukantaka	158
Veshika	188	Vishabhadra	191	Wari	52
Vetasa	77	Vishaghna	31	Yagnika	183
Vidanga	143	Vishaghni	145	Yaha	61
Vidara	231	Vishaghnika	83	Yajnanga	9
Vidari	131	Vishalatvaka	29	Yajnika	158
Vidari	168	Vishalya	166	Yakshaves	155
Vidari	186	Vishamushti	297	Yakshmaghni	323
Vidari	265	Vishani	171	Yamalapatrakah	62
Vidari	291	Vishari	257	Yamapriya	155
Vidula	77	Vishnukranta	154	Yashtika	103
Vijapura	100	Vishopaha	275	Yava	179
Vijaya	81	Vita	10	Yavaja	118

Contd..

Appendix–VIII–*Contd...*

Sanskrit Names	*Page No.*	*Sanskrit Names*	*Page No.*	*Sanskrit Names*	*Page No.*
Yavaneshta	25	Yojanaphala	157	Yushtika	103
Yavani	310	Yugaphala	242	Zandu	302
Yavania	48	Yugmapatra	62	Zergul	77
Yawani	310	Yugmaphalika	242		

Plate 1A
Acacia catechu

Plate 1B
Ageratum conyzoides

Plate 1C
Allium cepa

Plate 1D

Anethum graveolens

Plate 2A

Argemone mexicana

Plate 2B

Asparagus racemosus

Plate 2C
Bauhinia variegata

Plate 2D
Berberis aristata

Plate 3A
Bergenia ciliata

Plate 3B

Bidens pilosa

Plate 3C

Calendula officinalis

Plate 3D

Calotropis procera

Plate 4A

Canna edulis

Plate 4B

Cannabis sativa

Plate 4C

Capsicum annum

Carica papaya

Catharanthus roseus

Cissampelos pareira

Coriandrum sativum

Cuscuta reflexa

Cynodon dactylon

Plate 6B

Ficus hispida

Plate 6C

Ficus religiosa

Plate 6D

Foeniculum vulgare

Plate 7A
Hedera helix

Plate 7B
Indigofera cassioides

Plate 7C
Iris germanica

Plate 7D

Jasminum grandiflorum

Plate 8A

Juglans regia

Plate 8B

Justicia adhatoda

Lantana camara

Lepidium didymum

Mangifera indica

Plate 9B

Mentha piperita

Plate 9C

Momordica charantia

Plate 9D

Murraya koenigii

Plate 10A

Nicotiana tabacum

Plate 10B

Ocimum sanctum

Plate 10C

Opuntia dillenii

Plate 10D
Phyllanthus emblica

Plate 11A
Pinus roxburghii

Plate 11B
Prinsepia utilis

Plate 11C

Prunus persica

Plate 11D

Psidium guajava

Plate 12A

Punica granatum

Plate 12B

Rhododendron arboreum

Plate 12C

Ricinus communis

Plate 12D

Rosa damascena

Plate 13A
Rubus niveus

Plate 13B
Solanum nigrum

Plate 13C
Solanum surattense

Plate 13D
Tagetes minuta

Plate 14A
Toona ciliata

Plate 14B
Trigonella foenum-graecum

Plate 14C

Verbascum thapsus

Plate 14D

Viola pilosa

Plate 15A

Vitex negundo

Plate 15B

Vitis vinifera

Plate 15C

Zanthoxylum armatum

Plate 15D

Zingiber officinale

www.ingramcontent.com/pod-product-compliance
Ingram Content Group UK Ltd.
Pitfield, Milton Keynes, MK11 3LW, UK
UKHW022217051026
15624UKWH00001B/202

9 789351 306559